John Fox, Jr., Appalachian Author

CONTRIBUTIONS TO SOUTHERN APPALACHIAN STUDIES

1. *Memoirs of Grassy Creek:*
Growing Up in the Mountains on the Virginia–North Carolina Line.
Zetta Barker Hamby. 1997

2. *The Pond Mountain Chronicle:*
Self-Portrait of a Southern Appalachian Community.
Leland R. Cooper and Mary Lee Cooper. 1997

3. *Traditional Musicians of the Central Blue Ridge:*
Old Time, Early Country, Folk and
Bluegrass Label Recording Artists, with Discographies.
Marty McGee. 2000

4. *W.R. Trivett, Appalachian Pictureman:*
Photographs of a Bygone Time.
Ralph E. Lentz. 2001

5. *The People of the New River:*
Oral Histories from the Ashe, Alleghany and
Watauga Counties of North Carolina.
Leland R. Cooper and Mary Lee Cooper. 2001

6. *John Fox, Jr., Appalachian Author.*
Bill York. 2003

7. *The Thistle and the Brier:*
Historical Links and Cultural Parallels Between Scotland and Appalachia.
Richard Blaustein. [2003]

8. *Tales from Sacred Wind:*
Coming of Age in Appalachia. The Cratis Williams Chronicles.
Cratis Williams. [2003]

John Fox, Jr., Appalachian Author

by Bill York

CONTRIBUTIONS TO SOUTHERN APPALACHIAN STUDIES, 6

McFarland & Company, Inc., Publishers
Jefferson, North Carolina, and London

Library of Congress Cataloguing-in-Publication Data

York, Bill, 1948–
    John Fox, Jr.: Appalachian author / by Bill York.
      p.   cm. — (Contributions to southern Appalachian studies ; 6)
    Includes bibliographical references and index.

    ISBN 0-7864-1372-7 (softcover : 50# alkaline paper) ∞

    1. Fox, John, 1863–1919. 2. Authors, American — 19th century — Biography. 3. Authors, American — 20th century — Biography.
4. Appalachian Region, Southern — In literature. 5. Appalachian Region, Southern — Biography. 6. Coal miners — United States — Biography. 7. Journalists — United States — Biography. 8. Virginia — Biography. 9. Kentucky — Biography. I. Title.
PS1703 .Y67   2003
813'.4 — dc21   2002015411

British Library cataloguing data are available

©2003 Bill York. All rights reserved

No part of this book may be reproduced or transmitted in any form or by any means, electronic or mechanical, including photocopying or recording, or by any information storage and retrieval system, without permission in writing from the publisher.

On the cover: John Fox, Jr. Taken around 1894, this photograph appeared in many of the promotional brochures for the Lyceum Bureau; Courtesy of the University of Kentucky. Background image ©2002 Art Today.

Manufactured in the United States of America

McFarland & Company, Inc., Publishers
  Box 611, Jefferson, North Carolina 28640
    www.mcfarlandpub.com

For Judy

There is a tide in the affairs of men
Which taken at the flood leads on to fortune;
Omitted, all the voyage of their life
Is bound in shallows and in miseries.
On such a full sea are we now afloat,
And we must take the current when it serves,
Or lose our ventures.

William Shakespeare
Julius Caesar

Acknowledgments

It is certainly not original to say so, but it bears repeating that no work of this sort is ever done by one person. Dozens of others contributed to this work — some without even knowing they were doing so. To all those friends who listened to those funny and unusual stories that surround the life of John Fox, Jr., and made their own critical comments and interpretations — sometimes without knowing there was anything more than a casual interest on my part behind the telling of those stories — a sincere thank you for listening.

No one can even consider the life of John Fox, Jr., without taking into account the collection of Fox family papers at the University of Kentucky's Special Collections and Archives Service Center in Lexington. Under the capable and professional guidance of Claire McCann, literally hundreds of letters and manuscripts — both published and unpublished — were made available to me, time and time again, and always with a smile.

Dozens of librarians at the University of Virginia, the University of Tennessee, the Kentucky Historical Society in Frankfort, the Filson Club in Louisville, the Tampa Historical Society, the Wise County Historical Society in Wise, Virginia, and Duncan Tavern in Paris, Kentucky, were all most gracious, dropping whatever they were doing to help me find what must have seemed to them the most insignificant of materials, and giving me hints as to where something else of interest might be found.

Special thanks must go to those individuals who went out of their way to be kind to a stranger, most of whom were met only through letters and phone calls: Bettie Tuttle of Lexington, Kentucky, a cousin of John Fox, Jr., Hope Meade of the June Tolliver House in Big Stone Gap, Virginia, and so many others. Special thanks are also due to Frank Cabot, a grandson of Currie Duke, who more than once pointed me in the right direction, sent copies of letters John had written to his grandmother, and generally did more than anyone had any right to expect him to do. I suspect I would still be wandering in the wilderness, searching for elusive facts, were it not for his help.

Acknowledgments

Mrs. Kyle Dean Fible-Epperson of Lexington, Kentucky, provided me with enough information about her kinsman Micajah Fible to allow me to finally bring that chapter of John's life to a close, although the fate of her ancestor and John Fox's best friend still remains a mystery to everyone.

Over a hundred years ago, John Fox, Jr., all aglow with his initial successes in writing, took to the stage, reading his works and those of other authors to a public that was starved for nearly any kind of entertainment. He had not spent that much time away from home, and was meeting people unlike those he had known before, in towns and cities far removed from his home in Stony Point, Kentucky, and Big Stone Gap, Virginia. Once, he wrote home to his mother from one of those faraway cities, that "everybody is kind to me out here." Having traveled many of the same roads that John did so long ago, I couldn't agree more.

Table of Contents

Acknowledgments   ix

Introduction   1

1. Early Spring—1916   9
2. A School Teacher's Son   12
3. Off to School   18
4. Boom and Bust, Part I   26
5. New York Reporter   35
6. A Tale of Two Cities   49
7. Boom and Bust, Part II: The Gap   58
8. A Knight of the Cumberland   76
9. Feuds and Romance   97
10. The Whirl-Wind   106
11. The Spanish-American War   119
12. The Horizontal Hail-Storm   139
13. The Road to Recovery   155
14. The Crook of the Shepherd   170
15. Cherry Blossom Correspondents   179
16. The Backward Trail   199
17. A Noble Profession   208
18. An Imperfect Union   218
19. The Eyes of a Father   237

20. West and East — North and South ... 247
21. Last Dance ... 260
22. Surviving ... 266

Afterword ... 269

Chronology ... 277

Notes ... 281

Bibliography ... 304

Index ... 311

Introduction

SOMEWHERE, AT SOME TIME, the research of another's life must come to an end either through exhaustion of material or through the realization that there will always be something to do. The realization dawns that it is not possible to know everything there is to know about another person, nor perhaps even desirable to know everything. Someone who researches another's life eventually reaches a point where he must stop and say enough is enough. There will always be details we cannot know, places seen, things said and thought by John Fox, Jr., and the great fear is that something of importance has been overlooked or forgotten, that every clue has not been sufficiently dealt with, but finally one must accept that no matter how long and how meticulous the search, there will always be something left to discover. That is both the attraction and the exasperation of such things.

John Fox, Jr., would become one of the first writers to use the mountains of southwestern Virginia and eastern Kentucky as a backdrop for stories and novels of a time and people that were fast approaching extinction of their culture, but it was not a profession that he chose quickly nor painlessly. He would be well into middle age before he made that decision to become a writer, and it was a decision that he struggled with, even after that decision had been made.

After graduating from Harvard, and under the protective wing of his half-brother James, John Fox, Jr., first tried to make his fortune in the hills around Jellico, Tennessee (or Kentucky depending upon which end of that small town one was on), and failed. Newspaper work in New York after graduation suited him no better, and ventures into his father's vocation of teaching bored him. It was only with the family's move to Big Stone Gap, Virginia, in 1890 that his prospects for getting rich became, he thought, a certain thing, but even in the midst of money, financial success carefully avoided him, and them.

The entire Fox family would move to Big Stone Gap, Virginia, in 1890, and although John is nearly always referred to as a Kentucky author, he spent most of his adult life in Virginia, and did most of his writing there in a little out-of-the-way spot called Big Stone Gap. His first acquaintance with Big Stone

Gap in the late 1880s was not a promising one, he would later relate. He and James had ridden in a wagon over roads that went from bad to worse for four days, leaving Jellico and going through Harlan, Kentucky, and across Black Mountain and on into Virginia. "We had been four days in that wagon," he wrote, "and when we finally emerged from the woods at the southern mouth of the gap, there was a beautiful little valley quite surrounded by lofty hills and from 10 to 20 feet above the level of the two streams that followed the circle of the hills. It was 'the city'— the Gap. There was a frame hotel, a blacksmith shop and a woolen mill and three or four farm-houses and that was all." [1]

The Gap was where he did his best writing, and, though not strictly the truth, he often said he could write nowhere else. An acquaintance of John's, still living in Paris, Kentucky, in 1975, remembered that he "felt he could write [only] in his own room at Big Stone Gap. He was a careful writer, she said, going over each line time and time again, often reading it aloud to other members of the family to get their impressions and to make certain it conveyed the picture he had in mind. He wrote a lot at night … but any hour of the day or night he would go into his father's room and consult with him. They would discuss the choice of words and phrases for hours at a time." [2] Once he had decided on writing as a career, he often wrote in bursts of activity and wasted time with no less enthusiasm, but even so, from the time he began writing seriously in 1892 until his death (still writing) in 1919, his outpouring of published material was prodigious and respectable. His novels— some admittedly too short to really deserve the name — numbered nine volumes, not counting *Following the Sun Flag*, a compilation of his Russo-Japanese War dispatches. His stories numbered 25, not counting 12 nonfiction sketches, several unpublished sketches, unpublished stories and rough drafts.

He married once, and rather badly, and that marriage ended in divorce just a little over four years later. Fritzi Scheff was an Austrian opera singer who had been married once before and would marry again after she and John divorced in 1913, but in the beginning they were happy but busy, perhaps too busy with her singing and his writing to really get to know each other well enough. When they did become better acquainted, it was not always for the best. They had no children, and after their separation and divorce, John gave no further serious thought, so far as we know, of marriage. It was too late, he reasoned, and he was too old to start that sort of thing over again.

He tried to be a true Southern gentleman, and although he slipped from time to time, he tried hard to live up to the image he had of himself, and envisioned himself to be. At times, this gentlemanly persona that he had adopted for himself seemed affected, though it is difficult, if that is true, to say for certain where reality of character lay and affectation began. "I remember an episode of his early visits to New York," wrote Thomas Nelson Page after John's death in 1919. "There was an entertainment one snowy night at the house of one of his acquaintances and Fox was invited. Among the belles of the occasion was a beautiful foreigner, to whom Fox was presented. When the entertainment broke up, this lady was shown to her carriage by a number of gallants, one or two of them men of distinction in New York society. As they stood about the door after handing her in, a young man with a 'Beg pardon,' stepped into the carriage, closed the door, and at the sound the horses pranced away through the snow. In great surprise one of the gallants on the sidewalk turned to the other: 'Who is that?'

"The answer was: 'John Fox, a young Kentuckian.'

"'Well,' said the other, 'by heaven! Fox knows his business.'

"The simple fact was that Fox, finding the lady unattended, had, according to the Southern custom, asked permission to see her home to her door." [3]

The incident that Page described was not rare nor peculiar in the life of John Fox, Jr., and so, characteristically, his dedication of *The Little Shepherd of Kingdom Come* in 1903 to Currie Duke — then Currie Mathews, and an old girlfriend of his — troubled him. It did not seem gentlemanly for him to dedicate the book — no matter his innocent intentions — to a married woman. Again, characteristically, he could not decide what to do, especially in light of the fact that he and Currie had a flirtatious relationship some years before, and that relationship had continued, apparently, right up until a few months before her marriage to her then present husband Wilbur Mathews. They had been close friends for years, and remained so after her marriage, and perhaps with the exception of his mother and sisters, and Tom Page's wife Florence Page, Currie Duke was his closest female friend.

John wrote Currie in July of 1903, telling her that "When I started the story in the spring of '98, I made up my mind that I would dedicate it, in memory of a life-long friendship. I imagine, though, that if I were a husband, I might seriously object to such a dedication, but if Wilbur is broader-minded than I should perhaps be, and doesn't object, I should let the dedication stand since it was then a pet purpose of mine." [4]

He really should not have worried so much about it; apparently neither Currie nor Wilbur did, but even with their permission, he was still a little uncomfortable about the whole thing. In a little over a week, however, he had his permission from Currie and Wilbur — as if there was any doubt about it — and by November his resolve to keep the dedication as planned, had stiffened. "I have been mildly questioned by one or two intimates," he wrote her near the first of November, "as to that dedication and my answer has been that it was the business of three people only — you, Wilbur and myself." [5]

His absent-mindedness was legendary. More than once, he misplaced (his euphemism for "lost") a manuscript, and either had to rewrite the whole thing or wait until someone found it and returned it to him. He almost did not graduate from Harvard simply because he had neglected to sign his name to some essential paper and had to be hunted up by his school mates who found him completely oblivious to his difficulty. His absent-mindedness occasionally reached a point that it became almost debilitating. "He used to find and give much pleasure," remembered Page, "in descanting on his inability to make a final decision or, at least, hold to it when made — about going to visit anywhere — and especially about leaving a place where he might be. He declared that he had stayed in a little hotel in Bardstown [Kentucky] once for a week because he could not summon the resolution to match his socks which the laundress had sent them back mismatched. And he rarely arrived without having lost his baggage or some part of it. I recall his arriving once and being met at the station, when his first words after his greeting were: 'Of course, I have lost my valise. But' (Cheerfully) 'the conductor will send it on. He knows me.' Just then the station-master, to whom the check for his trunk had been handed, returned with the information that his trunk was not on the train. 'That lost, too?' said he with a laugh. 'Well, thank God for that! Now I can stay as long as I like. I knew I would lose it; but was afraid I'd lose it going somewhere else.'" [6]

Page also related another time that the offending trunk found its way to England, leaving John somewhere, stranded but happy. The most insignificant of decisions might find him wanting, but all his friends knew it of him, and he said that should he suddenly become punctual, and more decisive, his friends might think something seriously wrong had happened to him, and worry about him unnecessarily. At any rate, he found it easier, if a little inconvenient, to remain the way he was.

He loved to laugh, and sing, and play the guitar and piano. He had once considered taking to the stage as an actor, and once took singing lessons toward another career that never materialized. He loved people, and traveling, and his family, and always had a special place in his heart for the numerous dogs that roamed the house and yard at the Gap. His family would often find him on a winter's night, sitting close to the fireplace with one of the dogs asleep in his lap, or lying close by on the floor, both the picture of contentment. Children also held a special affection for him though he had no children of his own. He often went out of his way to buy the children of his relatives presents to send home from New York or wherever he happened to be, and the one will that he wrote prior to his death made special mention of his brother Horace's adopted daughter to whom he wanted to leave a substantial amount of money, notably for her education. There was sadness in his life. He longed for the security and pleasure that only a family of his own could have given him, but it was not to be. The things he desired perhaps more than anything else — wife, home, children of his own — were denied him. Perhaps he always thought there would be time enough, and perhaps, too, as he grew steadily older, he tricked himself into believing there was time enough when, subconsciously, he already knew it was already too late. There is some suggestion, however, that the Fox family was too close to allow any but the most determined outsiders into their family circle, married or not.

After his failed marriage to Fritzi Scheff, he resigned himself to bachelorhood. By that time, he was nearly 50 years old, and any family he would have had earlier would have been nearly grown by then, and no doubt he thought of the circumstances that denied him those most important things.

The mountains of Kentucky and southwestern Virginia was his salvation in a life of uncertainty and disappointment. In writing about those mountains, he carved out a niche for himself and, in the process, brought to the forefront of the American consciousness a strange and fascinating people. "I never saw a mountain," he once wrote, "until I started for Harvard in my seventeenth year and through those I passed at night, but ... [I] ... cannot recall the time when the sound of the word 'mountaineer' and the look of it in print did not have a singular fascination for me." [7] It was in the writing of those stories of those mountains and mountain people that eventually brought him the fame he strived for all his life.

He was not born of those mountains and those mountain people of which he wrote, and neither did he gain any creditable information about those people and places from books. All his information that found its way into the pages of his books and stories came from first-hand gleaning of mountain customs and mannerisms. There is more than one tale told by himself and others of his staying overnight, or several days and nights with mountain families, dressed in their rough garb, speaking their careless language, and eating their plain and sometimes sparse food, all the while soaking in their dialect, speech patterns, and stories to be used later in his own stories that made him and his semi-fictional characters famous.

Family histories of Letcher, Perry, Leslie and other eastern Kentucky mountain counties are replete with stories of his staying with grandfathers and grandmothers, sometimes for days at a time. It is not rare to find families who "know" that Chad of *The Little Shepherd of Kingdom Come* is based on great-uncle Bob or that great-grandmother Lucy is the basis for the character of June in *The Trail of the Lonesome Pine*. The simple fact is that most of those family histories and legends may be closer to the truth than even the descendants of Bob and Lucy ever dreamed of.

In 1898, or thereabouts, he stayed with Solomon Frazier at Kingdom Come in Letcher County, Kentucky, gathering — or so family tradition goes — background for his *Little Shepherd of Kingdom Come* that would not be finished for some years to come. Years later, after the publication of that book, and also that of *Lonesome Pine*, he would return to the Frazier cabin and be startled by the resemblance of the Fraziers' daughter Allie to his characterization of Melissa in *The Little Shepherd*. A son who played the banjo after their dinner of fried pork and buttermilk he liked so well would seem to John to be more like Chad, another character in the novel, than he could have imagined. Those characters,

At the train station, probably in Lexington, Kentucky, and probably wondering where his luggage has gotten off to (courtesy University of Kentucky).

and others like them, he had created from his own imagination, but a good deal of those characters were taken bit by bit from real people that he had known or seen in the mountains years before. They were composites, and he knew that, but he was still a little startled when he met those composites face to face. It had an almost surreal quality about it.

Years later, in old age, Allie Frazier — who had been 16 years old when John had ridden into her father's yard in 1910 — was still a beautiful woman who remembered a small, bespectacled man astride a big red horse who hitched that horse to their porch railing and ate dinner with them. Even then, she knew he was famous, but womanlike, she tried not to show that she was impressed with him. But he was impressed with her. "Her throat was a proud column of alabaster," he would write of that meeting in 1910, "and her eyes were frank and straight-forward, big, and violet blue. Melissa I had never imagined lovelier than this mountain girl...." [8] And when Allie Frazier died on November 12, 1975, in Whitesburg, Kentucky, close by her beloved Kingdom Come, she had not forgotten John Fox, Jr., and the impression he made upon her, and she upon him.

From 1892, when his first novel *A Mountain Europa* was published, until 1919, and the unfinished *Erskine Dale — Pioneer*, he never abandoned those mountains and those people and their stories. His reputation as an author and chronicler of a vanishing part of Americana was made and sustained upon those hills and valleys and the "quaint" peoples and their customs, and he would not be able to bring himself to change. Even when twentieth century taste in literature supplanted the romantic meanderings of the nineteenth, he stubbornly continued to write mountain tales long after the demand had vanished for them, and his popularity with readers of fiction began to suffer.

He had his faults. He could become insanely jealous in an instant, and just as quickly, regain his composure as if nothing had happened. He once threatened to shoot an innocent admirer of his wife, and in a matter of minutes, all was forgotten and he was buying his potential victim a drink, as if they were long lost friends. His temper could get the best of him at times. Once, when he found that the editor of the Big Stone Gap *Post* planned to publish an article that John objected to for some reason, John marched down to the newspaper's office and threatened to thrash the editor if it was published. Apparently, the article never appeared.

Off and on for all his life, he was troubled with one ailment or another. He seemed to be almost constantly sick. What those ailments were is still something of a mystery, and one is tempted to suspect hypochondria. He had more than one operation in New York hospitals, but there is no evidence what those operations were for, except that they apparently were not too serious, since they did not involve extensive bed rest on his part. His teeth bothered him as did his eyes. Little or nothing he did was able to alleviate his ills for very long. Eventually, he began to doctor himself with popular home remedies of the day, which seemed to work just about as well as those he received when he placed himself under a doctor's care.

He could become obsessed about his age and would often avow that to deny, or to at least ignore, his age would prolong his life. Both his friend Tom Page and John's wife Fritzi commented upon his adamant decision not to discuss how old he was. He was, early in his life, careless in giving the exact date of his birth, and it varied in days and weeks and even years from his actual birth date. "As to the years," he told an Atlanta *Constitution* reporter in 1898, "I reckon them in blocks of five, and I have a theory that the world could be happier and people would live longer if nobody knew his age. Therefore I have forgotten how old I am. I know, however, that I am somewhere in the seventh block." [9] At his death, for better or worse, his reputation as an author had been made, and that reputation was on the wane. Tastes in literature had changed, and John and other romantic writers like him were unwilling to change, or perhaps they were unable to change. Some, like Thomas Nelson Page, stopped writing altogether and started other careers. Over the years, however, his books continued to sell, though modestly, and though from time to time there is a resurgence of interest in his writing, it does not last very long.

He was born in a time that was so different from the times in which he died. He saw war at its worst; he saw vigilante justice and public hangings; he saw airplanes when they were so new they flew only at Fourth of July celebrations. He knew presidents and governors and senators, and counted them as his friends, and they counted him as one of theirs. He knew famous authors, and he became one himself. He seemed to go everywhere and see everything.

For most of his life, he was uncertain of who he was and what his place in the world should be. He searched for a profession that would assure him a substantial place in society and allow him the financial freedom he needed to pursue the freespending lifestyle he loved. He searched for that avocation that was respected and noble, and in the end, perhaps he was successful at last. The road to that end was often trying on both himself and his family.

The times he lived in were like no others before or since. He once wrote that those "times were lively from the start," and so they were.

CHAPTER 1

Early Spring — 1916

IT WAS SPRING IN VIRGINIA and it reminded him of the Kentucky springs of his youth. The air was damp with rain and the promise of more rain, and he wondered if this time of year this season was the same no matter where a person happened to be. But this time, spring was different, he thought, and unlike any that he had known before, and with the way that time passed so fast these days, the coming of spring and summer did not have the attraction for him that those seasons once did.

But it was not only those everyday changes of spring that reminded him of the change that one season brings when it replaces another. He had learned that Richard Harding Davis, his old friend, was dead, and John Fox, Jr., could hardly believe it. They had been friends for more than 20 years. They had caroused together, traveled together, been with Roosevelt's Rough Riders in Cuba together, and had argued with, and was exasperated by, the Japanese in 1905. One had been an usher at the other's wedding. They had seen each other through divorce, and one remarriage. They had covered both popular and obscure wars together in strange countries populated by even stranger peoples, speaking languages that even Fox himself — the master of dialect — could not decipher. They both knew famous people, and were both famous themselves. They knew politicians and people of the plainest sort, and both were welcomed into the genteel and not so genteel society affairs of turn-of-the-century America. Both had been successful writers, but above all, and through all, they had been friends. Now one was dead. And for John Fox, Jr., spring had lost its luster.

Richard Harding Davis had been in Europe covering a new war, and Davis had no longer been young, but he had gone anyway. Davis's health and stamina, once stout enough for people to refer to him as "Richard the Lion Harding," had begun to fail. He had returned home in February 1916. He had been ill a few weeks previous, but seemed better. A few days later, though, his wife Bessy found him lying near his telephone where he had gone to call a friend.

With the exception of Thomas Nelson Page, Richard Harding Davis was John's best and most influential friend. Davis's death in 1916 affected John more than any other with the possible exception of John's father (courtesy University of Kentucky).

Fox would soon write of Davis, "He was so intensely alive that I cannot think of him as dead — and I do not. He is just away on another of those trips and it really seems queer that I shall not hear him tell about it." [1] It must have been the height of irony to John Fox, Jr., that a man so seemingly strong and healthy should die so soon while he, who had been ill off and on most of his adult life, should still live.

Fox, when receiving news of Davis's death, could not help but place him on a mental list of those who had gone on before. Madison Cawein, his poet friend from Louisville, who had once worked in a pool hall while writing poetry at the same time, had died in December almost three years ago, and Robert Burns Wilson, the artist, with whom he had stayed in Frankfort those bright summer days so long ago, had died only a month ago, in New York, too poor to afford a hospital room where he could peacefully die.

His brother James had been dead just over a year now, and although James was only a half-brother, John had never thought of him in that way. James was getting close to his tenth birthday when John Jr. was born in 1862, and John had sought out his brother's advice almost as much as his father's, and he missed them both terribly, and was often at a loss when he felt he needed advice. He remembered regretfully that in later years, he resented his brother's advice, and now wished he had been a little more understanding when he had told James to mind his own business.

The pain of his father's death still lingered four years after the senior Fox had died in June 1912. They had taken his father back home to Kentucky, and buried him in the family plot in the Paris Cemetery. John Jr. had some years before purchased a number of plots there, wanting them all to be together, even in death. John Jr. expected to join him there, and he had always made it very

1. Early Spring—1916

plain to his family that no matter where he was when he died, he was to be taken to that spot in Kentucky, on a low hill, and buried near his father. Now, thinking of friends and family who had passed, his thoughts turned also toward his boyhood home at Stony Point, Kentucky. It had always seemed like summer in those days when his brother James was still alive, and his mother and father were still young, younger than he was now.

The family had always been a close one, and John had depended, perhaps more than he knew, upon the support and advice of them all, and they, in turn, depended upon him, perhaps more than they should. Now, they had grown older, and had begun to leave him. He missed them, missed their guiding hands and encouragement — friends and family alike — and even though he knew that they were no more, he wished that it was not so.

Most of all, he wished for the times of long ago, when it seemed that summer and youth would never end. When bare feet burned on dirt roads, and the smell of hay and cut grass, and flowers and all the other smells and feelings that go with childhood and summers everywhere. He remembered the smells of the old home at Stony Point, his father's school in Paris, the smell of the mines near Jellico, the smell of the heat and sand and death of Cuba, and the smell of the mountains he loved so well that in the end he could not, and would not, leave.

He stopped and turned back toward town — the "Gap" as most, including himself, lovingly referred to it — and looked at it for a while. It had grown, not nearly as fast or large or rich as they had once thought, but he was more comfortable with it now the way it was. It was to be the new Pittsburgh, the new Detroit, the new whatever, but the expected and anticipated boom did not happen, and the hopes of many were dashed, even his. The crash that destroyed any chance for a real boom in the Gap nearly destroyed them as a family, but they survived, and he survived, and for a while learned to adapt with less. He loved the Gap, especially in spring when winter finally let go of its grip on the mountains, and the waters from thawing snow and ice flowed downward, washing everything clean again.

And as he turned toward the house, he wondered if spring had come to the old farm at Stony Point where he had been born, and where he had been happy.

Chapter 2

A School Teacher's Son

> Changeable, cloudy this evening with storms of snow. Minnie was complaining along during the evening ... and our son was born at 9½ P.M.
>
> John Fox, Sr.'s Diary,
> Tuesday, December 16, 1862

JOHN FOX, SR., WAS A LOCAL SCHOOLMASTER who had taken as his first wife his cousin Catherine Hill Rice, a native of nearby Bath County, Kentucky, who was two years younger than her new husband. Both husband and wife had lived relatively close to one another all their lives, living at the extreme eastern edge of the famed bluegrass region of Kentucky. They married on August 3, 1852, and the happy couple had set up housekeeping in Clark County, where their first son, James, was born on June 11, 1853. During their short-lived marriage, the couple moved several times, following John Sr.'s work as school teacher, which brought in little in the way of finances. Although their moves were perhaps more than common, those changes of homes were not usually of any great distance from one school assignment to another, at least in modern terms of distance.

They found themselves moving to Montgomery County, near Mt. Sterling, Kentucky, in the fall of 1854. They could not settle down to any one place for very long, however, since John Sr.'s duties as a roving school master moved him around a great deal, but in the first year or so of their marriage, they hardly had time to build a fire in the fireplace before packing up and moving again. [1]

He called her "Kitty," and when she died, he thought his whole world and future had died with her. They had three sons: James, Sidney and Everett, the last of whom was just shy of his third birthday when his mother was first stricken. Early in the morning of June 11, 1860, Monday, she arose not feeling very well, and found that she was bleeding. The bed clothes were soaked through. A frightened and worried John Fox called for the doctor who came and stayed all day and into the night, but the bleeding would not stop, though they slowed its pace from time to time. Kitty's mother came as quickly as she could from her Bath County home, and things seemed to take a turn for the better on Wednesday, and Kitty felt quite a bit better, she said, all day. The respite was short lived, however, and by night fall she was having cramps again and in "a good deal of pain." On Thursday morning, June 14, 1860, an exhausted and disoriented John Fox wrote in his diary, "Before this day closed, my poor children were motherless, and my happiness broken and destroyed. My poor wife rested well last night, and this morning ... she grew worse between 11 and 12 and after the most sever suffering, died at 11¼ P.M." [2]

The next day they took Catherine back to Bethel, Kentucky, back to her home in Bath County, and there, just north of the small town, with few in attendance, John Fox stood the thing he thought he could not stand. Later, he would write, "It seems as if to day I buried all my hopes in this world for the future." [3]

He was grieved and thought his soul would certainly die within him. He had always thought of himself as a strong man, but now he felt that no hand could guide nor sustain him when he needed guidance and assurance the most. He continued to grieve through the succeeding weeks and months after Kitty's death, but time and the three sons he had at home began to effect a sort of healing within him. He was never entirely the same afterwards, but he began to accept what he could not change, and his life — not the same, and for a few years, not as happy — went on.

He could not handle it all by himself. His boys were young and needed a mother, and when he asked his wife's sister, Carrie Rice, to help him with his children, she came, and he was grateful for her help. Forty years later, when Carrie Rice died in 1900, his gratitude showed through his entry into his diary that praised her unselfish devotion to himself and his children in those dark days after Kitty's death. All those years later, he did not forget that she helped him when he needed it the most.

Now, in 1862, the widowed school teacher married for the second and what was to be the final time to Minerva Worth Carr. Change had come to the household of John Fox. He had been alone for too long, and had done something that two years ago he would have sworn could not happen. He had fallen in love again, and felt young again, although he was anything but old. He hurried around the old house cleaning here and straightening there, putting down new carpet and trying in his hesitant way to make his home one that his new wife would be proud to make her own. But he did not neglect his precious diary, even on his wedding day. "A happy day indeed," he wrote for January 28, 1862, "for to-day I exchanged a life of lonesomeness for one of bliss and happiness, in leading Miss Minnie Carr to the altar, and plighting our vows to love each other." [4]

They spent their first night together in Paris, and traveled over the muddy roads in rainy weather to his father's home the next day. He was impressed that his new wife took the hardships of travel over Kentucky's notorious winter roads so well. The next day, snow delayed their homeward journey, and they spent a second night at his parents, but the weather cleared enough on January 31 for them to start home. "I think," he wrote, "I have the best and most affectionate wife in the world. She is an angel diffusing life and sunshine in a once lonely home, and making happy one of the most unhappy of men." [5]

Minerva was pleased with her new home and the preparations he had made for her. She now had a husband and three sons where a few days before she had neither. She was young and inexperienced in the ways of motherhood, but she did not shirk from her responsibilities, not in those first days and weeks, nor in the years that followed. Happily for them both, there would be many of those years to come. John Fox Sr. would continue to move, though certainly not as much as he did when married to Kitty and first establishing himself as a capable and trusted school teacher, and things would become more settled as his reputation for teaching grew right along with the size of his family. By the first of April, Minnie knew she was pregnant, and John, remembering the death of his first wife, was worried more than he showed. On the sixteenth of December, nine days before Christmas, a first son was born to a second marriage, and named for his father, John William, and the somewhat uniquely American appendage of Junior was added to differentiate between the two. [6]

Childbirth was never easy nor taken for granted in those times. In 1862 the cemeteries all over the country, not just Bourbon County, were holding too many children and mothers who did not survive the trauma of childbirth. There was a myriad of diseases and illnesses that could make the future of a newly born infant and a newly delivered mother quite uncertain, and many newborns were laid to rest alongside their mothers who did not survive either. John Fox, Sr., was pleased to have another son to go along with the three from his previous marriage and pleased that his new wife had survived. The now senior Fox could not be more pleased with the birth of his son, but it would not do for a school master to show too much enthusiasm, and he did not. In his diary for that date, he wrote almost as much of the weather as he did of his new son's birth. "Changeable, cloudy this evening with storms of snow," he wrote on December the 16th, "Minnie was complaining along during the evening, towards night sent for Mrs. Talbott and Dr. Ray and our son was born at 9½ P.M." [7] He did, however, close his school the next day so he could stay home with his wife and new son who were, he wrote, "doing very well." [8]

The successful birth ended months of worry and concern. When John Fox, Sr., wrote to his father, Boaz Fox, the day after the birth, everyone in that extended family breathed a collective sigh of relief. Boaz did not receive his son's letter for ten days, owing to the condition of the roads. They lived only 20 miles apart, but winter travel was extremely difficult, even for mail carriers, and people did not travel much during the winter months from November to March. In any event, Boaz Fox's return letter of the 27th of December echoed the relief of all the family members. "Let me congratulate you," he said, "on your good luck in regard to Mrs. Minerva as I feel much relieved myself, and would thank Providence, that your suspense is at an end." [9]

Soon, John Jr. would be joined by several additional siblings that included two sisters—Minerva, named for her mother and called "Minnie," and Elizabeth (Lizzie)—and four brothers: Rector, Horace, Oliver, and Richard. These new additions joined the eldest half-brothers James, Sidney and Everett, and the Fox household was soon in a delightful turmoil that only children can bring.

The elder Fox had moved considerably in his first years of following his profession, but now, his family having grown beyond his own imagination, he found it more convenient to stay in one place. At the small community of Stony Point just outside of Paris, Kentucky, where they now lived, he established his own private school, calling it the Stony Point Academy. The school acted as a boarding school as well as an institution of learning. Students who lived close enough to do so went home in the evenings, but there were always a few who stayed with the Foxes most of the school term. As in most schools of the time, the classics in literature, mathematics, and history were the subjects in which students were drilled. Expectations were high and discipline problems appeared to have been handled sternly and swiftly.

John Fox, Sr., was a handsome man, although not of imposing stature. He was somewhat slim of build, and certainly not tall by anyone's standards, with a thick shock of dark hair that did not turn gray until far into his senior years. His son John Jr. inherited many of his characteristics, although he tended to resemble his mother in outward appearances. John Jr.'s mother was a strikingly pretty woman, and that beauty lasted all her life. [10]

John Jr. loved his mother as he loved no other woman in his lifetime. His affection for her surpassed all others in his family, even his father for whom he had the utmost of respect and devotion. It was his father, however, that taught John Jr. and kept him and his brothers at their studies long after the regular students of his academy had gone home or to bed. Elizabeth Fox Moore, John Jr.'s youngest sister, later wrote that their father "was equally as strict if not more so with his own sons than with any of his other pupils. They often in later life alluded to this fact and also to his habit both at Stony Point and at Paris of keeping them at their studies around the dining table every night until each one could recite his next day's lessons, sometimes keeping them up until 2 or 3 o'clock in the morning." [11]

His father was strict with all the boys, and Elizabeth remembered in later years that their father often gave the boys whippings for not studying as hard as they might have. One day, says Elizabeth, her mother exasperated at a boyish prank or boyish insurrection, told her youngest son that his father did not whip him nor his brothers nearly enough. "'You should be down in the old woodhouse sometime,'" answered John Jr. [12]

Teaching was not exceptionally lucrative in those days, and though not as successful financially as it might have been, teaching was the financial mainstay of the Fox household, and they managed to maintain a comfortable, if tenuous, standard of living from it. There would never be, by any stretch of imagination, a surplus of any material thing in the Fox home, especially money, but they remained comfortable in their circumstances.

None of the Fox children suspected that they might not be rich, but for the time being, while John Jr. grew into his teenage years, the Fox household at Stony Point was a island of peace where the rest of the world, if not totally ignored, was held at arm's length, and the lack of those material things did not trouble them overly much. None of their neighbors had more, and many had less.

Like most teachers, the senior Fox found it necessary to supplement his meager salary with any extra work he could find. In addition to teaching, he was a surveyor, which in Kentucky offered its own brand of troubles with landowners and those who thought they were landowners but found out differently. As they grew older, the boys, including John Jr., worked alongside their father, in whatever venture he entered into to make a few extra dollars.

Perhaps one of the most uncommon outside activities that the Fox household employed to make additional monies was in the gathering of grass seed around the bluegrass. During the summer hiatus from school, John Sr. and the boys would travel, gather seeds from the fields, bag them up and sell them in Lexington. It was slow, tedious work, but not unpleasant. They were outdoors, and the labor was not overly strenuous, but the income from those ventures could not have been substantial. Money was tight, as always, and John Sr.'s letters reflect this and his loneliness while working away from home. On June 27, 1875, he wrote to Minnie from Georgetown, Kentucky, where he and the boys had gone to gather seed:

> Dear Minnie,
> I could not leave the boys to come home yesterday or to day, as we have just come to a place, and are boarding ourselves for a few days, and as much as I should like to see you all, I felt it my duty to remain here with the boys, besides I could not borrow a horse, and hate to hire one as we are at a heavy expense anyway.
> Some of us may come home in a few days, but if we do not, you must not feel disappointed." [13]

John Fox, Sr., was not by any definition an idle man. Even when not working at his school or gathering grass seeds in Georgetown, or surveying for his neighbors, he kept busy with a variety of interests that were the mark of an educated man, and one who never thought himself too old or too accomplished to forgo further learning. Neither did he neglect his community, being active in several organizations in and around Paris. His daughter Elizabeth wrote of him in 1947 that "he was also interested in various other activities. He collected botanical specimens and mounted and classified them and exhibited them at the Bourbon County fair; he also collected and exhibited geological specimens and Indian relics; he was one of the organizers of the Bourbon Historical Society [1879], and served as one of its Vice Presidents; he was one of the examiners of school teachers for the Paris Public School.... He studied when he could, rising at 5 for this purpose, and read much...." [14]

He kept diaries and recorded his biological and botanical findings around Stoner and Houston Creeks in the pages of those diaries. "The blossoming of the fruit trees, cherry and pear and apple," were noted in his diary. He watched the unfolding of the seasons. One October he wrote: "It looks much like fall since the cold and frost of last night. It is sad to see the change but there is no life without death, and spring returns." [15]

John Jr. learned from his father to appreciate learning and curiosity, but perhaps the most valuable lesson he learned was that a person's family can be the center that does not move in a person's life. As John Jr. moved from childhood into his teenage years, he knew that he had a place in the life of his family, and that his family was also a part of his life. Always, there was the knowledge that whatever he did, wherever he went, whatever he became, his family would be a net that would catch him when he fell, and a net through which he would not fall.

When John Fox, Jr., left his family and the agrarian community of Stony Point for the "big" city of Lexington and school at Transylvania University, he took with him the valuable lessons of his childhood that had been imparted to him, both inside and outside the school rooms of his father. The world stood ahead of him, and in his youthful exuberance he was prepared to offer it battle. With the unshakable confidence of his youth, John Jr. was soon off to school, carrying everything he owned in one bag, in one hand, and feeling older and more mature than he actually was, unconsciously feeling his own importance and a determination that there was something special that lay ahead of him.

He dove into his studies in Lexington and was determined to do well at the University. At least at first had little thought for anything else. On March 21, 1880, he wrote his brother Horace from Transylvania: "An education is everything. Give me an education and I ask a nickel of nobody, and that I am determined to get if patience and perseverance can accomplish anything." [16]

CHAPTER 3

Off to School

This is the most expensive place I ever saw.
John Fox, Jr.

WHEN JOHN FOX, JR., ENTERED COLLEGE that fall of 1878, Transylvania University (or Kentucky University as it was then known) was one of the premier institutions of learning in the United States and in Kentucky. Located now in Lexington, it is a small college with a well-established reputation for learning. At the time John Jr. enrolled, the curriculum was well founded in a classical education with the usual emphasis on Latin, Greek, and those who wrote in those languages. With the educational background he had gotten from his father, John Jr. no doubt felt quite at home there. John would often say that when he left college — both Transylvania and later Harvard — he knew more Latin and Greek than he knew English.

It was not easy. Those who enrolled were expected to already have a good foundation in the classics and related subjects. There was no place here for "catching up." The curriculum was basic, and yet it went far beyond simple "readin', writin' and arithmetic." The faculty was small and youth was a rarity among them.

There was the usual list of prohibitions for students (and probably the faculty as well) against "attending the races, theaters, circuses or barrooms, drinking intoxicating beverages, or keeping deadly weapons in the rooms." [1] When Sundays rolled around, all students were compelled to attend church services; there were no exceptions. There were, however, few serious discipline problems at the school. One of Transylvania's presidents, perhaps only half jokingly, said that "only two rules were necessary: no student shall burn down any of the college buildings, and no student should murder any member of the faculty." [2] For the most part, however, the student body was made up of those who were sometimes older and studying for the ministry, and although that group is not immune from pranks and horseplay, it is certainly easy to assume that they did not contribute much to the isolated incidences of rambunctiousness that did occur. All in all, with notable exceptions, lack of discipline was not a factor on the campus at Lexington. It seemed the perfect school for John Jr. to attend. [3]

John had wanted to go to Harvard from the very beginning, and had passed Harvard's entrance examinations in 1878, but the family's old nemesis of money, or rather the lack of it, had prevented him from going. Instead, he had enrolled at Transylvania. His sights, however, did not venture very far from Harvard, and his time on the campus in Lexington was spent in preparation for bigger things. James wrote to his father from Louisville in September 1879 that "John, Jr. has his studies at Lexington arranged and is doing good work. He will be well prepared to enter Harvard's sophomore class next September. John is a sharp boy, diligent student and I believe is the coming member of the family, for he will certainly make his mark." [4] They all had high hopes for him, even though he exasperated them from time to time with his constant requests for money. Still, they never relaxed in their belief that he would "make his mark."

This photograph of John Fox, Jr., was taken when he was 15 or 16 years old and a student at the old Kentucky University, now Transylvania University, in Lexington, Kentucky (courtesy of Duncan Tavern).

It was not, however, a free ride for John. Despite the temptation of not taking his studies seriously enough, knowing that he was destined for another school far away, John bore down and did not slacken from his family's admonition that he should study hard and learn as much as possible as quickly as possible. He wrote home to his father soon after arriving at Lexington: "I am studying harder now than I have done this year and am going to keep it up until June as I have a great deal to do." [5] College agreed with him. His health improved, which in and of itself was worth the trip. Never really robust, he had already begun a cycle of illnesses that would remain with him for the rest of his life. He now attributed his "splendid health" away from home and at college to the fact that he got plenty of exercise walking from one college building to another.

In 1880, John took the examinations for Harvard again, passed, bid farewell to Kentucky, and arrived in Cambridge on September 25, 1880. Three days later, he took the examinations for the sophomore class, passed, and was admitted at the wizened old age of seventeen.

Things seemed to be happening quickly for him now. He had entered Transylvania at the age of fifteen, and now two years later, found himself attending another, larger and more sophisticated school. Harvard was a completely new experience for him, as was all he saw on his way to Cambridge. He had passed through New York on his way, had seen Niagara Falls, toured the Hudson River on the *Albany*, was impressed, and had met his half-brother Sidney in New York where that brother had ventured into medical practice, walked forty blocks and had lunch, and with the supreme confidence of youth, determined that he "soon got the run of the streets and could go any where by myself." [6] He walked up and down the famous "Bowery" several times, and tourist that he was, he spent ten cents to see a trained pig, a Circassian girl, an albino, and other sideshows and was, in his own words, "completely taken in." "I walked down Broadway," he wrote, "and ran across the street several times to see if I could perform that feat without getting run over." [7] New York was already a busy, fascinating place.

John Jr. may have been away at school, but his heart and mind were not far distanced from Stony Point and his family. Letters flowed back and forth constantly describing the most mundane of occurrences, and he hungered for more. His Sundays were spent mostly writing letters home to his mother or father or brothers and sisters. It almost seemed a game to them, the rules of which depended upon who wrote last, and who owed whom the next letter, and pleadings to those who did not write often enough to try harder to write more. He and two other students managed to find themselves a room at $100 per month. He complained in a letter to his father soon after, "This is the cheapest board I can get. The regular boarding price is $12.00 a week." Echoing college students and parents of college students the world over, he said, "This is the most expensive place I ever saw." [8] But he suspected that he would be satisfied with his accommodations — price and all — once he had become settled in.

The complications of finding the right courses, the required courses, and in John's case, the courses he could afford to take, became a jumble of confusion to him, and being so far from home and his father's and James's guidance and advice, he found that for the first time he had to act on his own initiative and own judgment. That was a new experience for him and not a pleasant one. He was, all his life, indecisive. Making any decision, no matter how trivial, became something of a phobia with him early in his life, and the older he became, the worse his decision making skills became. He was notorious among his friends and acquaintances for being forgetful, and vacillating to the point that he could hardly make a decision of even the most ordinary of problems, but in the end he managed. The lack of money limited his social contacts. Attending parties and other events cost money, and there were several times he did not go because he did not have proper clothes to wear, or there was an admission cost that he lacked, or there was a price he felt he could not afford to pay.

Not all was study and classes, however. It was thought to be a good thing to exercise the body as well at the mind, and athletics and sports were considered important to a student's education. John loved sports and any kind of physical exertions. He was not, to be sure, of the physical stature to play "hard ball" (he only weighed 133 pounds when he first arrived at Cambridge, but had gained about 15 pounds by the end of November), but he took part whenever he could. Other times, he stood on the sidelines and cheered until his throat was sore and tight. And he exercised and grew stronger. Writing to James in October of 1880, he said that he was exercising every day and it was doing him "a great deal of good. My muscles," he said, "are getting larger and hard as a rock. Yesterday I jumped nearly 5 ft. high... Am now learning to turn somersaults in the air, handsprings and all sorts of didoes. The students play foot-ball, base-ball, lacrosse, cricket and lawn-tennis, and it is a fine sight to see Holmes' and Jarvis' fields covered with a hundred boys, dressing in various colored suits and playing the games above mentioned." His running war with finances again reared its ugly head when he continued the letter, saying: "As the ticket for admission was fifty cents I did not attend." [9]

A day or two later he was asked to attend a Republican political rally that would cost 25 cents, but he declined, he said, because he was a member of another party, but again, it may have been because of the money or a combination of both reasons. His health was better, but by no means was he always healthy. Catarrh, which may have been chronic sinus infections, plagued him from time to time and would continue to do so for years. He had seen a doctor in New York and was getting some relief, however. He mentioned in a letter to James that the climate in and around Cambridge and New York was "a fearful climate for catarrh." [10] But he was determined to remain as healthy as possible, and in a letter dated October 2, 1880, to his brother James he said he would rather be healthy than to maintain a high scholarly standard.

By late November, his illnesses hardly troubled him at all, and his headaches had virtually disappeared. The weather was turning quite cool, and he bought a used overcoat from a classmate for $10, and thought it a good bargain. It was not long before he became acclimated to the Cambridge weather as well as its social structure. He had at first thought that the students there had divided themselves into the "cliques" but was pleased to find out later that he had been mistaken. "There are very few swell heads in college at which I am much surprised and greatly pleased," he wrote after nearly three months at Harvard. [11] Many of the students there came from the best and richest families in the nation, and it is not unlikely that he felt somewhat ill at ease because of his lack of funds. Money continued to be a problem from time to time, and with the encouragement of his brother James, John set his sights on an academic scholarship, and was disappointed when he did not receive it. John passed his fall examinations. He stayed in Cambridge for Thanksgiving, the first he had missed away from home, and, because of the expense involved, neither did he go home for Christmas that year.

By April of 1881, he was well into his second semester, and was looking forward to seeing home for the first time in almost ten months. He was busy with preparations for his examinations, and expected to finish on the 23rd and start for home on the 25th. His only complaint with Harvard was the cooking, or rather the lack of good cooking in the North in general. "I shall be very glad," he wrote his mother, "to get some of your fried-chicken for I haven't had any since I left Ky. People don't live here as well as we do in Kentucky. They don't know how to cook." [12] Things being what they were financially, James sent him $50, and John replied the middle of May that he thought he could "land in Kentucky on $50 more." [13] By the middle of June, with examinations over and the money sent to him by James in hand, John started home at last. He was always glad to see Kentucky, but in his absence, he had forgotten what a dreary place small towns can be, and so, after the first few days at home, he found himself almost anxious for the new school year to begin. But for the time being, he could surround himself with family and what few friends still remained in Stony Point and nearby Paris and Lexington.

The old house at Stony Point was gone, having burned three years before in 1879 and the new one still had about it the smell of new paint and plaster. In time, he knew, that smell would be replaced by more familiar ones, and the house might eventually become a home. But deep inside, he knew that the place of his birth and youth would never be his home again, and he often wrote his friends at this time that Paris and Stony Point held nothing for him anymore—no friends, no jobs that would pay a living wage, nothing but his family to whom he remained strongly attached and devoted, and they to him.

It was especially hot that summer in Kentucky, and John's father suffered from the heat perhaps more that summer than any other before. John Sr. would not allow himself to let his outside work go if there was something to be done, no matter how dangerous or uncomfortable the weather happened to be. The elder Fox was often sick that summer, and when John Jr. arrived home from school in the company of James on the first day of July, their father was glad to see them both. John Jr. stayed home on the 2nd, enjoying the sights and smells of the old home place, but the next day he was off to Paris, and then to Lexington, and did not return until the 19th. He should have stayed away longer, for on the day after the 4th of July—a hot, sultry, humid day—his father introduced him and his brother Horace to the charms of forking hay into a barn loft. That evening, in the relative cool of the house, the elder Fox wrote in his ever-present diary that it had been "a very warm day. Have been putting hay in barn loft, and Johnnie and Horace both gave out from the heat." [14] The old man could still outwork his boys. It secretly pleased him, while at the same time it galled both James and John Jr. to death. By the first days of August, John Jr. was sick, and remained ill for nearly a week, but soon felt well enough to visit Paris with his father. When September finally rolled around, he was more than ready for the fall semester at Harvard to begin. He was tanned, had lost a little weight, and except for being sick for a week in August, he felt fine. On the morning of September 22, he was on his way back to Cambridge.

John's return to Harvard in 1881 did not have the confusion about it as did his first exposure to those hallowed halls. Now, he was a veteran, and fell into his studies like an old hand, and without the concerns that he had previously shown about his studies. His sophomore year had been one of some loneliness, which was to be expected, and even though he had made a few friends, he still imagined himself then as an outsider, and someone from the South who did not quite fit in. But now he was considered an old hand, not only by others at Harvard, but his opinion of himself was gradually undergoing a change, and he began to feel a new confidence. At times, he became unsure of himself, but usually managed to cope with his newfound independence. The one thing that school away from home did for him was to offer the realization that, for the first time, he could survive by his own wits, but not without a concentrated effort on his part.

In previous months at Harvard, he had boarded wherever he could find a room to let, and got plenty of exercise walking to his classes on campus. In the fall of 1881 and through his senior year, he lived in Thayer Hall on the college campus. Naturally, living in closer proximity to the other students brought him greater social opportunities. Always a lover of music, and something of an amateur musician himself (although he eventually had to sell his banjo for five dollars for pocket money), he now found himself among others with the same interests. He soon found himself acting in college dramas, and because of his smooth features and youthful appearance, even for a college student, he was cast in women's roles. He did well at acting, and for a while he even considered taking up the stage as a career, but the whim was not a lasting one.

In the meantime, however, he and his fellow actors at Harvard had taken a stage play on the road into New England. They visited towns like Portland, Bangor and Augusta, and although they made no money on their dramatic venture, the excitement of travel and the meeting of new people in new towns made the effort worthwhile. John loved it.

At any rate, the fall term came and went, as did the spring term and eventual final exams, and in the early summer of 1882 he was headed back to Ken-

3. Off to School

tucky once again. But this time he would encounter that which he had long been searching, though at the time he did not realize it, nor did he suspect that he was even searching for something. That summer, he would go to work with his brothers James and Horace in the mines near Jellico, Tennessee, and for the first time come face to face with the mountaineer of Kentucky, and would be fascinated and changed forever by what he saw and heard there. There he would be thrown up against a people who also held family and friends in deepest regard, but for education, they had none to speak of, and as for cultured conversation, they had never heard of Homer or Plato or algebra or Greek or Harvard. Their world ended at their doorstep, sometimes literally, and as far as culture, they hardly knew the meaning of the word, and John Fox would be taken in completely by the strangeness of all he saw and heard — of what seemed to be the simple life and charm of those mountain people.

The oldest of the Fox children, James would prove to be the most adventurous of them all when it came to money making opportunities, and seemed more willing to take financial chances than most of the other members of the Fox family. He was moderately successful, though never in the way he would have liked. He left teaching early and became a businessman, never straying from the pattern that he set for himself of making money.

In 1881, James had ventured into southeastern Kentucky near the little coal mining community of Jellico, Tennessee, that straddled the Kentucky border. Coal had been discovered, or perhaps rediscovered would be a better word. Transportation in and out of the Kentucky mountains was only of the roughest sort. There were few or no railroads, and as a result, little, including coal and lumber, could be transported out of the mountains. But the railroad was coming, and anyone with the foresight to get in quick was apt to make his fortune. Land could be bought for a pittance from the mountain people who owned it in those days.

James and Lexington businessman John Proctor, backed by their own money and that from any other investors they could persuade to finance them, jumped in and attempted to make a fortune. Before long, Horace Fox was in Jellico helping his brother, and in the summer of 1882, younger brother John Jr. joined them there. It would prove to be the experience of his life.

Opposite: John Jr. (right, with hand on hip) and friends likely dressed for one of the many Harvard theatricals in which he participated. Most of the time, because of his clean-shaven and youthful appearance, he was given female roles (courtesy University of Kentucky).

CHAPTER 4

Boom and Bust, Part I

> John works very well, but I think he doesn't like it extra well.
>
> Horace Fox

FOR THE FOX FAMILY, THE FIRST GREAT venture into land and coal speculation began near a small town in southeastern Kentucky straddling the border with Tennessee. Originally, the town was called Smithburg, but by the time John Fox, Jr., and his brothers lived there in expectation of making their fortunes it had changed its name to Jellico, and soon the town began to lose its innocence. The town itself lies on the border between Kentucky and Tennessee, largely buttressed up against an imaginary line, with most of the town situated south of the border, in Tennessee, and laid out in simple streets and narrow alleys. The main concern of the Fox family lay north of the border, solidly in Kentucky proper, but with strong Tennessee ties. Although there was a Jellico, Kentucky, for a short time in the 1930s, for the most part when one talked of Jellico, then or now, they were referring to the small Tennessee town that remains, in many ways, much today as it was in the late 1880s.

When the coal and timber boom of the 1880s hit, sleepy little communities all over the southeastern part of Kentucky and northeastern Tennessee, and on into western Virginia, were inundated with people the like of whom the local inhabitants had never seen before. Many who lived far back in the hills and hollows might assume that the world stopped somewhere just out of sight of their cabin. The influx of fortune seeking "furriners" was something of a culture shock for them. They were used to a lifestyle that avoided haste like the plague, and these nonworldly citizens of towns like Jellico and Big Stone Gap, Virginia, sat in their doorways and marveled that anyone would be in such a hurry about anything. They saw and heard strange looking people with even stranger accents, or completely foreign tongues altogether. But these new people were workers. You could certainly say that of them.

Almost overnight, it seemed, new buildings sprang up without even the hint of paint or whitewash to soften their harsh, rough exteriors. The smell of sawdust and raw, green wood filled the air, and the smell was not all that unpleasant. Things sure were changing, many a mountaineer remarked, and the changes were so fast they seldom thought such changes might not bode well for them. Mostly, the mountain inhabitants were left bewildered and not a little mystified at all that was happening around them. Middlesboro, Kentucky, is a good example of what happened to sleepy little towns all over Appalachia in the late 1880s and 1890s. "In May 1889, there were fifty inhabitants," wrote Maury Klein. "Four months later the L & N's extension reached the town, and by August 1890, the population reached 6,200 and soared rapidly toward 15,000. Watts Iron and Steel was busily constructing two large iron furnaces and had contracted for the construction of a steel plant. Another iron furnace was going up, and several New Yorkers were building a tannery. The Middlesboro Water Works and several other industries were also under way. Lacking adequate facilities, the men flocking to the boom town camped in tents while construction projects rose around them. An incredible mélange of humanity poured into its crude streets: Englishmen with impeccable Oxford accents, engineers, metallurgists, geologists, miners, farmers, lank, rifle-toting mountaineers, gamblers, bar keeps, hostellers, and the entire range of boom-town camp followers." [1]

There was money to be made, and the Fox family, principally James, were no exception to the fever. Early on, they saw the opportunities for making money in the area and soon began to concentrate most of their energies and interest toward that end. Unfortunately, their return of investment never grew substantially beyond those dreams, and the ups and downs of their financial situation troubled them all, and especially John Fox, Jr., for the rest of their lives. But for the time being, the smell of a fortune was in the clean mountain air of Kentucky and Tennessee, and it was not from want of trying on the part of the Fox family that the venture was something less than a complete success for them personally.

For John Fox, Jr., the rewards that did not involve riches would be more immediate and longer lasting than the fortunes they did not realize. For it was here that John first came into intimate contact with the mountaineer and their culture, and formed the idea of writing about those people and their circumstances. His and the rest of the family's sojourn in Jellico, and later Big Stone Gap, Virginia, however, came with a not insubstantial price. The debts that began during this period in their lives would haunt them almost until the end of their lives.

Mines began to open everywhere, and small mining communities grew up around those mines. Sometimes the best business was the money that could be made selling necessities to the miners and their families. The "company store" to which legendary John Henry owed his soul became an integral part of the mining communities. These mining camps or towns often took the name of their principal investor, or his wife, or possibly his girlfriend or sister, but mostly, they were just casual names for temporary towns. Towns with the names of Proctor, Red Ash, Gatliff, Bon Jellico, Vicco, and Wofford grew up almost overnight. There was money to be made, quick riches for those who were not afraid of hard work and responsibility. Those with money enough to invest could hire the work done for them and seemed to accumulate the most riches. But in the 1880s to the end of the century and beyond, there seemed to be enough money and financial risk for everyone.

By the time John Fox, Jr., first visited the area that would influence his writing so much, his older brother James was already well on his way to being established there. Land could then be had dirt cheap, and the timber, mostly virgin, that covered the hills and valleys of the mountaineer were of little value to their owners and could be bought for practically nothing. James and his partners were fast accumulating mineral rights to vast areas of land for what was even then pocket change. Soon, the enormous amounts of capital from English investors arrived, and small time entrepreneurs like the Fox brothers had little chance of making much more than a living wage, and that was something they were never satisfied with. If nothing else, the Fox brothers were ambitious, but sometimes ambition is not enough.

John's first visit to the area came while he was on summer vacation from Harvard in 1882. During the hot and notoriously humid Kentucky months of July and August, he worked, despite his infamous ill health, beside his brothers with pick and shovel in the heat and humidity that could take your breath away. Torrents of sweat poured from his body as he worked. Calluses grew on his hands. His face tanned, and he never felt better in his life. He did not like the work, but during the summer of 1882 he kept at it and did not complain, at least not much.

In his spare time, he walked the primitive roads and trails around the workings, where he met and talked with the strange people who stared at him with occasionally suspicious eyes and demeanor. But for all that, they were often friendly enough, and when they found he was a good listener, they opened up, and told him things he never forgot. At night, back in James's office or his own bedroom, he wrote down what he saw and kept notes of the people he had met. Perhaps he did not know it yet, but his writing career had begun.

The physical work was hard and the weather hot. Roads were being built to and from the mines, and with pick and shovel. Everything was hand labor. Concessions had to be made on account of the oppressive heat. The Victorian notion of "the more clothes the better," soon fell by the wayside, and the brothers were soon working without shirts in the hot sun. "John and I are working the Tram Road; get so much per cubic yard (20 cts)," wrote Horace to his brother Oliver in August 1882. "The road runs through a thick woods & we are in the shade all day. We don't work in any thing but our pants & shoes. We take off our shirts and hats. Our man works in his drawers." [2] The work was there, if the rain would stop long enough, and he wrote that he and John could "make $10 or $12 a week apiece" if the weather would only cooperate. [3] And he added, "John works very well, but I think he doesn't like it extra well." [4]

Two weeks later, Horace wrote his mother, and sensitive son that he was, neglected to mention that he and John were working, at least by those omnipresent Victorian standards of the day, practically naked. He did, however, ask that she send some old clothes if she had any lying about that might fit either him or John. "Rail roading," he said, "is hard on pants." [5] Horace wrote that he wished they could all move to Jellico so they could all be together, or if not that, could they come down for a visit before the summer was out? But the main thing he wanted right now was pants, and if his brother Oliver should come down for a visit, then he could bring what clothes they could spare with him. And, in a manner that was sure to assuage any concerns a mother might have, Horace wrote offhandedly, "Tell Ol [Oliver] to bring the Derringer with him. Everett says he put it in a pocket of some of the pants." [6] Near the end of August, John was on his way back to Paris, leaving his summer job in preparation for returning to Harvard for his senior year.

John Jr. had seen what hard, manual labor could be like during his summer vacation, and he was not overly fond of it, although he appreciated the effect that kind of daily exercise had upon his health. He had come into close contact with a "race" of people he had only heard of in vague rumors, and in bloodthirsty tales of feuds and feudists, and found them not at all like what he had imagined or had been told. But perhaps, a seed, a small one to be sure, had been planted, and one might wax a little philosophical and say that the fruit of that seed would stand him in good stead in years to come. But for the time being he was soon back to Cambridge and books and studies and school chums, and he was glad of it. There, among the leisure of student life, there was no one to ask him to build a road with practically his bare hands and bare back, and absolutely no work that involved a pick and shovel.

The lack of money again made its presence known once he was away from his family. Back at school he again had to provide for himself what he had taken for granted while at home. Food, clothing, shelter — everything seemed to cost more and more, and his funds grew less and less. His parents were in dire financial straits themselves and could not afford to send much, though his mother, concerned that he was not taking good care of himself and concerned that he lacked warm underclothes, offered to send what she could. [7] He applied to the Garth Fund, a local scholarship administered by Paris businessmen, and was successful in receiving a modest amount, though it would be some time before he would be able to repay the loan. And James, immersed in his own troubles in Tennessee, could not be as cooperative, money-wise, as he had been in the past.

All in all, things went as well as could be expected. He was little sick that fall, and soon adjusted himself again to living on limited finances. Before he knew it, 1882 was drawing to a close. He had not been able to go home for Thanksgiving nor Christmas again that year, and he missed that, but despite his protestations to the contrary, he did not really miss home all that much. Paris and Bourbon County, Kentucky, was a far cry from New York and Cambridge, and later, when he had to be at home for a lengthy stay because of sickness or to help his father at the Paris school, he was not hesitant in writing that the old town practically bored him to distraction.

At Harvard, he was surrounded by boys, young men really, of his own age, with all manner of schemes and plans for fun and games. There was always something afoot, and the absence of classes during Thanksgiving and Christmas gave them more time for mischief. And, of course, there were girls. He occasionally wrote to his brothers about the charms he had fallen for. His brothers were sympathetic, as was his father, and occasionally, they were even enthusiastic, but these dalliances seldom lasted from one letter to the next. His letters home or to his brothers seldom mentioned the "fun" he was having, and when they did, his part was convincingly played down, and one suspects they were given in third person to obscure John's own part in those festivities. Fellow Kentuckian and friend, Micajah Fible of Louisville, wrote his sister just after Christmas in 1882 that "all the Kentucky boys—Berryman, Drane, Fox & Davis are here, & Christmas is by no means not lively. I had expected, in fact had hoped, to spend the time quietly. But it is impossible. I am overrun with callers who have all sorts of schemes for fun on hand." [8]

"At college," wrote Charles Wingate in 1897, "he was one of the most delightful companions a man could have. He had more friends than mere acquaintances, for the latter very quickly joined the ranks of the former. He was not a man who sought prominence in his class," continued Wingate, "In his quiet, easy way he went along through the college course, making things pleasant for those around him without attempting to extend his sphere. I do not think that he had any idea, at that time, of taking up literature as a profession. He was a very clever amateur actor, and on account of his good-looking smooth face was called upon to interpret women's characters in the college theatricals of one of the chief societies. Before me now lies a picture showing him in quaint, old-fashioned woman's garb, with odd little ringlets hanging down all over his head, and a most absurd bonnet perched upon the top." [9]

The fun, however, would soon be over. June and the end of his senior year were fast approaching, and he came face to face with the prospect of leaving school without the slightest idea of what he would do. Every important decision of his life had been determined by someone else, up to that time, at least, and so his indecision in the latter months he spent at Harvard was natural. He simply did not know what to do with himself, and there was no one who could now advise him. He gave quite a bit of thought to entering Columbia Law School; the sons of anyone who was *anyone* always went to law school, and not only sons of the South. John's principal hope was that he would receive a scholarship that would enable him to do that very thing. His friend from Harvard, Micajah Fible, had determined to become a lawyer himself, and Fible would be successful, though that success would be short lived. The profession of law attracted John greatly. He would be respected, have a good paying career, be his own boss, and perhaps most important for him, have plenty of leisure time to pursue whatever caught his fancy on any particular day. The work could not be that hard, he reasoned, and he thought it would be something in which he could be a success.

He had his misgivings and his enthusiasm was not overwhelming. He wrote his mother in February 1883, "I suppose I shall soon begin to look around for something to do after I get out of school. I shall try to get into some lawyer's office in New York." [10] He had not given up the idea that he might try acting as a career, even though his only experience had been in a less than successful rendition of *Papa Perrichon* with his college pals. His brother Everett, wanting or needing to be helpful to his indecisive brother, wrote him a rare letter in May of 1883: "Jim tells me you have a fondness for the stage. Now let me advise you a little. If you have any talent that way, follow it. Make something out of it, let Law go to hell. There is more money in the 'profession' than in law... It is a sorry time I'll tell you for a young fellow starting into a profession, (law or medicine), without no funds, & a dam few friends." His brother went on to say that money was everything, and that he should find a job that paid a lot of it. Nothing else was of any great importance, Everett said. "Let me tell you," Everett continued, warming to his subject, "money is what everybody wants. You are aware of the fact that a fellow that has no money 'ain't much'. Money is what 'does things'. A fellow without money ain't worth a dam. He had as well be in hell." [11] Everett's advice was certainly questionable, but James's advice was hardly any better: He suggested that John consider a career teaching athletics in a gymnasium while he studied law. "I think," he wrote, "that avenue is not crowded like others: the pay would probably be ample, if too much time was not taken from your studies, and law duties next fall." [12]

Of course, James may have had other motives in suggesting his younger brother find a job to help support his way through law school. James had subsidized John Jr.'s academics and personal expenses ever since John had left home for Transylvania University, and then Harvard. The strain, both mentally and financially, was beginning to show, and to be truthful, John was never very backward about asking for money. On May 17, 1883, James wrote from Knoxville about John's constant request for more money: "I sent you," he began, "a week ago $40 P.O. order for a suit, etc. I hardly understand from your letter if you want money for other purposes, or, if so, how much. You must go slow on expenses; very slow." He also asked that John send him an "itemized list" of his needs, and promised to send the money once the list was in his hand. [13]

John's indecision showed through in nearly all his letters home and to his brothers. One letter would mention law, and then another would mention the stage, and another teaching or tutoring, and then back to law again. Soon he would become almost frantic about trying to choose what he wanted to do the rest of his life. He had not thought much about it at all until now, and he was beginning to show signs of panic. Throughout his life, anytime he was hurt, sick or uncertain of what to do, his eyes and his yearning turned toward home. "I want to come home very much this summer," he wrote his mother near the end of his senior year, "but I don't know for certain that I can. You see that I must strike out for myself as soon as I graduate and if I can get a position in New York I may take it immediately. I shall always come home every year at least if I possibly can." [14]

His mother would not be put off so easily, however, and wrote him about the middle of May that she knew he would be glad to finish college. It had been hard on him, and she worried about him getting too tired with his examinations and entering law school in the fall without getting enough rest. "We all want you to come home this summer," she wrote. "It is too hard for you to go into a law office before fall. So come home and rest up, eat peaches and fried chicken." [15] James added his own incentive by writing that he had been in Lexington, and he was not sure if it was because he had been in the mountains too long, but he did not think he had ever seen so many pretty girls in his life, and that "It really does a fellow good to gaze." [16] True, they all had advice for him to follow, but in the end, he knew it would be his choice and his choice alone. He could, he knew, be able to count on the family's support. That certainly was not the problem. The problem was, he had to decide, and soon. The end of the school year was upon him almost before he realized what was happening. The hustle and bustle of last minute studies, testing, packing for home and the realization that he would be seeing some of his Harvard friends for the last time was something of a shock to him.

The end came in June, and he came very near to not graduating. He had neglected to sign his name to some relatively important document, and had simply forgotten it. A group of his classmates finally hunted him down, and found him completely oblivious to his predicament. The paper was signed, of course, and his graduation from Harvard—cum laude, no less—went off without a hitch. It was, however, a disappointment that none of his family was able to attend the ceremonies. He sat alone, although dozens of other graduates and their parents and family sat around him. His school mates were distracted with their families and friends who had come to see the great day, but on this one day when he needed his family the most, there was no one. His father felt bad about it, and wrote on June 24 that the whole family was glad he had finished his "long and arduous course of study at Harvard, and I regretted very much indeed that none of us could avail ourselves of your kind invitation to be present." [17] The letter from his father did not do a great deal to salve his

The Harvard graduate. His forgetfulness was already becoming an accepted part of his life, and because of a paper that went unsigned until the last minute, he almost did not graduate (courtesy University of Kentucky).

hurt feelings. Unconsciously, no doubt, John Sr. helped none at all when he mentioned it no more in his letter, but wrote another page and a half about gathering grass seed, the progress of the railroad in reaching Paris, and the possibility of John Jr. helping him in his surveying that summer. John Jr. felt himself cast adrift, a long way from home, and in his youth he could be forgiven if he felt a little lonesome and sorry for himself.

After graduation was over, he did not immediately go home, nor did he go anywhere else. He remained at Cambridge, still unsure of what he should do. He wrote in his diary: "I am now brought face to face with what I am to do and how I am to do it. Away with all nonsense now. I must begin to look at myself carefully and direct my life with a better purpose than heretofore." [18] He was giving some thought to writing, though he seldom mentioned this in his letters to his family, fearing, perhaps, what they might think of his chances of success should he choose such an uncertain career. No doubt, he had his own misgivings about it. He confessed that he did not want to do just anything, nor be tied to a profession that offered him no excitement or satisfaction. "I want to be an unprejudiced observer," he said of a writing career, "capable of receiving a pure impression from any source and recording it truthfully. Now I am getting material which will assist me if I ever may write anything." [19] And, he was having second thoughts about becoming a lawyer, concerned with dry briefs and shuffling papers all day, not to mention being trapped inside an office. "I wonder if life in a lawyer's office will not make me too practical; grind down my thoughts to one channel, and a commonplace one, too?" [20] But, he resolved to try. In the meantime, law school did not start until the fall, and he needed something to do until he decided whether or not to apply for admission. He went to Brooklyn, New York, arriving there on July 21 by boat in search of a job.

He was sorely disappointed in what he found, or rather, what he could not find. He could not find a tutoring job. A promised position in a law office did not materialize. The city seemed bigger than he remembered and more hostile now that he was looking for something from it rather than being an occasional tourist who would accept anything the city offered. Soon he became miserable and dejected. "My dreams are passing away," he wrote in his diary soon after arriving in New York, "Verily, day dreams must yield to the demands of the needy stomach. Came here with almost childlike trust in Providence. That childlike trust is passing away." [21]

He did not, however, have long to search for something to do. On July 9, he began work as a reporter for the New York *Sun*, and was fascinated with his new job. He reported on runaways, strikes, women who took poison thinking their husbands were unfaithful, biting dogs, sweat-shops, domestic disputes, and all manner of strange and unusual — at least for him — happenings of the big city, and the newspaper printed them all. But it was a lonely life. He missed his school chums, and the periods of inactivity between racing for the "story" grated on his nerves. But it paid him $15 per week, and was his first "real" job.

Back home, his father was enthusiastic, and so were the other members of the family. "It affords me great pleasure indeed," his father wrote him on June 25, 1883, "to hear of your success in Journalism in which all the family share with me, and your friends here in Paris, too… All think your Salary alone is wonderful for a beginner, to say nothing of being on so celebrated a paper as the Sun." [22] His father encouraged him to stay with journalism, and to forget law, if it so suited him.

He fell into his new job with a vengeance and learned quickly the ways of a New York reporter. The job was new, and for a while everything about it absorbed his interest, and he learned quickly. He would also soon see that he was embarking upon the loneliest time of his life.

CHAPTER 5

New York Reporter

> I imagined I was quite brave.
>
> John Fox, Jr.

 NEW YORK DREW JOHN FOX, JR., ALL HIS LIFE. Despite his sometime loneliness there, once he saw it and all the lights and excitement the city had — even in the late nineteenth century — he was hooked for life. He loved Kentucky and in later years had the same affection for Virginia, but the great city of the North had a special attraction for him. That fascination never wavered.

After John's graduation from Harvard, there was no other place he really cared to begin his career than the beloved and much maligned city of New York. His father often asked him to come home to Paris and find something to do there, but invariably John refused. Even now, while working at the *Sun*, and despite all the encouragement he received from home and others, he still was not fixed upon a permanent career. It was still the law that he was half-heartedly determined to study. His choice of a career remained unfixed in his mind, and his determination wavered from day to day, from one career choice to another.

His father and mother read between the lines of his letters home and wanted him to come back home to Paris, Kentucky, and teach school, and study law in his spare time but he continued to decline their offers. Despite the advice he received from the other members of his family and his friends, John seemed determined to follow his brother James's advice that he enter law school in the fall of 1883.

He would work at the *Sun* during the summer and enter Columbia Law School in the fall. It was a tentative plan at best, and certainly had no permanent connotations about it, but reasoning that any plan was better than no plan at all, John embarked upon his career as a journalist, believing it was only for a short time, and only for the money he would earn that summer between his graduation from Harvard and his entrance into Columbia Law School in the fall. For the time being, however, he found himself in the middle of the exciting world of a nineteenth century newspaperman. The excitement of a new job and new responsibilities drew him, and the freewheeling methods of newspapers and reporters of the time excited him. It was a profession that held many attractions for him, and despite the hours and conditions under which he was required to work and live, he found that he was beginning to enjoy it more than he had thought that he would. But like most things, the life of a reporter looked attractive, exciting, and wonderful from the outside looking in. On the inside looking out, one could have an entirely different perspective.

He had arrived in New York full of ambition and enthusiasm that spring after graduation, but those initial feelings swiftly descended into discouragement and despair. He wrote to one of his college pals soon after arriving that he had a "pretty hard time" after getting to the city. "I had a series of disappointments at first...," he wrote, "Was discouraged at every point but I persisted and after tramping all over New York on Saturday a week ago I got a position on the 'Daily Sun' at $15 per week and here I am. The work is as you know hard and my hours extend anywhere from 11 A.M. until 3 A.M. next morning, but I like it. The excitement in it makes it very fascinating to me. If I succeed well, I shall take to journalism ... and let law go to the devil... Gad! its dangerous as well as exciting to rush around here in the worst part of New York after mid-night. One night I wandered around in dives and beer-saloons from 12 to 1, then climbed up a pair of dark, rickety stairs and pounded up the proprietor of an old tenement house, to find out where a noted thief and murderer lived. Gad! What a commotion I stirred up! Broken English, children's cries and shrill female tones, formed a harmonious chord. 'Dunder und Blitzen! vat is dat?' the old Deutcher yelled. 'Only a reporter', I answered. Presently he opened the door and there was a scowling little Dutchman, holding up his breeches with one hand, his huge mountain of a wife with a shawl disclosing her massive upper half, and the dirty shirt tails of some bare-legged children just escaping through a side door. The little fellow was pretty good natured after all and I got my news. I got away as soon as possible, though I imagined I was quite brave." Later, in the same letter, in a revealing passage, John wrote, "It makes me homesick to think about going to Ky." [1]

It was a time of wild, almost uncontrolled journalism with stories concocted of less than half truths, and occasional outright fabrications despite the assertion of one New York newspaper publisher who said that truth was to a newspaper as virtue was to a woman. That in and of itself, coupled with what is generally accepted about nineteenth century newspapers, offers an interesting conclusion about the status of virtue, at least in New York City. A reporter was duty bound to make his stories interesting. That was the bottom line, and nothing else would satisfy the editors, and if a little poetic license needed to be used, then no one objected too much. The goal of the editors was to make their papers interesting at all costs. A dull paper did not sell, and most of the reporters were only too happy to oblige their editors' wishes. "The revenge or suicide of an outraged young woman would be lovingly detailed. Brawls among immigrants, fires in tenements, freaks of nature, fits of madness—the [newspapers] played up the weird, the piteous, the startling." [2] It was, in effect, a war among all the papers with the prize being the largest circulation possible.

The reporters for the *Sun*, including John Fox, worked long hours for their $15 a week. Some even brought cots to open up at their desks in the hope they might get some sleep from time to time. It was a cutthroat business, and every reporter was expected to work ten to twelve or more hours a day, and be on call 24 hours should they be needed to cover a "fast breaking" story. Reporters who would not or could not keep up did not last very long in the newspaper business. There was just too much competition between rival publishers to retain any dead weight among their staff of reporters. It was a nerve-wracking and nerve-shattering business. John was, despite it all, a very good reporter, but his casual nature about work and his uncertain and sporadic health forbade him from ever thriving upon the hectic schedules and deadlines that are the twin sisters of a newspaperman's life.

At the *Sun*, a reporter could make enough money to live on. Rural newspapers and those of smaller cities paid next to nothing for a reporter's efforts, but the New York papers paid handsomely, at least for those times. "The salaries of first-rate editors and reporters in New York almost doubled in fifteen years. At least twenty reporters were earning $3,000 to $5,000 a year; one managing editor was paid $12,000, and five or more others were topping $6,000. No wonder that young men and women from all over the country applied to New York newspapers for jobs." [3] The New York *Sun* was an excellent place for a young reporter to start his career and learn his trade. The paper itself was widely regarded as being one of the best written papers in New York. Some disagreed, especially those whose reputations were pierced by the paper's keen edged writing and editorials. Those who had the societal power in New York, and had been gouged and embarrassed by the paper's editorials, succeeded in having the paper banned from some of the reading rooms of "respectable clubs," claiming that only lower sorts would deign to read such garbage as the paper published. As uncomfortable as it might be for some, the truth was that more people of all classes were reading the *Sun* than the high society types cared to admit. As a newspaper it was definitely a success, and John Fox did well to acquire a position there.

As time went on, however, John grew less and less enamored of his new profession. He was associating with the lowest forms—or so he thought—of people: murderers, robbers, prostitutes, pick-pockets, wife beaters, drunks, and policemen who were all too ready to use their clubs on criminals and reporters alike. None of his life previous to the summer of 1883 had prepared him for this. He was unaware, in his 20-year-old innocence, that such a world existed. His letters home soon began to reflect his doldrums that he could not completely hide from others. His father, seeing much of this in the letters he received from his son, again proposed that he return home and study law with an as yet unnamed local attorney. John Sr. was in the process of closing his Stony Point Academy and opening a new school in Paris, and he may have thought that his New York reporter son might be able to lend a hand teaching in the new school, but he was careful not to broach the subject too hastily. "You could be studying law," his father wrote on July 18, 1883, "with some of the Paris lawyers." He tempered his suggestion with an apology for interfering but could not resist a small dig at the newspaper profession. "Now, Johnnie," he continued, "I do not want to embarrass you about what it would be best to do by proposing this, or wish it to interfere in the least with any thing you are already doing, or may have in contemplation, but propose it in lieu of something better." [4]

He need not have bothered too much trying to convince John Jr. to return home to Paris to work and live. If John did not especially care for his present circumstances, he liked the thoughts of returning home to a more uncertain future even less. His old indecision about what to do was back again, if it had ever left. "As for me I know not what I want," he wrote in his diary, "I do not lean toward any profession. Sometimes I feel that I should like to dash out into the world with no ties behind me and let Fate carry me where she will. [But] I know that then I should not be happy." [5] In a letter to his old Harvard schoolmate, Micajah Fible, who would soon become a lawyer himself, John listed the reasons for leaving the *Sun* and taking up the law as a profession. But even as he listed the pros and cons for his friend, old doubts began to pick away at him.

In the meantime, unknown to him, his father was already making plans for a teaching profession for his son, trying to obtain a position teaching Greek at the University of Missouri, but the plan fell through when another was hired to fill the position. The senior Fox would not be put off entirely, however. In a letter to his oldest son James, an idea had "suddenly" occurred to him: "[I]t occurred to me that if Johnnie did not realize his expectations in N.Y. it would be a good plan for him and me to open a school together, as he is fresh from college, and has a great popularity about Paris, provided he would like to teach…Johnnie could be studying too with some one of the Paris lawyers. This seems to me the best thing that presents itself…" [6]

James was not as enthusiastic as his father about John Jr. returning to Paris. It was a dull little town, lacking entirely the prospects of New York. "There is no need to seriously consider the Paris proposition," he wrote to his homesick and confused brother from the mines in Jellico, "although I see it announced the father will be there next year. Why? Some less able man will for $600 or $800 fill the same place about as well. I don't think a young fellow with strong ambition and life before him would want to be in such a place even for a year or two. You want to be where your surroundings stimulate you to effort and where contact with abler men calls out the best in you. Paris means mere existence, stagnation." [7]

His father continued to pick away at John Jr.'s already weakened self-confidence. If he had no doubts about writing and journalism as a career, it only took a few lines from his father to crash any lofty ideals he had about the profession. "The popular taste and morals of the millions are not high," his father wrote, "and the newspaper writer finds it easier to descend to their level to please, than to lift them to his own plane to instruct." But, he said he knew that John knew his own mind more than anyone else, and that whatever he decided to do, it would be the right thing for him. He encouraged him not to be too hasty, however, as he was still young with his whole life ahead of him, and something proper and acceptable would soon come his way. [8]

In the end it was just too much for the confused budding reporter. He decided not to decide. He would, he thought, stay with the newspaper business, tutor if he could get students, and study law at the same time. It was a recipe for disaster, but he was too much like his father — making plans that looked and sounded so good on paper — to realize that some things just cannot work.

He wrote his friend Micajah Fible in Louisville, who had advised him otherwise, saying "I am going to Columbia Law School. I can teach and write and thus keep myself alive in the meanwhile. If I have a good opening in the law, I shall take it, if not, then I shall be where I am now with the advantage of a legal education. Besides many a fellow decides on his professions after he is 22 years old and I shall have the professions reduced to law and Journalism and it is not impossible to combine both to a certain degree." [9] John's father, surprisingly, thought it a capital plan, and encouraged him to stay in New York by all means, although he could not help but mention that he could take on more students at his new school in Paris if only he had an assistant to help him. Had John Jr. known what was to happen in regards to his father and his father's school, and the effect it would have on him in a few months' time, John might have studied harder at Columbia, for it would not be long before he was back in Paris, teaching in his father's school, and the words of his brother James would become prophetic.

John's stint at the *Sun* during the summer of 1883 was soon over, and in no way had he ever seriously intended his summer employment there to be permanent. In many ways, he had just been marking time until he could enter Columbia in the fall, although he was certainly indecisive about any career as a lawyer as well. Journalism was interesting work, although at times it became drudgery to him. Surprisingly, considering how well he performed at the New York *Sun*, and still later at the New York *Times*, he never seemed to consider seriously a permanent career as a newspaper reporter. Seemingly, it never really occurred to him that journalism might be where he belonged and was best suited for. He even considered joining the United States Civil Service Commission to get away from journalism, though it is uncertain whether he took the examination or not. And so, when the fall term at Columbia began, he was there, but only half-heartedly. His father continued to encourage him to do whatever he desired, but could not help but be more admiring of the profession of law. "I hope you will like the study of the law better as you get into its mysteries. It is a noble profession and worthy of the efforts of mastering and I trust you will like it well." [10]

John was not a chronic complainer, but his letters home and to his friend began to more and more mention his lack of money. It bothered him a great deal, and the experience of eating meals without end of thin soup and crackers because he could not afford anything more substantial never left him, and he would write about it years later. After graduating from Harvard, he would write, "It was not financially convenient for me to go home, and so before I should enter the Columbia Law School, I spent the vacation time as a reporter on the New York *Sun*" [11]. Of his time with the *Sun*, Fox had little to say except that, "As a reporter I was a failure — I hadn't the 'news sense'." [12] But he underestimated his worth as a reporter. He was a good reporter, as his later dispatches from the Spanish-American War and the Russo-Japanese War show. He had a good eye for detail, and when John finally left the *Sun* that fall of 1883 for Columbia, the night city editor of that paper praised him simply by saying "Fox can write." [13]

At any rate, Fox left the *Sun* and ventured into Columbia Law School which proved to be a relatively short-lived experience for him, and apparently not a pleasant one. He related that the only question he was asked in law school was one about money, and his answer did not satisfy his professor, who asked him, "Surely you know what a bank account is, don't you?" Without having to think about it, John answered, "Not from personal experience, Sir." [14]

In the meantime, he was living in New York with a family named Ryer now, and tutoring their children. [15] He had quit his job at the *Sun*, and was now entirely dependent upon any monies he would receive from his tutoring position, which was becoming tenuous at best. So far, his plan to juggle three separate professions had resulted in the temporary, as it turned out, loss of journalism, the entry into a law school that would not last as long as his journalistic venture, and the tutoring of the sons of rich people which had already, after only a few weeks, begun to look less and less attractive. The situation with the Ryer family was not working out very well. Its beginnings were auspicious, but quickly degenerated. He was not at ease in the household, and when his mother offered to send him a turkey for his birthday, he declined, being fearful that Mrs. Ryer might think he had written home saying he was not getting enough to eat.

He mentioned his situation in the Ryer household to his brother James in November, but said only that "unpleasant parts develop gradually, however, which may make my delicate footing in the household still more unstable than it now is. I hardly feel I have a right to explain more so I will reserve it until I see you in New York." [16] A few days before Christmas he left the Ryer home, although the details of his leaving are sketchy at best. His father wrote on December 23, 1883: "We have your letter stating the circumstances that led up to your leaving the Dr. Ryer family and think that you did right and that there was no other way. It was an unkind act since you had given up your place as a reporter on the 'N.Y. Sun' to go to their home... Keep up a brave heart, be not discouraged as it is a great world where there is much to do." [17]

But the matter of law, and whether or not he should stay with it was still foremost in his mind, even as he bade farewell to the school that December. Columbia had not suited him. Never completely enthusiastic about law as a career in the first place, he had barely been able to stick out his first semester there. His diary entry a few days after Christmas illustrated the usual dejection he felt when things did not work out the way he planned or thought they should. Christmas was an especially lonesome time for him, being away from home as he was. He had not been home for that holiday for nearly six years—since he was 15 and off to school at the old Kentucky University—and it did not look like he would be there this Christmas either, though his mother wished he would.

"Sunday last," he wrote in his diary, "I was 21. The day on which I became a man was different from what I expected. It found me worse off than when I graduated nearly six months ago—without money, without a position but, thank God, not without friends." [18]

One of those friends was Buckner Allen, with whom he visited several times each week. Allen was from Kentucky, and he and John had many things in common that they could talk about for hours on end. John grasped at his friends from home like a drowning man at straws. He was homesick most of the time now, and anyone he could call a friend was fair game for him in his attempt to alleviate that homesickness. He and Allen had started writing a play in November together. For the first time in several weeks, he was excited about what he was doing. He cautioned his mother that should she hear of "a new play to be brought out soon in New York, written by the talented young dramatists, Messrs. Allen and Fox, you may know that means your humble son and servant. We are going to do it." And, he said, "There will be fame yet in the Fox family." [19]

His friendship with Allen and Fible and others with whom he had graduated had a profound effect on him. They, but especially Allen, knew all sorts of society people, mostly rich, and when there were parties to go to, it was Allen who introduced him into those affairs. The thing that bothered him the most was that Allen and others seemed to have everything he desired and did not have. They already had positions both in business and society, and he had neither. They had the confidence of their decisions and convictions, and he vacillated from pillar to post. He admired them for what they were, and was grateful when he was introduced into the society affairs with a confidence that seemed so effortless and casual. He insisted to his mother that he did not "think much of these fashionable people." [20]

But that did not keep him from attending as many parties and society affairs as time and invitations and finances would allow him. Despite his protestations to the contrary, he loved it, and always would. "We must go in full dress," he wrote his mother of one such occasion, "swallow-tail coat, high standing, dude collar with a little white tie, expansive white shirt front with a single diamond glistening in the centre, etc. and be extremely polite and dignified. Everything will be in grand style ... and there will be probably a dozen courses. It will, as it were, be my first introduction into fashionable society and I am looking forward to it with considerable anxiety." [21]

The year 1883 ended for him much the same way it had begun. He was still waxing and waning between this career or that, and as yet made no conscious decision about what he wanted to do. Perhaps, he thought, 1884 might prove to be more productive, though he would not make New Year's resolutions to that effect like other people. To do so, he thought, evidenced a lack of character and determination. At any rate, the new year was upon him. He was now 21 years old, and he entered that next year thinking surely his situation could not be much worse. "Do not be at all disturbed about me," he wrote to his mother in January. "Cultivate papa's faith in me and follow his example, and believe that I will come out all right. I feel considerable confidence in myself and opportunities arise frequently and I think I can seize one which will be satisfactory to all of us." [22] But his mother *was* concerned, and she worried about him. The weather had been cold in Paris, and she knew it was supposed to be colder in New York. They had no money to spare, but wished he could come home.

In the early part of 1884, he managed to secure a position with the venerable New York *Times*, and although he kept up the fiction of studying law, it was only a pretense. He began, in what little spare time that he had, to spread his wings a little wider into literature, and tried his hand at writing poetry, which went much the same way as the promised play he and Buckner Allen had been so excited about. He found that in his short absence from newspaper work, little had changed. He was covering the dregs of society and their doings for the *Times*, and did not find very much in his work that impressed him about the condition of the human race, at least not in New York City.

On the next to the last day in January 1884, he wrote his friend Micajah Fible in Louisville, Kentucky, a vivid description of his duties as court reporter for the *Times* and the people he saw pass through the halls of justice he was sent to cover: "Nowhere in the world will you see more varied types and nationalities of criminals than are daily collected right here," he wrote. "They file along in a line up in front of a bald-headed justice who whacks a poor devil off the end every minute. The court room is crowded with morbidly curious sight-seers who look little better than the criminals themselves. Swarthy skinned Italians dressed in filthy remnants of what were once probably picturesque costumes, innocent looking China men, big, good-natured Germans, two Hindoos with smooth, oily olive faces and the blackest hair you ever saw in your life, an idiotic looking Swede and several representatives of that striking product of 19th Century civilization known as the 'Bowery tough.' Describe one and you describe all… This is all interesting enough if you can just accustom yourself to the foul odor which you can almost feel with your fingers and look calmly at instances of depravity which must be seen before believed to exist. Here comes one! An officer is pulling a woman into the court-room. She has on clothes which it would be an euphonism to term rags, her gray hair matted and filthy straggles over her shoulders and around her wrinkled dissipated face and she is drunk. He gets her in and leans her against the railing. Her eyes are blurred, her face red and bloated and she is slobbering and working her lips like a gibbering idiot. A dirty old shawl has fallen away from her shoulders and leaves exposed her sunken breast. An old black bonnet with bright artificial flowers placed jauntily on the side of her head presents an inconsistency which is simply frightful. Her time comes and she is hustled away yelling out a string of curses which would have Andrew Jackson ashamed of himself." [23]

He toured the courtrooms looking for tidbits of crime and gossip to turn into columns for the *Times* that might interest its readers. But he also showed a marked ability of description that would become his stock in trade in later years. Readers over a hundred years later can see, hear, and smell the "foul odor which you can almost feel with your fingers" when reading of his courtroom scenarios.

He was, whether he realized it or not at the time, a good reporter, and one who could make his readers see and feel what he saw and felt. It was, however, terribly hard work. That much had not changed in his move from the *Sun* to the *Times*, but he seldom mentioned how hard a life it was to his family, not even to James to whom he confided the most. But to his friends like Fible he was more candid. The work week was seven days long, and Sundays accounted for no more notice than did Wednesdays or any other day. "The work as you know is devilish hard and the hours are almost inhuman," he wrote to Fible. "To-night I shall get to bed at 3 o'clock, my day's work will amount to almost sixteen hours. Besides Sunday is like every other day, no holiday at all." But he was optimistic that he was receiving a good training that would eventually bring out the best in him. "If there is any good stuff in me," he continued, "I suppose I can pull out of this some day." [24]

He had not been sick for some time, but soon his health started to decline.

He was beginning to be, at about this time, regularly afflicted with a mysterious illness that has never been fully explained nor diagnosed. He personally thought it was the result of an old injury he suffered while at school in a gymnastics class, and he apparently underwent surgery to relieve the symptoms of that injury in April of 1884. Other times, he referred to it as "catarrh" which is characterized by an inflammation of the nose and throat that refuses to go away. It was a recurrent illness that sapped his strength and put him to bed for days, and at least once placed him on what he thought was his death bed.

His salary on the *Times* had risen to $26 per week, and he was beginning to think that perhaps, just perhaps, he had found his calling. He realized now that he wanted to write, and always had, and it was something of a revelation to him and a surprise that he had not really thought of it too seriously before. To be sure, he had often toyed with the idea, and had even from time to time written down sketches of conversation and descriptions for possible later use, but had given it only casual thought. What should have been obvious to him from the start now became his new goal in life, and he entered into it with the enthusiasm of his past ventures. His new job at the New York *Times* suddenly took on a new meaning for him. He might not always be a journalist, he reasoned, but the training he would receive while making a few dollars would serve him later. For a few weeks, he was enthusiastic in his newfound profession. Again, he wrote to his friend Micajah Fible in Louisville about his renewed interest in journalism: "[I]f I did not hope that it were a stepping-stone to something higher, I would break rock rather than continue. I am depending on something which I am not certain that I possess. I would rather be a writer than anything on earth but it yet remains to be proven that I have the elements of one." [25]

But he had enough of his father in him to be practical about it, almost to a fault. He would not, he told Fible, depend entirely upon his talents as a writer for his living. He did not think that was a wise course to pursue. "I have always thought that a literary man should have some other profession or business and that his writings should be the expression of the best part of his life." [26] He would not leave the *Times*, at least not yet, but he confided in his friend that he did not think being a "hack" writer for any newspaper was much of an honorable profession. He continued trying his hand at writing poems and stories that came to naught. His duties at the *Times* under the guidance of the city editor who was "the meanest Jew I ever saw" [27] prevented him, or so he thought, from concentrating on other work. His working conditions were terrible, and he was soon looking for something better to do, even at a reduction in salary from what he was making at the *Times*. He thought perhaps that *Harper's* or Scribner's might find a place for him, and was keeping an eye in that direction. He longed for peace and quiet, and a few hours to himself in a comfortable room with pen and paper, and without anyone shouting at him to do this or do that, or to run here or there. He was sick of it, and began to think that a trip back home to Kentucky would cleanse him of the city and the unwelcome cares of newspaper work, and he began saving toward that end. He wrote his mother the first part of June, "I am longing for even a week's quiet rest at home where I can see you all, eat and sleep to my heart's content. I am beginning to save up already, and I am going to get me a little box and drop spare nickels and pennies (which I would otherwise spend) into it. I want to leave here," he continued, "about the middle of August. If I don't come you may know that the lack of funds is my only reason." [28]

The nickels and pennies soon added up sufficiently to buy passage home on a southbound train. The railroad had entered Paris finally, to nearly everyone's delight, but it was still nearly eight miles farther to Stony Point, and the old home place. He walked from Paris, and the train station there, carrying all he brought with him, uphill and down, finally breathing in the air that he had longed for. In his diary, he wrote, "What a glorious walk I had over the pastures with the glorious sunset behind me. Then as the shadows lengthened and twilight came on with its cool, fresh fragrance I took off my hat and could hardly restrain myself. I crossed the hill and looked down on my home thrusting its roof up through the trees. I looked back at the sunset, gave a loud halloo and started down the hill. My sister heard me and recognized my voice and as I climbed the fence the whole family met me in the yard." [29] He never dreamed that this would be the last time he would come home to Stony Point and the old farmhouse of his youth.

His father's school in Paris was doing quite well, but the 16 miles or so John Sr. had to travel each day from Stony Point to Paris and back was just too much, especially in winter. Then, too, Minerva was lonely at Stony Point. There were no school children nor boarders there anymore since the school had closed in favor of the one in Paris, and she was home with little to do except wait for what family there was left in and around Stony Point, to come home. So, on August 26, 1884, the entire household moved into Paris, into a rented home, and although Minerva was pleased to be closer to the school and her husband, she felt dejected that they now, after so many years, did not even have a home they could truly call their own.

After John returned to New York from his August visit and his mother had become somewhat more settled in her new home in Paris, she wrote him of her feelings: "...at times I feel like one forsaken [but] I am more satisfied here than at Stony Point if I could only keep away the thought that we own no home and have only the income from the school.... I should be more thankful than I am for the blessings bestowed on us." [30] But even as strong a woman as she was, even she could reach a point when she could not fight off the dejection she felt. In her letter to him in December 1884, she told him that she knew she looked "on the dark side too much [but] if there is any bright ones in the future for me I wish they would hurry on." [31]

Neither was John Sr. entirely happy without his old home. He had lived at Stony Point for a long time, seen the old house burn to the ground, and seen another built on the old foundation to take its place. He had walked every square foot of the old farm, knew every tree, every dip in the ground, knew where the breeze blew coolest in the summer, and the harshest in the winter. It was hard for him, and later, in Paris, he confided to Minerva that if he had the old farm back again, nothing on earth would make him leave. But it was too late for that. He and they would spend the rest of their lives in Paris and Big Stone Gap, Virginia, and except for their last years, they would always hold out a thought and a vague hope that someday they might go back to Kentucky, and even back to Stony Point.

By September, John Jr. was back in New York continuing to work on stories and poems in his spare time that were seldom finished, and never published. He was visited by his brothers James and Everett early in January, and Sidney, who lived in Brooklyn by this time. Their time together made the transition of his visit home and his return to New York somewhat easier for him. But too soon they were all gone back to their respective homes and jobs, leaving him alone, and things began to get lonely again for him. He was soon sick again, and although he did not think it was anything serious, it quickly put him in bed, unable to work. His parents learned of his illness through his brother Horace to whom John had written and mentioned he was very sick, and Horace passed the news on to Paris.

Whenever John was sick, both his father and mother suggested the best cure would be for him to come home, and this time it was no different. "Your Ma and I," wrote his father when he learned he had been seriously ill, "both think you should come home, and stay awhile, and rest up till you get entirely well and strong: for if you do not get so now at your age, a foundation may be laid for good deal of bad health and trouble hereafter. If you will come," his father added, "we will send you the money." [32] He did not improve as quickly as he at first thought he would, and his father continued to admonish him to come home. In a letter dated February 13, 1885, John Sr. wrote his errant son, "Our house is large and roomy ... and we would be glad to do all we could to make you feel pleasant, and improve your health." [33] And, he added, it might be best for him to give up his reporting job at the *Times*, since he was sure the hard work and long, uncertain hours had a great deal to do with his deteriorating health. The same day, his mother wrote in much the same vein.

His illness, whatever it was, was serious. Once again, he wrote details to Micajah Fible that he would not dare burden his parents nor any others of his family with. "I have had a siege of illness for nearly a month," he wrote Fible on February 3, 1885. "Inflammation of the bladder came on me about two months ago and I had to have a painful operation performed which kept me in doors for two weeks. I have begun work again but I don't get well very fast and the physician says I may have trouble with it for some time. I was in quite a dangerous state at one time." [34]

His work at the *Times*, naturally, suffered, and he did not receive much sympathy from his editor, "...that d----d Jew of whom I spoke to you, has actually become angry with me for getting sick. 'Well you're pretty useless' says he with that proud and humane sympathy which characterizes him." [35] Things were so bad, he thought he would be fired almost any day now, but it did not bother him too much. He was too ill to care. Then, too, his doctor was advising him to go home and recuperate, but in the meantime, he returned to work, finding things there at their worst.

On Valentine's Day he wrote: "With the advice of the doctor I went back to work on Thursday until Monday next when he would decide whether it will be necessary for me to go home. At the office that Jew of whom I spoke to you, informed me that he had been obliged to take on three more men during my illness and would not, consequently, encourage me to continue [and] When I reminded him of his promise to reserve my position he denied it up and down." [36] John had exhausted all his meager savings while he was sick, and was placed in the humiliating circumstance of asking Micajah Fible for a loan of $50, if his friend could spare it.

Another series of tests changed the physician's mind about John's returning to work, and although John thought he was improving, though slowly to be sure, his doctor ordered him home. Under the circumstances, it was an order John was comfortable with and welcomed. He would go, even though he preferred to stay and fight it out with his editor and retain his position on the *Times*, but he saw the uselessness of that battle—the cause of which, he felt, was personal and in no way reflected upon his performance as a reporter and writer. And so, by the end of February he was again home in Kentucky, determined to get well, return to New York, and fight the good fight against what he was certain had been slights against him personally. But he would not return to New York quickly, and not in the capacity in which he left.

He found, upon returning home, to his utter amazement, that his doctor in New York had secretly written his father informing him that his returning son could not have long to live, and that there was little or nothing anyone could do for him. John immediately took to his sick bed, convinced that he was dying. "This news," he said, "stretched me out on my bed, sick at heart and stomach as well, but in a few hours I was quite resigned, for only the young, I find, are willing to die." [37] He did not die, however, but it was six weeks before a local doctor got him on his feet again. He continued to be weak and ill, however, for some months to come.

He was glad to be home again, even though sickness and the idleness of not working exasperated him. "It is pleasant to get back among hearty wholesouled people whom you understand and who understand you, people who never change in their affection for you and whom you can trust with your life." [38]

And so, after only a short time, his career as a reporter was put on hold.

His journalism career was not finished, however, and he would fall back upon reporting from time to time, never really leaving the profession until his success as a writer of fiction was assured. But for now, sick in body from a recurring illness that would not leave him alone, he had arrived home to Kentucky to regain his health, if he could.

CHAPTER 6

A Tale of Two Cities

I felt like a fraud.

John Fox, Jr.

EXCEPTING HIS SENIOR YEAR AT HARVARD, and the one summer afterwards working for the New York *Sun*, John had spent his summers at his older brother James's mining concerns on the Kentucky border with Tennessee. Now, at home in Paris, "miraculously" retrieved from death's door, and bored to distraction, he began to seriously reconsider his brother's invitations—which began as early as February 1885 when he first came home from New York—to join him in Jellico. "Come up," James wrote him, "when you are tired of the Blue Grass, and try our mountain air and water." [1] By the middle of May, however, John was bedridden again, and he went nowhere. His father, of course, wished him to stay in Paris, and help out at his school there, and for a short time, when his health permitted, he did. It was just enough to remind him why he left law school in the first place. He found he did not like teaching all that much, but he felt guilty enough, however, about his father's need for help, so that he said little about it. His brother James, however, knew John and their father better than anyone else. "I will not attempt to advise you," James wrote on August 25, 1885, "in the matter of teaching: you must decide that for yourself ... you may find it as nerve-wearing as journalism." [2]

John was writing a little between bouts of sickness. He sent a small story he had written a year before to *Frank Leslie's Illustrated Weekly* entitled "The Betrothal Ring." Another, "Deceiver's Ever" was sent to *New York Life*. They were his first attempts at "serious" fiction, and he received $8 for the former, and apparently nothing for the latter. "It don't amount to anything," he wrote Micajah Fible about "A Betrothal Ring," "except a very agreeable little check which came this morning." [3]

By September, he was too sick to begin the new school term with his father. His illnesses came and went, but between times abed, he was strong enough to visit Micajah Fible in Louisville and another old Harvard school mate, Upshaw Berryman, in Lexington, but invariably these trips, no matter how enjoyable to him, nearly always resulted in a relapse. A New York doctor suggested a "dangerous operation" to correct his affliction, but there was, cautioned the doctor, no guarantee of success, or even of survival. Another doctor suggested the life of a farmer, preferably in the West, should cure him, but nothing that he tried within reason was of any use. In his personal narrative, written in 1908, John wrote of this mysterious ailment that his "trouble whatever it was kept laying me out periodically for years until a decade ago [1888]." [4]

He was not physically working, and that galled him. His brother Sidney wrote from New York that he had a position for him there, but he was held up by the weather and could not leave home for four days past when he was supposed to leave. As a result, John was too late in reaching Brooklyn, and the position was already filled when he got there. He decided to see about his illness while there. The doctors looked him up and down and made an astonishing decision: He would get well in time. His father thought he would do just as well, recuperating in Paris, and should come home.

He resisted, however, and in March 1886 he landed a position as tutor to another wealthy New York family. His doctors had advised him against ever going back to a journalistic career which, he said, "narrowed my field of employment very much." [5] He had tried to find a job without success, and finally, almost desperately, he had placed his name with an agency for teachers, though his previous experiences as a tutor had not ended well. In any event, he was soon hired "to a millionaire who lives in a veritable palace on 5th Avenue. This millionaire," he wrote Fible, "has just had a private car constructed for the purpose of taking himself and family on a 5 week trip through California." [6] The millionaire had a 15 year old son, and John was to tutor him along the way with a "good salary and all expenses paid." [7] It did his self-esteem good to know that he had beaten out at least six other competitors for the job, and he reasoned that his success was based primarily upon his Southern breeding and manners, and not necessarily his Harvard education. At any rate, his duties as tutor would not be severe, and he thought he would have plenty of spare time to write along the way.

It was a special railway car, and John was impressed. In all his life, he had never met anyone who could afford to rent an entire train car for six weeks for no other reason than to take a vacation with his family, and be thoughtful enough to take along a tutor for his children. He had never been so pleasantly situated, and never so excited by his prospects.

There were eight passengers, not counting John. There was Mr. and Mrs. George Kemp, and Mrs. Kemp's sister, Virginia Kent, the Kemps' two daughters Marie and Julie, and their two friends Adele Turnure and Mary Clark, and Master Arthur Kemp, whom John would attempt to tutor throughout the trip. [8]

"Up to Albany we went," wrote John in *The Wanderers*, "across New York State to Buffalo, and by early morning we were dashing along Lake Erie and catching occasional glimpses of its blue water and broken ice through the openings in its wooded shores. Erie, Cleveland and Indianapolis were points in the journey, but not until we reached St. Louis was the dividing line between east and west sharply marked in the face of the country and in the appearance of the people." [9]

In Kansas City, he saw his first cowboy, and was not too impressed. By the time they had reached Colorado, he had seen a country more "bleak and desolate" than anything he had ever seen before. The people at the stations where they stopped seemed to match the landscape. "At every station there was a crowd of rough-looking men, unshaven and unshorn, with the wide-brimmed white hat of a cow-boy here and there among them." [10]

They reached California on the first of April. It was still early spring, even late winter in New York, but here it seemed summer had started a long time ago and had not left. "Trees appeared here in full foliage and on each side the stretches of wild flowers in full bloom were a delight to the eye. The air was soft and mild with that particular deliciousness that only Spring-time gives." [11]

After more than three weeks in California, the group turned back east into Utah. Through Ogden and Salt Lake City they went, stopping only long enough to see what little there was to see, but nothing came up to the standards that the California towns had set for them. On the 31st of April, they struck Council Bluffs, Iowa, switched trains, and on to Chicago, Buffalo, Niagara and New York by the end of the first week in May. On the 8th, they arrived back in New York City, worn out as only long traveling can do, but happy to have gone. In their absence, spring had arrived in New York.

When he returned that May of 1886, he was ecstatic. This was more like it, he thought. Never before in his life had he experienced anything like the trip he had just made. He knew that there were people to whom money meant very little, and he knew that there were people who had plenty of it, but he had never really been associated with anyone who spent money as carelessly as the Kemps did. He was surprised, and pleasantly so. He had gone from eating 15 cent lunches and a cot beside his desk, to the lap of luxury, and he would never forget the difference. And, although he vowed to Micajah Fible that he cared nothing for money, only enough to keep himself alive and "scribbling," he was not being honest with his old friend. Money, or the quest of it, guided him as surely as a compass guides an explorer in the wilderness.

Now, he was back in New York with some time on his hands until the middle of June, when he would again pick up his tutoring duties. Against his doctor's orders, he was again working as a reporter, but hated it more than ever.

He had gradually become convinced that journalism, and those who practiced it, destroyed their capabilities for producing "pure literature," [12] but he had not worked at any real job in the past two years, and had gotten himself into debt again. It was the worst of all possible situations for him. His health was not back to normal, and in New York, away from family and most of his friends, he had to work to live. He was trying to find a career that would allow him to work and write at the same time, but he was not having much success and was beginning to see the return of his old doubts about his capabilities, and his old thoughts of an uncertain future returned to haunt him. Perhaps he wondered if he was not making the wrong decisions, and no doubt he wondered if he would ever be the writer he wanted to be. Perhaps, too, he was wasting his time — time that could be better spent in some other profession before it was too late. "I have no personal complaint to make," he said about his situation. "I think I get about what I deserve, I would be very well content with just a competency and enough leisure to follow out this craze (maybe foolish) for writing and I mean to have both." [13]

To complicate matters, he had found a new girl whose beauty and personality he found quite distracting, and though he had never really been in love before, he somehow found the experience at the same time pleasant and disturbing. Currie Duke was the daughter of Basil Duke, a powerful, rich and respected member of Kentucky's highest social strata. Additionally, Currie's father had been a one-time lieutenant to John Hunt Morgan, a famous Kentucky Civil War character and general in the Confederacy, which could not help but be an asset in Kentucky politics and social circles.

John and Currie's relationship was an association that never ventured, apparently, very far past the platonic. To begin with, she came from one of the premier families of the Blue Grass. Her father had money, power, and respect. Her mother was the sister of the aforementioned John Hunt Morgan. Currie was beautiful. John had nothing — not ever a job that could remotely be called a profession. His friend Fible encouraged him to go ahead anyway. "Great God!" John replied, "on what grounds could I storm the heart of any woman much less of a beautiful, delicate creature the very thought of whom in any other circumstances than luxury or certain ease and comfort, would give me exquisite pain? Sometimes I wish she were married, and I am glad, very glad that she has a mother who has the reputation of possessing high matrimonial ambitions for her children." [14] He could not blame any mother or father, he said, from keeping him, especially in his present circumstances, away from their eligible daughters. He had nothing to offer, and his prospects were not getting any better as time went along. By November, the Louisville papers would be announcing Miss Duke's engagement to be married, but not to John Fox, Jr. [15] Their friendship would, however, last as long as either of them lived, and they both cultivated a real mutual affection but it never realized anything more than friendship, and there is not a lot of evidence that Currie wanted it to go any further. The stagnation of his relationship with Currie only served to remind him of his uncertain future and lack of prospects. He was disappointed in himself, unsure of what to do, and not a little sorry for himself.

As usual, when things were darkest for him, he went home to Kentucky. He often seemed to put it off as long as he could, thinking the expense of travel was too much for his purse, but when he could stand his isolation from his friends and family no longer, when his homesickness finally overcame him, nothing but Kentucky would suit him. By the last of June 1886, he was in Frankfort, had stopped in Louisville for a time, and would stay overnight in Lexington before arriving in Paris, finding it duller than he remembered. "There is not one congenial companion for me in this small town," he wrote. "No, not one, and the sense of dissatisfaction is keener than ever." [16]

This portrait of Currie Duke shows John's penchant for being drawn to beautiful women. When he left to cover the Spanish-American War for *Harper's* in 1898, Currie married Wilbur Mathews (Frank Leslie's *Illustrated Weekly*).

Small town life had a discouraging effect on him, especially after his time in New York. He would visualize himself trapped for the rest of his life in this small town or another just like it. It made him restless and nervous to think that a life at the end of a dirt street was all that was waiting for him. There were no theaters here. Nothing like ballrooms and restaurants and businesses and smells and colors and people here like there was in New York, or even Lexington and Louisville. To make matters worse, he had received a letter releasing him from his obligation as tutor for the summer. Even the Kemps did not want him anymore, it seemed.

Now, he had no prospects. His doctor had informed him that to return to journalism, the only profession he felt he was capable of doing successfully, would cause a relapse of his old ailment. And so, he found himself dejected and discouraged in a small town. It was one of the lowest points of his life. The signs of failure were all around him: the Kemps, unpublished stories, failure at law school, a wasted Harvard education, doctors who could not find out what his mysterious illness was (even a sort of death sentence from one of those doctors), a failure at winning the hand of someone he could have truly loved, and now back home with fewer prospects for what he really wanted in life, and more miserable than he had ever been in his life.

He had plenty of the peace and quiet that he thought he must have to write, but he found he had neither the inclination nor the will to make himself sit down with pen and paper. His mind was too full of his own problems, both physical and monetary, to allow him the concentration he felt he needed, though he produced a modest amount. He spent a great deal of his time out of doors, taking long walks, sometimes in the company of his father who loved the outdoors as much as John did.

Despite his lack of success and accomplishment, writing was not, however, very far from his mind during those long days and nights in Paris. By the middle of August he had finished a short story, and wrote his friend Fible in Louisville that he had "material for two or three more." [17] And he had been writing for the local newspapers, submitting small articles of local happenings, historical pieces and similar meaningless articles that filled the local presses in those days.

By the middle of September, he was back in New York, feeling that his illness and the threat of that sickness—whatever it was—had abated enough for him to resume his journalism career somewhat, but not at the pace he had worked it previously. He was determined, this time, to find enough tutoring to allow himself to write when and what he wanted. He would write enough articles for the *Sun* and the *Times*, he said, to stave off starvation, but no more.

His father was pleased that John had gotten back to New York safely, and whether or not a diphtheria epidemic that broke out in Paris just after John left had anything to do with it, he cautioned his son to take especial care of his own health. "We are extremely glad to know that your health continues good," John Sr. wrote, "and whatever you may engage in, be sure to make the preservation and improvement of your health of primary importance, and ever thing else secondary, for every thing depends upon health, without which, even life itself may become a burden." [18] And as far as his work in New York was concerned, his father could not resist but to offer a small bit of advice concerning what his son was doing in that faraway city: "Don't forget that if you fail to secure some congenial employment, that you will still have a home to fall back upon, and pursue your literary tastes for authorship." [19] John was discouraged enough as it was with his prospects, and although it is very unlikely that his father intentionally made his discouragement worse by perhaps subconsciously suggesting his son could not make it in the big city and should come home where he "belonged," there can be little doubt that such "encouragements" from his father did little to raise his son's spirits when they needed raising the most.

A few days after writing to his mother from New York not to be too discouraged about her son's progress, John was able to secure enough students to tutor privately to enable him to ease his mind and pocketbook, and he was happier still. He continued to write stories, and finished a play entitled *A Stage Tragedy* that was named appropriately since nothing ever came of the effort, but he was not discouraged. His letters home were full of excitement and confidence. As usual, however, those letters seldom mentioned anything bad that happened to him. He had been sick again.

By the middle of December he was facing another birthday, and he was characteristically philosophical about it. "This is my birthday," he wrote in his diary for December 16, 1886. "I had forgotten it entirely when a glance at the Calendar reminded me of it. It sobered me at once. There is something very serious in the coming of a birthday after the age of manhood is reached. What have I done? Am I making proper use of the years that are slipping by? Am I rising or sinking? These are some of the questions that crowd into one's mind. Three years and a half out of college, nearly two years illness— a good part of which I must attribute to carelessness, I suppose. No, my expectations are not realized. I have done nothing upon which I can congratulate myself. And yet, my hope is just as keen, my aspirations as high and my confidence even growing a little… I suppose there is some excuse for me for having done so little though I might have done better… I know that there is work in the world for which I am fitted and in doing which I shall be contented. I long for such work that will never cause me the slightest feeling of a loss of self-respect, work every phase of which is independent and honorable to the world. Alas, all honest work is not honorable." [20]

In the meantime, he had been hearing from James, who reminded him of the offer of employment with himself and Horace in Jellico, and distance and time had softened his memory of the hard work he had done there. It seemed like an attractive opportunity for him. Perhaps, he thought to himself, he might be able to do something there, perhaps divide his time and work between New York and Kentucky, and make enough money to free him from any dependence upon his friends and family, and not incidentally to write. He quickly saw the advantages, and made up his mind. "I want to be in Kentucky… I don't want to write about anything else than Kentucky," he wrote to Fible in May of 1887, "and I feel now that I am losing. I want to be steeped in its history, have its people, their characters, their personalities, their modes of life & thought in my brain, and feed my senses in the beauty & grace of its women, its soft climate … and its wondrous surface of grass & foliage. This is my plan." [21]

The month of June 1887 found him in the hills around Jellico, working again for his brother James, not digging roads nor laying railway tracks to the mines as he had done in his previous summer there, but working in the Proctor Mining Company offices.

It was not hard work, was enjoyable really, and his penchant for working with pen and paper, studying and correcting invoices, and generally doing the day-to-day business with payrolls and those other duties inherent with running a business anywhere appealed to him a great deal. Not incidentally, he had plenty of leisure time as well, and this too appeared to him.

As usual, he wrote Fible from the Jellico mines all that he felt about himself and the position he found himself in: "I am drinking in like a sponge the peculiar life, the peculiar ideas of this peculiar mountain-race and its beautiful natural environment. I rise at 5, dine at 12, sup at 5:30 and go to bed before 9. I work in the office with the books, study all the departments during the day, write letters, scribble conversational bits, vernacular, idiosyncrasies, in my note-book and work a little, a very little, on a hopeless & unsatisfactory story." [22]

Despite the allure the mountains had for him that summer of 1887, by the first of September, when the mountains were at their most beautiful, he had returned to New York. Now, his capacity was as a representative and as a sales agent for his brother, trying to find investors, rich ones, in the North who would turn loose of a few dollars. The prospect was the making of millions, and he assured those potential investors that their money would be quickly and easily made. He was not oblivious to the absurdity of his situation. "Jim," he wrote to his brother from New York, "it requires nothing this side of monumental cheek to sit up & talk to a man about investing $10,000 when you have but $10 in your pocket. It makes me feel like a fraud." [23] It was a feeling he would have to get used to, and quickly, but the thoughts of what business was doing to his sensibilities, worried him a great deal that fall.

He did not like the business that he found himself in, and furthermore, his eyes began to bother him — from too much reading while he was younger, he was sure — and his old "ailment" was back again, forcing him to place himself under the care of his brother Sidney, a doctor, in Brooklyn.

By the first of January 1888, he was improving but wanting to go home so badly that he wished half-jokingly that he was sicker than he was so he would be forced to go home. But he was soon better, and back from his brother's home in Brooklyn where he had been staying, and living along again in New York, picking up a few dollars tutoring for $2 per hour, attending to James's interests, and working haphazardly on what he was sure was to be a worthless story of the mountains.

He was alone and lonely, more so than he had ever been before in New York. He wrote his friend Fible on January 7, 1888: "I am lonely here. I have a room in a house where I know no one & I eat alone, anywhere, in my room, in restaurants or not at all. I am as solitary most of the hours as I ever was last Summer in the sacred silence of those mountain forests." [24]

He was not so revealing to his mother when he wrote her the next day. "I am back in New York again, you see, in a new place and very much pleased I am. I have a pretty little room facing the South & I get the sun through three small windows. I have a little oil stove and a brass tea-kettle. Every morning before I get up, I reach out and light it, heat some water & have a cup of beef tea with some crackers in bed. Then I work writing, correcting themes etc., until noon when I take breakfast." [25] He added that he ate in restaurants or anywhere else he felt like eating, and gave the impression he was enjoying the life of an independent a great deal. The truth, however, was not so cheery as he would have had his mother believe, and by the first of April he could stand it no more, and was back home in Paris, Kentucky. "I am going to the mountains & I mean to stay there till I get well, if I ever do," he wrote to Micajah Fible. "I long for absolute solitude. I want the beautiful, clear, tender, luminous atmosphere of those wooded mountains. I want to get into the heart of those lonely forests where there is no suggestion of humanity & dream & think & be thrilled through & through by the rustling of leaves, the quivering of shadows, the hum of insect life, & the cool, faint, flute-like notes of a wood-thrust." [26] By the end of the month, he had again returned to Kentucky, and not only that, he had returned to the mines in Whitley County, Kentucky, at a little coal camp called Red Ash, hard against the mines of Jellico.

CHAPTER 7

Boom and Bust, Part II: The Gap

...the light of hope was in every man's eye.
John Fox, Jr.

They have tied me to a stake; I cannot fly, but bear-like I must fight the course.
Macbeth

THE DOZEN OR SO YEARS from the middle 1880s to the latter part of the 1890s were the most troublesome for the Fox family. Ventures into coal and land, and the resultant headaches and monetary responsibilities that such ventures incurred, frayed the nerves of every family member. It was not necessarily bad judgment on their part that caused them the most trouble. More than anything else, it was simply bad luck, and perhaps not enough monetary capital to see them through the difficult times such investments required.

John Jr.'s sojourn at the mines in Jellico had not been an entirely satisfactory one from a business standpoint. Petty troubles and squabbles and jealousies among the leadership in the mines had caused no end of headaches. There had been suspicion among some of the partners—not especially by the Fox brothers—that one partner would undercut the other. When that happened, no one was above suspicion real or imagined.

The Proctor Coal Company at Red Ash, Kentucky, on the northern edge of Jellico, was a typical coal camp of the 1880s and 1890s. James—John Jr.'s half-brother—was a director and sometimes salesman for the company along with about six others. John worked in the offices, taking care of the paper end of the coal business, and he was young, and it is not unlikely that he felt the importance of his position all out of proportion to his actual status, and although John did not appear on the company's letterhead in any capacity, he felt responsible for his brother's business when James was not there, which was often. In any event, trouble, possibly brewing long before John Jr. arrived in Jellico, broke out between himself and the store keeper, Joe Wilson, who, by the way, *did* appear on the letterhead of the company, which in the pecking order of such things, seemed to give Mr. Wilson a little more status.

One thing led to another in the absence of James, the upshot of which was that Wilson closed and locked up the boarding house, packed up all the dishes from the dining room, "took the key to the store & gave Aunt Mary orders not to cook for anyone but herself & Ben." [1] When John and the others refused to move to another dining room, Wilson threatened to fire them. Whether he had the authority to do so was not important by this time. Wilson had gotten his dander up, and was going full throttle. John was reduced to making his meals of crackers and beef tea, which, he lamented, he had to prepare for himself. He did not want to be the cause of any trouble, he wrote to James, and he did not want to be put in the position of having to thrash an old man. So, he said, he would go home, and he did. By the middle of January 1889, he was back home in Paris, with a sour taste in his mouth for the Jellico mines, and once again without prospects, and by January of 1889, he was no more settled on a career or livelihood than he had been upon his graduation from Harvard nearly six years ago. But, of course, he was harder on himself than others were.

The move to Big Stone Gap, Virginia, to cash in on the expected boom in real estate there would come next year, but for the time being, preparations had to be made to make their family's move more accommodating for everyone, especially their mother and father. So, in the meantime, things went on much the same as they had before in and around Paris. John found himself once more teaching with his father, and hated it, not just for the teaching, but for the sheer boredom of classroom work. He hated it with such a passion, that he was unable to keep his feelings from his brother James, but he apparently was able to conceal those feelings from his father.

The elder Fox was pleased to have the help, and thought that John was doing very well in the school. John, however, did not have the same high opinion of his father in the classroom. "Papa simply hasn't the life to stir boys up to explain things," he wrote to James, "to hold them to a certain mark & by George it is almost pitiful to hear him talk in the school-room. How in Heaven's name he has stood this work 40 years is without my comprehension. In a year I would be a moral & physical wreck." [2] In another 30 years, when he was the age that his father was now, he would look back and remember, a little more kindly, his father's efforts.

He could not, however, keep himself from feeling guilty about his father's situation. John Sr. had worked hard all his life, and as far as anyone could see, he had gained very little through those efforts. One of the reasons they had not immediately moved to Virginia was the debts his father owed. They could not, in good conscience, just move off and leave those debts unpaid, small though they were. The elder Fox also was not well and was beginning to be dogged by sickness, especially in winter when he tried to supplement his income by surveying no matter the weather. "Papa is not well," John wrote to James that February. "He has been killing himself by inches. He is 58 yrs. old & looks 70. This is the 40th year he has taught & worried — 40 years of incessant toil & self-denial & thrift...& to-day he is in debt & without a home & six grown men in the family!" [3]

Having been away from home so much in the last few years, he had not thought of how desperate things could get in any small town, and it was a shock to him that, upon returning home, he found his parents poor to the point that they could not pay the most trivial of debts. He was so concerned that he thought if something was not done soon his father would not live to see spring come again.

Soon, John's eyes, which had been giving him problems off and on for some time, necessitated a trip to Cincinnati, for which he had to borrow from his father the sum of $75. It was like rubbing salt into an open wound to do so, but he salved his conscience by telling himself he was working for nothing teaching school, and had no money of his own. Also, he reasoned, he could not continue teaching if he could not see. In a letter to James toward the end of February 1889, he tried to justify the loan by saying, "I deny myself all forms of social entertainment here because I consider it selfish to go [to parties] alone & because I can't risk the possible expense of a carriage." [4] He could not help but remember James's advice to him earlier that teaching would drive him mad in short order, and he thought that it just might do so before the June recess.

Though the drudgery of school teaching weighed upon his mind up until the last day of classes, he continued to work on the story he had begun at the mines in Jellico, and in his most dejected hours thought he would never finish. He found himself in Louisville in March trying to interest investors in Big Stone Gap. He found himself in Stanford, Kentucky, in June, doing much the same thing, and in Lancaster, Kentucky, selling coal from the Jellico mines in July, and back to Paris by the middle of the month. He was never still that summer of 1889, running here and there trying to interest the uninterested in land and coal schemes around a practically nonexistent town. By September, he was back in Paris, teaching school, and this time it was his father who was becoming dejected at their plight.

Hearing his sons talk of fortunes to be made so easily in Big Stone Gap captivated the imagination of the old man. It would mean moving and leaving everything he had known for most of his life. Most of his friends would stay behind, and he did not fool himself into thinking that he would always be able to see them after the move to Virginia. It was a long way from the people and places he knew and was comfortable with. His and his wife's financial condition, not to mention their prospects for a better financial future, seemed dark indeed these days. His school was not going nearly as well as he would have liked, and his age was beginning to bother him. John Jr.'s assessment of his father's teaching abilities—though certainly not mentioned in his father's hearing—certainly did not go unnoticed by the old man himself. He knew he was not doing as well as he would have liked, and more and more he was thinking—these days—of quitting the profession he had loved all his life.

In the latter half of 1889, while he was up to his ears in his brother's mining ventures in Jellico, John received his last letter from his old friend, Micajah Fible. Somewhat suddenly, Fible had taken leave of his law business in Louisville and moved to Chicago where he went to work for the *Chicago Tribune*. The break between the two friends does not seem to have been one simply of distance. The letters and family history hint otherwise, but what those reasons were, if any, may never be known. [5]

In December of 1889, John wrote to Fible for the last time. It is a letter sad in tone and scarcely hidden regrets and melancholy. "Christmas is near at hand & I always want to know where my best friends are," he wrote, "[and] what they are doing about this time of the year." [6] John had gone to Louisville to visit Fible's family, and found them also gone to Chicago. "I hope with all my heart that you are in good health & spirits," he wrote, "that you are doing well, climbing where I know you can. I am at home preventing my father from working himself to death & writing when my eyes permit me. Give my love to your mother, father & sisters & remember, old fellow, that I am your friend." [7] As far as is known, he never received a reply as to what his friends were doing.

Later, long after his friend had vanished, John would hear rumors and innuendo about his friend, and guesses about what had happened to him. Fible had been sent to Seattle, Washington, to cover a sex-scandal story that began in England with members of the royal family involved. He arrived, sent back encouraging telegrams to his paper, spent his expense money, and promptly dropped off the face of the earth, never to be seen nor heard from by his friends and family. No one knew if it was foul play or not. No one could find out anything, and it would be months before John even knew his old friend was missing.

Fible was gone, and for a while, after the news of the disappearance of his friend reached him, John wondered almost daily at the mystery, and half expected to hear from "Dear Old M." one more time. Finally, time and his own problems and concerns pushed Micajah Fible out of his mind, but from time to time he would see or hear something that reminded him of Micajah Fible, and the whole of those times when they were both fresh out of school and struggling, and those thoughts would rush back to haunt his memory of his friend once again. [8]

In the meantime, events in the Gap and at Paris were rapidly coming to a head. Horace was already in Big Stone Gap working as a surveyor, and John wrote him that the rest of the family were coming that spring, house or no house, and they could live in a tent, if necessary, until a house could be built. Apparently, he had no idea, nor did any of the others in Paris, of the time and trouble involved in building a house, and a large one at that, big enough to hold them all in a new town where lumber and carpenters were at a premium.

Horace received John's letter early in January and sought out a house that he assured them would hold the family until a better one could be built. It needed a few minor finishing touches, but if they were coming, he was sure it would be ready for them. He also noted that land was now selling in the Gap for around $100 per acre (land could be bought as cheaply as 50 cents an acre nearly everywhere else). It seemed that the boom was indeed taking shape and that land prices were destined to go up.

By the third week in January, John had waited as long as his anxious temperament would allow him to wait, and he joined Horace in Big Stone Gap and had begun to dive headfirst into real estate investing. His spirits concerning the family's move to the Gap were dampened somewhat by his own difficult trip there. He found the town nothing like what he had envisioned it to be. There were few buildings, and those were green and rough. The streets were pure mud, and practically impassable. The Gap had all the markings of a frontier town at its most raw, and he immediately had second thoughts about the rest of the family moving there any time soon. "It w'd be ridiculous," he wrote to James, "for the family to attempt the journey here now. They will have to wait perhaps several weeks." [9] He was excited almost to the point of giddiness at the prospects he saw beyond the Gap's first impression, however. He had caught "gold fever" where there was no gold, but the prospect of riches, no matter if it be land or gold or timber or coal, is an all consuming thing, and he found himself caught in its snare. It would be small comfort later when he realized that he was not the only one so entangled or so blind. Others, more knowledgeable veterans of "booms and busts," found themselves in much the same circumstances. The elder Fox found himself somewhat bewildered by what was taking place around him. "When I hear of your enterprises involving such enormous sums," he would write in March of 1890, "I rub my eyes to see if I am not dreaming, or reading some fairy tale." [10]

Before it was all over, all of them, including the elder Fox, would think of the whole of the Gap as a dream, and at least in a business sense, a bad dream at that. But in the meantime, the Gap was being touted by nearly everyone as the next Pittsburgh, and would in time, everyone seemed to be saying, become one of the greatest of American cities.

The plan was a simple one, and time honored in business circles: buy cheap and sell high. They would buy land and options on coal, timber and more land, and when the "boom" hit like everyone assured them that it would, when huge factories were built, when railroading terminals criss-crossed the valley, they would be able to sell the same cheaply purchased lands and options for a fortune. With James's expertise in the business of speculation, they reasoned they could not lose.

Years later, John would write of the enthusiasm that captured every man's imagination at the Gap. "The in-sweep of the outside world was broadening its current now. The improvement company had been formed to encourage the growth of the town. A safe was put in the back part of a furniture store behind a wooden partition and a bank was started. Up through the Gap and toward Kentucky, more entries were driven into the coal, and on the Virginia side were signs of stripping for iron ore. A furnace was coming in just as soon as the railroad could bring it in, and the railroad was pushing ahead with genuine vigor. Speculators were trooping in and the town had been divided off into lots—a few of which had already changed hands. One agent had brought in a big steel safe and a tent and was buying coal lands right and left. More young men drifted in from all points of the compass. A tent-hotel was put at the foot of Imboden Hill, and of nights there were under it much poker and song. The lilt of a definite optimism was in every man's step and the light of hope was in every man's eye." [11] Everyone was caught in the maelstrom, and all, regardless of station or status, thought they would become rich. "The labour and capital question was instantly solved," John would write in 1908, "for everybody became a capitalist—carpenter, bricklayer, blacksmith, singing teacher and preacher. There is no difference between the shrewdest business man and a fool in a boom, for the boom levels all grades of intelligence and produces as distinct a form of insanity as you can find within the walls of an asylum." [12]

Big Stone Gap, Virginia, then as now, is a community hemmed in by mountains of the steepest sort. John would write incessantly about the isolation of the mountaineer and what that isolation denied those people he wrote about. This time, he neglected to imagine that a town could become as isolated as the people who lived there before the Foxes and all the rest of the newcomers had arrived. He had forgotten, or refused to think about, the first three universal rules of business: location, location and location. The simple truth was that the Gap was not in a good, accessible place. It was hard to get to. Roads were few and primitive. More than once, John would write of seeing mules trapped in the mud holes in the street, and drowning there. The railroads would eventually build to the town, but even that could not negate the fact that Big Stone Gap, despite all the promotional brochures, was simply a mountain town, and was almost certainly predestined for trouble and failure. But they all tried hard to bring civilization—at least their conception of civilization—to the Gap, and for a time, they succeeded.

With the advent of the first railroad to reach the town, life became somewhat easier for them all: "supplies came from New York, eight o'clock dinners were in vogue and everybody was happy. Every man had two or three good horses and nothing to do. The place was full of visiting girls. They rode in parties to High Knob, and the ring of hoof and the laughter of youth and maid made every dusk resonant with joy. On Popular Hill houses sprang up like magic and weddings came. The passing stranger was stunned to find out in the wilderness such a spot; gaiety, prodigal hospitality, a police force of gentlemen — nearly all of whom were college graduates— and a club, where poker flourished in the smoke of Havana cigars, and a barrel of whiskey stood in one corner with a faucet waiting for the turn of any hand." [13]

The Fox brothers— James, Horace and John Jr.— opened an office in Big Stone Gap, and by the end of January 1890, John was able to write to James that lots within the city limits, when they could be found for sale, were selling for as much as $2,500 each, and that overnight they had made $250 on one lot. "If I had been here three months ago," he wrote, "I should have had a bank account of $10,000 by this time." [14]

Somehow, in all the confusion and helter-skelter scurrying that he had done in the past few months, the story he had been working on for so long was finally finished. He sent it to *Century* magazine in New York, but instead of accepting it with the praise he thought it deserved, they wrote him that they thought "the general situation of your story 'A Mountain Europa' is somewhat hackneyed," but, they said, it interested them quite a bit, if there were some changes made. [15] It was his first association with an editor, and he did not like it very much. "The greatest fault, to our minds," they continued, "is a certain 'wordiness' which pervades almost the entire narrative. The style is dignified," they said, possibly trying not to be too hard on him, "but it is

James Wallace Fox was the oldest of John's siblings and perhaps the most daring in his financial schemes. It was primarily James who brought the family into the coal boom of the 1880s and 1890s, and was instrumental in bringing the family to Big Stone Gap, Virginia, in 1890 (courtesy John Fox, Jr., Museum).

7. Boom and Bust, Part II: The Gap

rather conventional and over-stately in expression." [16] They suggested that he go over the entire story with care, and condense every page down to its most important points, and if you would "send the story back to us, we should be glad to use it." [17]

He made the changes they wanted, and sent it off as soon as he could. "I sent the story in early one autumn," he wrote, "and one black night about six months later I was wading back from the post office through rain and mud by the light of a lantern with a note of acceptance and a checque for $262.00 in my pocket. I had that checque photographed, but by that time the 'boom' had started. I was a budding Napoleon of finance, and I would have paid the *Century* twice that amount to publish the story." [18]

The "Napoleon of Finance" had much to do. His parents were still waiting impatiently in Kentucky for their move to the Gap, and the responsibilities of the land office he was now running along with his brother Horace was beginning to take up more and more of his time. He thought little of writing, in fact, he had little time to think of anything but business, the price of land, and the availability of capital. His two younger brothers, Rector and Richard, soon joined him, as did James, and the younger two proved to be a handful. At ages 17 and 19, respectively, they could hardly be controlled in that wild country, and ran pretty much as they pleased without any parental direction and took complete advantage of their older brothers' time-consuming work habits. By the time James had been in the Gap only a few days, the three older brothers had gotten their heads together and were thinking about sending north for a tutor to keep Rector and Richard under control, if only slightly.

The next three years would be spent in dizzying business schemes and the ever-present search for investors and more money to keep them from bankruptcy. Principally, James, John and Horace were involved. Sidney, having married and having a successful medical practice in New York, had little actual involvement, excepting a few dollars' investment, and the others, Oliver and Everett, were off on their own in other parts of the country, pursuing their own interests. Ultimately, the weight of the "boom" fell upon the shoulders of the three brothers.

By the middle of April, their father and mother and the girls could not be put off any longer. "We are packing today," John Sr. wrote on April 20, 1890, "and will load car tomorrow, and start it off at once." The tales of money being made and spent that made the elder Fox "rub his eyes to see if I am not dreaming," [19] did nothing to encourage patience in the old man, and he was bound to go. It is unlikely that anything short of physical infirmity could have stopped him.

By the middle of May 1890, the stresses of the new business in the Gap was beginning to make itself evident. John's ability to hold his temper began to deteriorate while he tried to keep track of the day-to-day operations of their land office, and even his father was beginning to get on his nerves, only a few weeks after his arrival. "Papa gets up at 6 o'clock, takes a cold breakfast & makes himself sick and miserable by dark," he wrote. "Yesterday morning, he actually left me a list of miserable little trifles — hauling wood — paying butcher, etc., etc. to perform while he hurried away to his important duties." [20] And that was only the half of it. Nothing seemed to be going as he would have liked it. Perhaps ominously, nothing was selling, and next month, June, $26,000 in loans would come due at the bank, and they had not enough money on hand to pay. Somewhat sarcastically, he said, "Now with a batch of correspondence, the books to overhaul, study and complete, fitting up the office, building fence, kitchen altering interior, painting ... $2000 or $3000 behind in bank — some $26,000 in notes due at bank in June, everything dull, selling nothing & expenses big ... I have an abundance of leisure, cheerfulness and charity." [21]

Everyone, he complained, was caught up in the get-rich-quick schemes, and everyone wanted to get rich without working, even carpenters, painters, and ordinary laborers. He would later write that while the "boom" was going on, you could not even get a carpenter to hang a door. And to top it all, his brother Sidney, who showed little interest in the Gap thus far, was bold enough to send an unemployed, and unemployable, friend down from New York with a promise of a good job when he arrived. John did not know what on earth he would do for this friend of Sid's. "He can't do physical labor, he can't keep books, or even set the type." [22] He would do what he could. After all, he had plenty of "leisure" and "charity," or so he said.

Very few things are as they are advertised, and when the expectations of quick fortunes did not live up to reality, some of those erstwhile enthusiastic investors — of more faint heart than others — began to sell their properties at any price they could get. Some of the later investors in the Gap, including the Fox brothers, had bought at a high price with the expectation of prices going higher, and making a good profit. It did not happen, and to sell now would mean the loss of hundreds, if not thousands, of dollars. The Gap had not "boomed" quite as much, nor as quickly as it was thought that it should. The lucky ones were those who got out early. The Fox brothers, including John Jr., were not among those fortunate few, and there would be years of convoluted and confusing financial dealings.

One letter written to James on June 23, 1890, their first year of their venture into Gap real estate, is indicative of all that followed, and John's exasperation:

Dear James,
   I wrote you but tore the letter up in view of more developments. Howard had to have some of the money on that $15,000 & I was compelled to sell 15 of the 35 bonds (held by Bullit, Bryant & me) in Louisville for $9000. I thought I had the right to sell all or certainly my third $11,666⅔ and I expected to replace the rest $3,333⅓ either with

bonds or with money at the rate at which Bullit & Bryant should sell their own.

According to the contract I learn to-day I had no right to sell at all without the consent of either Bullit or Bryant and accordingly I must replace the bonds—fifteen. Bullit offered the $9000 to the $21,000 note, but I consider it better to replace the bonds, for this reason.

I should still be liable for the rest of the $21,000 note. Moreover Bullit says he can borrow $21,000 on the 35 bonds in Louisville (with the endorsement of Bryant and me). If he can, as much money can be received on the $11,333 ⅓ bonds in that block as you or I could borrow on $15,000 of bonds alone (as a matter of fact, it seems, we can't borrow at all — I have tried all the banks in Louisville, Lexington, Knoxville and Gainesville).

So that I applied $10,000 ($1000 of the app. loan) to the $15,000 note and got Howard to carry the remaining $5000 as long as he can. Horace had overdrawn in my absence. Sid had written that "we must sell his and Smith's and Lair's lots at once" and that he must have $1000. And to-day came a telegram from him saying he must draw on us. The father's notes came due in July — signed by Horace and me — some $600 or $700 & we must watch for them. There are my reasons for holding back the rest of the insufficient (I forgot there is still another note of Horace's $1000 due July 13). Now you can consider yourself as owning $11,666 ⅔ or those 35 bonds and Horace and me owing you $3,333 ⅓ in bonds. [23]

Later, in the same letter, he bemoaned the fact that money seemed to be tight and in short supply everywhere he asked for more loans. The Fox brothers, early on, made the serious financial mistake of trying to, and sometimes succeeding in, borrowing money to pay off interest on loans they had already received elsewhere. Everything hinged on whether or not the expected boom would soon come. Should that happen, they fully expected land prices to shoot skyward. They would be able to sell their investment lands at a huge profit, pay off the loans they had incurred to purchase those lands, and still have a fortune left over. It was a risky business to be in, and for the inexperienced, without substantial capital to sustain them through hard times now and in the future, it was especially fraught with pitfalls of the most serious kind.

Money was not only just tight in the banking business. Things in the Gap had quickly gotten to the point that money needed to finish James's house could not be had. They needed every dollar they could scrape together just to stay in business. Somehow, they managed to keep their financial heads above water, and just when things could not get any worse, they did.

Sanitation in the town was almost nonexistent, or of the most rudimentary kind, and with the influx of people, outbreaks of disease was only a matter of time. The result was malaria and typhoid, and it could not have come at a worse time. Investors, already leery of investing their money in a place that seemed to have come to a standstill, ran backwards at the idea of even coming to a place that was under the attack of pestilence. John played down the significance of it, however, saying that, "As I expected there is a small epidemic of malaria here — due to the drainage — with occasionally cases of typhoid." [24] One cannot help but wonder at the characterization of a "small epidemic," and it would not be long before the news of typhoid and malaria were in the newspapers, and the situation quickly became major.

Building in the Gap quickly ground to a standstill. The hotel, which was to have already been built, stopped completely with little more than the foundation in place, and the "little epidemic" became worse. "This malarial epidemic, the collapse of the hotel, ... the utter lack of building ... and a general feeling that somehow the place is being deserted have made this place 'bluer' than I have ever known it," he wrote to James. [25] By the end of the month, nothing had changed. "Our enemies tell many lies about our fever & people come in having heard all around us that the sick are dying at the rate of 3 or 4 per day," he wrote James. "All these rumors & the general distressing stagnation hurt us fearfully." [26] Their only salvation, he supposed, was the railroad which was yet to make it into the Gap, and they waited impatiently for it to arrive.

His letters to James continued in much the vein all that summer. John Sr. had been sick most of the time with malaria, it was supposed, but was soon better. Their brother Sidney, who had invested $1,000, was planning a trip abroad, and had demanded his money to pay for the trip. They had grudgingly sent it to him, and James Lane Allen, friend to both James and John Jr., had been convinced to invest, but being a writer was often without funds and had asked for $200 of his investment back. They felt obligated to send it, and did. But for every dollar that went out, none was coming in, and the bank account of Fox Brothers Real Estate Brokers, Big Stone Gap, Virginia, was often, as John Jr. said, "minus zero." [27] He could not help but see the dark humor in their situation, however, intentional or not, and despite all their problems, sickness and debt, he ended his letter of July 25, 1890, to James by saying, "Papa is getting better & we are all quite cheerful." [28] But his father was homesick for Kentucky, and was beginning to think that he had made a bad decision in moving away; he had never been quite the same since leaving Stony Point. The state line was only eight miles or so away — you could almost see Kentucky from the Gap — but it may as well have been 8,000 miles to the elder, homesick Fox.

Stress almost always made John Jr. physically sick, and so it was only a matter of time, under the circumstances, that he would fall ill. By the end of August 1890 he had a touch of malaria, and noticed his skin had become a pale yellow in color. It did not seem as if he was accomplishing anything in the Gap, and so it was not a hard decision for him to leave. He wound up in West Virginia at White Sulphur Springs searching for a doctor. The trip made his con-

A diary entry for August 1890 mentions Robert Burns Wilson's visit to the Gap to see John. This photograph may have been made then. Left to right: H. Clay Horland (?); Robert Burns Wilson; Miss Berryman of Lexington, Kentucky, a relative, no doubt of Upsur Berryman, John's roommate at Harvard; Elizabeth Fox; John Fox, Jr.; and Minnie Fox (courtesy of Duncan Tavern).

dition worse instead of better, but he found a doctor, though by that time he was "as yellow as a pumpkin in September." [29]

The change of scenery did him more good than the doctor's prescriptions, and he could not bring himself to return to the Gap anytime soon. He had grown to hate the office work in the Gap and the day-to-day aggravations and worry that went along with that job. By September, feeling better in all ways, he found himself in Boston. His mind was not far from business, however, and he occupied himself with trying to gather up new investors for the boom that was sure to come to the Gap in the fall. In this he had some modest success, but he would no longer ask his friends to invest, and told James that if it had not been for the others in the family — naturally referring to James — he would never have involved his friends in the Fox family ventures in the first place. In any event, he vowed, he would do it no more.

The *Century* company had not published "A Mountain Europa" as yet, but that did not keep him from beginning another tale. He wrote to James from

Boston the middle of September that he was "Improving right along, living cheaply," and had "knocked out 40 pages of my novel." [30] That "novel" would later turn out to be *A Cumberland Vendetta*. Then, out of the blue, a letter from *Century* confirmed that publication of "A Mountain Europa" would be soon, and they were sending him "unfinished proofs" of drawings they had commissioned to illustrate the story. At about the time he was hearing the good news from *Century*, they received the most unwelcome news from New York.

Sidney, living in Brooklyn, was the first of the Fox boys to marry, and living as far away as he did, he and his wife, Polly, had little contact with any of the rest of the family. Except for occasional visits by the other children to New York when they sometimes stayed with Sidney and his new wife, none of the family ever saw him, nor heard from him except through those who had visited him. Letters from him, unlike the rest of the Fox family, were practically nonexistent. Toward the end of November, Polly, Sidney's wife of only three years, became sick with what was diagnosed as typhoid, or at least, a sickness with typhoid-like symptoms. But as serious as her illness was, Sidney was a doctor, and no one in the family was worried too much about it, feeling that she was in very capable hands.

The elder Fox wrote in his diary for December 9 that they had received "two dispatches to day from Jimmie in N.Y. saying that Sidney was very sick with pneumonia, and we are very uneasy about him as it is his second attack, but hope he will grow better soon." There was no indication that it would be more than a temporary illness, and John Sr. wrote nothing more about it that day, moving on to other thoughts in his diary. Somewhat forlornly, he added underneath those first few lines, "I am 60 years old today." [31]

And so it was with quite a shock that they received a telegram from James that, although Polly had recovered, Sidney had not, and had suddenly died. All were devastated. "It does look so hard," wrote the elder Fox to James who had written him of Sidney's death, "for one so young, and so devoted to suffering humanity, to be called away from his life work, and sometimes I feel it all so keenly, that a stifling smothering feeling almost overcomes me." [32] His letters, usually long and involved, became terse and abbreviated. He wrote in his diary, the day after Sidney's death, "This has been a sad long day to me." [33] He could not bring himself to talk or write very much about his son's death. Instead he busied himself with the minor details of having obituary notices published in newspapers back home in Kentucky, and walked the muddy streets of the Gap, hands thrust deep in his pockets, hunched from the cold of December, feeling his age as more of a burden than ever before, and thinking of another long dead for nearly forty years, buried far away in Kentucky. And now, he thought of the son of that long dead first wife, buried farther away, whose grave he would never see, whose funeral he could not attend, and whose life should have extended past his own.

He could not bear to think of it over long, and so, through those short, cold and bitter days before Christmas, he walked of the evenings up the hill from the house, and back again, wondering what he had gained and accomplished over his 60 years, and when he went inside, to warmth and family, he felt the darkness outside closer, thicker, and somehow more forbidding than before, and wondering about that unwritten law that says a man shall not outlive his sons.

Sidney's stepmother was not unaffected. She was aware, too, of that same adage that was troubling her husband those long winter days, but she held her grief close to her heart, letting little show. Sidney was not her own in flesh nor blood. She remembered those three little motherless boys, of 30 years before, now grown into men, and now one dead, and her mother's heart broke.

By the end of January 1891, John had returned to Big Stone Gap to find things much the same as when he left except for the possibility of business being worse. The same old problems assailed him, and it was not long before he wished he was somewhere, anywhere else but the Gap. Money continued to be in short or nonexistent supply. Loans and the interest on loans were falling due with a terrifying rapidity, and one could easily liken their situation to a snowball that gained girth and speed as it rolled down an inevitable hill of doom and destruction. It was often more than he felt he could handle.

James was off to England and New York and nearly everywhere he thought money and investors might be had, trying almost frantically to save them a little while longer, thinking all the while that prosperity was just around the corner. John was left alone at the Gap, trying to do the best he could against overwhelming odds, and it was not long before he began to feel abandoned, trapped and alone. He was not entirely without a businessman's instincts. The last few months had taught him well, perhaps better than any school could have done, and if he perhaps could not yet see the writing on the wall, he was perceptive enough to know that the wall might be close. He swore to himself, time and again, that if he ever survived the situation he was now in, he would never allow himself to be caught in such a predicament again for as long as he lived. Things had gotten to such a point that he felt confident in saying to James that "if somebody here should bring suit we should go to the wall in a hurry." [34] He did not know then, how prophetic his words would be.

Finally, things had gotten so bad, money-wise, that he had to make the suggestion to James that their younger brothers, Rector and Richard, come home from Harvard where they were only just beginning their studies. With things the way they were, they just could not, he thought, afford to keep them both in school. "I think we had better face the facts," he wrote, "and have these boys come home and go to work." [35] Once the crisis was over, perhaps next year, they could go back, but for now, he did not think they had "the right to send the boys away when we can't pay the grocer's bill at home." [36] The "boys" would have, no doubt, been agreeable since they had not done as well in school as their brothers before them, but fortunately, that decision was put on indefinite hold, and Rector and Richard stayed away at school.

Writing and his frequent trips north were his only escapes. He soon began to cultivate friends who were writers themselves. He had known James Lane Allen for years, and had become friendly with Madison Cawein, the Louisville poet, and Robert Burns Wilson of Frankfort, Kentucky. [37]

Burns, like Cawein, was a poet, songwriter, and artist. He and John became close friends and would, from time to time, share a room at the corner of Ann and Clinton Streets in Frankfort. The times he spent in Frankfort with Burns were some of the best of his life. Arm in arm, the two took Frankfort by storm. Burns was arguably the best-known personage in Frankfort — with the possible exception of the governor — at that time, and all doors were open to him. Parties and society to-dos on a grand and seemingly never ending scale were always at the top of the menu. They became fast friends, these two young and handsome and aspiring writers, and John treasured these times away from his more mundane responsibilities in the Gap.

But perhaps his best and most influential friend, especially during the years from 1891 until his death in 1919, was Virginia born writer Thomas Nelson Page. They had met for the first time while John was working for the *Sun* in New York, and although their friendship grew somewhat slowly for the next three years or so, by 1891 it was renewed, and never flagged. Page would become John's encouragement, his protector and his confidant. A successful lawyer and author himself, Page had everything that John Fox wanted and did not have, but it was never a one-way friendship, and there was never any jealousy involved in their friendship. They were good and best friends always. John would be a welcome addition to the Page homes in Richmond, Washington, and York Harbor, Maine, and over the years, he visited there often, sometimes for weeks at a time.

He thrived on the encouragement that his author friends, especially Page, gave him, and it is likely that without that encouragement, he would not have written as much as he did. By the first week in October, he had finished "Vendetta," and had sent it off to *Century* magazine who was still holding on to "A Mountain Europa," though they continued to promise publication any day now. "I have finished up another story of the mountain feuds," he wrote to Page, "and have sent it in to 'The Century'. Of course I am on the anxious seat, for it would pretty near break my heart to have it refused, and yet it is not to my satisfaction. That's what makes me uneasy... I did not do my full duty." [38]

He had not been in the Gap since March, and was having so much fun in New York and Boston, Lexington and Frankfort that he refused to allow himself to think about returning to the drudgery he had left. Along toward the end of October, he finally received long awaited good news about "Europa." The publication of the story had been held up because of the illustrations. The

7. Boom and Bust, Part II: The Gap

This photograph of Thomas Nelson Page was taken around 1917 while he was ambassador to Italy. Tom and his wife Florence would arrive home from that assignment on the very day that John died in Big Stone Gap (Charles Scribner's Sons).

magazine had not been satisfied with the initial drawings, and had them redrawn by another artist. But they were just as enthusiastic about "Vendetta," and he felt sure they were going to publish it also. He needed the money that story would bring. The $262 he had gotten for "Europa" was long gone, and a doctor had assured him that he needed an operation for his old injury. His hopes were high that he might get as much as $300 for the story, maybe even as soon as next month. By the time he had written James on October 30, his estimate of its value had risen to $400.

He had not entirely neglected the Gap's business while in the North, but the preoccupation he was having with the upcoming operation distracted him, and made him uneasy, wanting to get it done and over with as soon as possible. "I wish you could come on here," he wrote James, meaning the Gap though John was not there himself, "and let me go on to Boston and do all I can there for a week or ten days and then have my operation performed… I might get out of the hospital in 3 weeks and then I should be willing to go back. But where the money is to come from, I don't know." [39] He was down to his last

$30, and owed his landlady almost a third of that. He began to shop around for a cheaper doctor. [40]

On top of his worries about his health came the news from *The Century* that they were somewhat "disappointed" in his last story, "A Cumberland Vendetta." They wrote that they were "much pleased and much disappointed with the new story ... in construction, movement and style ... much disappointed." It seemed to them to have been "conceived in too sentimental a spirit, and written a little tritely with phrases that are out-worn." [41] He was hurt by their rejection, but upon reflection, he remembered telling Page that he was not satisfied with it in the first place, and did not think he had done his best. He flew into the revision of the story with a vengeance, and found himself enjoying the rewriting. "Am working every day on my story," he wrote James from Louisville, "and am in a glow of enthusiasm over it." [42]

Thomas Nelson Page was already a nationally recognized and respected author and would continue to be so for so long as he cared to write. In recent years, he had been reading his stories to audiences all over the East. There was good money to be had doing this kind of work, and Page had encouraged John to do the same, but there was just one problem: John had not written enough, or so he thought, and had not had anything published, though there had been promises aplenty. And now, the idea struck him, that if he could devote more of his time to writing, he could take to the road, speaking in towns near and far, rubbing elbows with other famous people, and not incidentally, living the free and easy life he had always, perhaps subconsciously, longed for. Having immersed himself in the literary circles for the past year, he became determined more than ever to make writing his vocation.

Back in the Gap, the end was in sight. There were still those who insisted that the "boom" would surely come, that success was just around the corner. They had done everything the way it should have been done, and could not, would not, understand that location and pure luck could be the determining factors in whether or not they were successful.

Luck had turned against them, but still they held out hope. Maybe they were just too soon. Maybe the people, the times, were just not ready for a boom, and it they could just hold on a little longer. "Some day," he wrote years after all hope of a boom were gone, "the avalanche must sweep south, it must — it must." That he might be a quarter of a century too soon in his calculations never crossed his mind. "Some day it must come." [43]

But the boom was over, or rather the expectation of a boom was over. The capital they needed to build a town — and to interest outsiders into moving industry and themselves to a new town — had not materialized despite high hopes from nearly everyone involved. Finally, even the most optimistic could see the truth. It had been a good plan, but they simply ran out of luck.

In *The Trail of the Lonesome Pine*, his bitterness still showed through, years later. "Not suddenly did the boom drop down there," he wrote, "not like a falling star, but on the wings of hope — wings that ever fluttering upward, yet sank inexorably and slowly closed. The first crash came over the waters when certain big men over there went to pieces — men on whose shoulders rested the colossal figure of progress that the English were carving from the hills of Cumberland Gap. Still nobody saw why a hurt to the Lion should make the Eagle sore and so the American spirit at the other gaps and all up the Virginia valleys that skirt the Cumberland held faithful and dauntless — for a while. But in time as the huge steel plants grew noiseless, and the flaming throats of the furnaces were throttled, a sympathetic fire of dissolution spread slowly North and South and it was plain only to the wise outsider as merely a matter of time until, all up and down the Cumberland, the fox and the coon and the quail could come back to their old homes on corner lots, marked each by a pathetic little whitewashed post — tombstone over the graves of a myriad of buried human hopes. But it was the gap where Hale was that died last and hardest — and of the brave spirits there, his was the last and hardest to die." [44]

John Fox saw himself as Jack Hale, the hero of *Lonesome Pine*, in many respects, and years later, John would pour through that character his own thoughts and beliefs and feelings. When he had Jack speak of the Gap and what had happened there, it was John speaking just as surely. "It was the gap...that died last and hardest — and ... his [hopes were]...the last and hardest to die." [45]

The years following 1891 were not uneventful years for both the town of Big Stone Gap, Virginia, nor for the Fox family who had, for better or worse, tied their fortunes and their futures to the town in the mountains. The Gap did not die, but continued to grow, although at a much more sedate pace. Homes and businesses were still being built, but now they were being built because people wanted to build, not because they felt they had to or were obligated to build because of a contract. John, and the rest of the family, despite the money real and promised they had lost when the boom failed to materialize, found that they liked the new atmosphere around town, and except for what seemed to be a few minor loose ends of real estate to tie up, hoped they would come out of this whole thing in one piece.

CHAPTER 8

A Knight of the Cumberland

From the beginning, times were lively...
John Fox, Jr.

...for the boom is a shadow, and whatever
of light there be in it is as a flash of lightning...
John Fox, Jr.

THE STRESS OF HIS BUSINESS DEALINGS in the Gap had somewhat abated, but they would not completely leave him alone for years to come, and that after the first blush of the supposed "boom" had not materialized. John Fox wanted to get away from the Gap, and try something new. Every day he spent there reminded him of his failures as a businessman, and it seemed to him that no matter how hard he tried, he could not shake his bad luck.

He was upbeat about his new home, however, and in the middle of his dejection about his financial prospects, he at the same time began to see that the Gap might have other possibilities. Not, perhaps what they had at first envisioned for it, but it could be something, perhaps something even better than all of them had first supposed. He began to see the Gap as his permanent home now, and seeing that, he was determined to make it as liveable as he could for himself and his family. They had built themselves a home there, large enough to hold them all, and although some of them had or would have their own individual homes, it was important for them to have at least one where they could all be together, and just off Shawnee Avenue, a two story rambling structure was built for that purpose.

The new town of Big Stone Gap continued to have all the problems of any frontier town. Streets were muddy almost all winter, and were actually dangerous for man and beast, which did not make the Gap any different from many more "settled" towns of the time. One local paper in Kentucky wrote in January of 1896 a tongue-in-cheek description that applied to most rural towns of the period when they said that "water and mud are no longer a scarcity," especially in the winters. [1] Things had not gotten any better with the passage of the summer and repairs of the now relatively dry streets. The same paper wrote again that November that "the condition of the county roads in this county is a disgrace to civilization and should be inquired into by the grand jury," and three years later, "The streets of our town are in a most fearful condition, the mud in places being knee deep, and what work has been done on them this fall seems to have tended to make their condition worse." [2]

There was also violence of the meanest sort. When the president of Berea College in Kentucky characterized the mountaineer as being "much addicted to killing," he should not have so narrowly construed his statements nor the area to which he applied them. It seems that all Americans, mountaineer and otherwise, were "much addicted to killing," and general lawlessness. The Gap of the 1890s was no better, nor no worse than many places in America at that time, as far as streets and violence were concerned.

If there was a common denominator that characterized the violence and mayhem, it was the combination of whiskey, both store-bought and homemade, and the prevalence of firearms. Neither one by itself posed any real problems out of the ordinary, but when mixed together, they often led to disaster. "Another noticeable recreation of the day was the drinking," said James Lane Allen. "Indeed, the two pleasures went marvelously well together. The drinking led up to the fighting, and the fighting led up to the drinking; and this amiable co-operation might be prolonged at will.... It was not usually bought by the drink, but by the tickler. The tickler was a bottle of narrow shape, holding a half-pint — just enough to tickle. On a county court day well-nigh a whole town would be tickled." [3]

Newspaper accounts of the times are filled with instances of petty quarrels about the most insignificant of provocations turning deadly, if the combatants were not too drunk to hit their targets. Kentucky's reputation for marksmanship suffered mightily around the turn of the century, as did the marksmanship in those frontier towns of southwestern Virginia and other places in the mountains, and it is quite apparent from those newspaper accounts of the mayhem that most killings were little more than bad luck on the part of the deceased. Account after account mentions that "15 or 20 shots were exchanged ... but strange to say no one was hurt." [4] In another instance, a store clerk in London, Kentucky, was mortally wounded after seven shots were fired between himself and a customer. The incident started over the clerk throwing a pig's foot at the other, who was drunk. The courts, as usual, were little help, and juries often sympathized with, or were afraid of, the defendants. "The Highlander is ... a clannish person," wrote John C. Campbell in 1921, "and he does not like either to inform on his kinfolk or to witness against them in court.... As an individualist he has ... an instinctive sympathy with the person under arrest, unless that person has been guilty of an offense against himself. Evidence thought secure sometimes melts away in the publicity of the court room, and a verdict of 'not guilty' may be secured when knowledge of the offender's guilt is quite general throughout the community." [5]

That Big Stone Gap, Virginia, was essentially lawless has not been disputed by many. Weekdays were bad enough, but Saturday nights were the worst, except for election day which fostered its own brand of violence. The Gap became a meeting place for miles around, bringing in toughs from everywhere convenient to horseback and foot travel, and Kentucky, within shouting distance, contributed as much as possible. As bad as "normal" times could be, local elections often made things worse. "Every Saturday there had been local lawlessness to deal with," [6] Fox wrote. "Here they drank apple-jack and hard cider, chaffed and quarreled and fought fist and skull.... And always during the afternoon there were men who would try to prove themselves the best Democrats in the State of Virginia by resort to tooth, fist and eye-gouging thumb." [7] Sometimes, he said, they would completely subdue the entire town, taking it over with gunshot and general rowdiness. Storekeepers would close and lock their doors, exiting the back door, locking it on their way out, seeking cover in the nearby woods or private homes until the whiskey had its desired effect: rendering the horse backed shootist unconscious until the next day, when the clean up could commence, in preparation for the next weekend's festivities.

The local newspaper, not a year old in 1890, wrote that "men would quarrel, shoot and cut at will, and were seldom arrested, even to go through the form of a trial.... Drunken men would ride along the streets at a rapid rate, firing pistols, yelling like wild Indians, and no one dared even to attempt to detain them." [8]

Certainly, not everyone living in and around the Gap were lawless. Most were not, and even those who liked to drink a little and go to town with a pistol in his pocket had a respect for the law. "I venture the paradox," wrote Fox of the mountaineer, "that he still has at heart a vast respect for the law." [9] They respected the law most times, but they were more likely to respect it most when it suited them best.

With the influx of outsiders into the Gap, the prevalence of shooting, killing and general roughhousing — the absence of any real enforcement of the meager laws that were on the books — soon became a real problem. Those who had moved to the raw town from the North and other reportedly more civilized states did so with a long history of having peace and quiet and letting the law run its course. They were not used to lights being shot out, signs being riddled with bullet holes, and horse racing drunks careening down the middle of the streets, scattering men, women and children to the dubious safety of sidewalks and mud holes. Something had to be done, and whatever it was, had to be swift, sure and permanent. The result was the now famous "Guard."

There had been other police officers before, but they were outnumbered and woefully unsupported by the citizens who had elected or hired them. Many had resigned in disgust over the way courts and juries had coddled those arrested. One of the last constables before the formation of the "Guard," locked horns with a miscreant on the streets of the Gap, and was soon overwhelmed, shouting for help with every other breath he took. Not surprisingly, no one came to his aid. "'I've fit an' I've hollered fer help,'" he cried, "'an' I've hollered fer help an' I've fit — an I've fit agin. Now this town can go to h--l.' And he tore off his badge and threw it on the ground, and went off, weeping." [10]

Everyday citizens often winked at the violence; it had been going on so long, longer than many had lived in the Gap, and it had become something of a tradition. It practically never occurred to them that they might want to put a stop to it, or even that they should, but some of the newcomers had different ideas. They had come from lands that had more or less peace-loving communities, and now they wanted peace for the Gap. "They could not build a town without law and order," Fox would write later, long after peace was established, "they could not have law and order without taking part themselves, and even then they plainly would have their hands full...but they meant to hew to the strict line of town ordinance and common law and do the rough everyday work of the common policeman." [11] The Big Stone Gap *Post* was lavish in its praise of the new volunteer Guard, and it was right that it should be. "They proceeded to organize a volunteer police force," wrote the *Post* proudly after giving accounts of lawlessness, "and elected Mr. Joshua Bullitt their captain. Under the leadership of Captain Bullitt this untrained corps of police have developed into one of the best organized and most effective bodies of the kind in the country." [12]

Another place, another time, it might not have worked, but the mountaineer was so shocked and taken aback at the appearance of derby-hatted, suited, pistol-packing, whistle-blowing constables, that it is likely that they thought that it must be some kind of dangerous trick. By the time they realized that it was not a trick, it was too late. The ground work for law and order had been established, and the rowdies found themselves unsupported on their home court.

While their initial uneasiness was wearing off, others who had been hesitant at first were beginning to see that a little law and order and peace and quiet was not such a bad thing. They liked it, and whether or not they admired, trusted or even liked the outsiders posing as policemen, they did like the results. Whatever the circumstances, those results apparently were nothing short of miraculous. John, always liking adventure, soon joined them, as did his brothers, but by his own admission, most of the organizing and initial confrontations were over by that time, but not all. For lawmen in a frontier town, East or West, there is always plenty of work to be done.

John, his brothers, and other members of the vigilance committee broke a few heads when necessary, but generously did not hold any ill feelings toward their arrestees once the culprits were released from the local jail. More remarkably, those who had been arrested for one thing or another did not hold any grudges against the arrestors. The Guard always tried to work in twos and threes, more if possible, for a mountaineer would not be shamed if he could go back home and tell that it took three men to arrest him in town, but if he gave up to only one he could hardly live down the shame. It worked. After a period of about 18 months, the streets of Big Stone Gap had undergone a transformation that could hardly be believed by insider and outsider alike, and "a woman could even walk on the streets at night and not be insulted — a thing never dreamed of before in that region." [13]

Then, another strange thing happened. The rowdies who had previously caused so much trouble began to join the Guard themselves and became, if anything, more ardent enforcers of the law than the original members. "Indeed," said Fox, "the enthusiasm for the law was curiously contagious. Wild fellows, who would have been desperadoes themselves but for the vent that enforcing the law gave to their energies, became the most enthusiastic members of the guard." [14] It is not unlikely that some thought to use their positions to settle old feud scores, but it does not appear that this happened too much. That very little abuse of power occurred only goes to underline Fox's initial belief that the mountaineer is genuinely respectful of the law, and subconsciously hungers for the establishment of a fair justice.

"It was a unique experiment in civilization," [15] said Fox, and so it was. The wonder is, that the members of the Guard were not slaughtered where they stood. "In this town," he wrote, "certain young men — chiefly Virginians and blue-grass Kentuckians — simply formed a volunteer police-guard ... and each man armed himself — usually with a Winchester, a revolver, a billy, a belt, a badge, and a whistle — a most important detail of the accouterment, since it was used to call for help." [16] Apparently, they were not completely stupid, since they included a whistle in their equipment to call others of the Guard for help, and to make sure no mistakes could be made, they forbade anyone else from blowing a whistle within the town limits.

The Guard was adamant in their enforcement of the law, but even they had their limits, sometimes. "I have known members of the force to protect a Negro from a mob while he was on territory in which they were sworn to preserve the peace," John told a reporter in 1900, "and join the mob in lynching him after he was taken beyond our jurisdiction where the oath had no binding effect." [17]

They passed ordinances against shooting and racing up and down the muddy streets of the Gap. "The wild centaurs were not allowed to ride up and down the plank walks with their reins in their teeth and firing a pistol into the ground with either hand; they could punctuate the hotel-sign no more; they could not ride at a fast gallop through the streets of the town, and — Lost Spirit of American Liberty — they could not even yell!" [18] They cracked down on the illegal sale of whiskey and set up patrols to enforce the law. "They were lawyers, bankers, real-estate brokers, newspaper men, civil and mining engineers, geologists, speculators, and several men of leisure," John wrote, "and most of them were college graduates, representing Harvard, Yale, Princeton, the University of Virginia, and other Southern colleges." [19]

Once organized, the Guard bore down on the lawless, and to their credit, they would knock a townsman on the head as quickly as a drunken mountaineer, though that did not happen very often. But when it did, it was this evenhandedness that gained them respect from the rowdies outside the Gap as well as the citizens within. At first there was the usual resentment, from the mountaineers especially. It appeared to them, and with some foundation, that outsiders were coming in and telling them, at the point of a pistol and the business end of a billy-club, what they could and could not do inside their own town. They did not like it, and all manner of idle threats were made against the members of the Guard, but, in the words of Jack Hale, John's main protagonist in *The Trail of the Lonesome Pine*, "The law has come here and it has come to stay." [20]

The advent of law and order, and the slackening of the "boom," gave John more time to concentrate on his writing, which was getting off to a hesitant start; even *he* could admit that to himself. He found himself in the unique position of beginning to think of himself as a writer, but one who had, as yet, had nothing that amounted to anything published. Excepting a rather insignificant piece in *Frank Leslie's Illustrated Weekly* called "A Betrothal Ring," some six or seven years before, and another like piece entitled "Deceiver's Ever," that appeared in *New York Life* the same year, there had been nothing — if one did not count promises from the *Century*. So, even though he was beginning to think of himself as an author, he had yet to see anything of his published in any magazine, or anywhere else that fiction might be printed. Even the publication of his first story "A Mountain Europa" was still being held up by *Century* Magazine.

John was not too concerned with excuses from *Century*. He just wanted to see "Europa" in print, but was beginning to think he never would. In the meantime, there was enough excitement in the Gap and nearby towns to keep his mind occasionally from his literary problems. One of these distractions came about in the fall of 1892, in the little town of Gladesville, the county seat of Wise County, Virginia, a few miles north of Big Stone Gap.

It sat on a small hill in Gladesville. It stood practically windowless and gloomy even in the brightest of sunlight. It was an ugly building, but most jails are not built nor maintained for their beauty. Inside, looking a little lost and bewildered sat Talton Hall, waiting the day of his execution, taking little solace in the fact that he would be the first man ever hanged in Wise County, Virginia. Newspapers of the day made the somewhat dubious distinction that Hall would be the first man "legally" hanged in that part of the country.

What made the situation even more ridiculous was, in the next cell, awaiting his own execution was M.B. "Doc" Taylor, the man who had been most instrumental in bringing Hall to justice. Hall made life a living hell for his neighbor, and jail-mate, but if he ever thought of poetic justice, he never said anything about it to the volunteer guards from the Gap who kept a watch over him day and night.

How both men came to be where they were is a story as convoluted and mysterious as the men themselves. No one ever starts out to be hung, but if anyone had, Talton Hall would have been the one to have sought guidance from.

Talton Hall, a killer before the word "psychopath" was in vogue. Hall's thirteenth murder — that of Enos Hilton (sometimes "Hylton" in Norton, Virginia — proved to be his undoing (courtesy Wise County Historical Society).

By the summer of 1881, Hall had already killed at least 12 men, one of whom had been his uncle, which he denied having any part of. The number of his victims vary with the source, but it cannot be denied that, in a time of bad men, Talton Hall seemed to be one of the worst. In 1881, Hall killed his thirteenth and last man.

Hall was in Norton, Virginia, north of Big Stone Gap, with a friend named Bates, and not far from the Kentucky border. Details are scarce, as most such killings do not lend themselves to complete disclosure, but the upshot of it was that a policeman in Norton, Enos Hilton, arrested Bates for stealing a watch. When he turned his back with his new prisoner in tow, Hall shot him dead, and both he and Bates turned their feet toward Kentucky and safety.

One year later, captured and tried, and sentenced to hang, Talton Hall admitted to a smallish looking man who was his sometime guard while awaiting execution, that he knew when he killed Enos Hilton in Norton, he had done his last bad thing. That slightly built guard was John Fox, Jr., pulling his duty as part of the volunteer Guard from nearby Big Stone Gap. "He swore to me," remembered Fox, "that, the night of the murder, when he lay down to sleep, high on the mountain-side and under some rhododendren-bushes, a flock of little birds flew in on him like a gust of rain and perched over and around him, twittering at him all night long." [21] The "twittering birds" gave him what was likely the first cold stab of superstitutious fear Talt Hall had ever known.

The singing birds at dusk was a bad sign; there was no two ways about it. He would still run and hide, and do his best to escape, but deep in his heart, he knew it would be useless. Hall was a bad man. Neither he nor Doc Taylor, his captor, were good men. They were not misunderstood, nor were they products of any number of psychological tremors in their childhoods. They were not especially a product of, nor indicative of, their times and environment. They were both aberrations. They were killers, and to a people and a time that could, and did, condone a certain amount of killing, there was also a point past which those people were not prepared to go in their justifications, and both men, Talton Hall and Doc Taylor, had passed beyond the pale.

There is no doubt that the stranger of the two men who occupied the Wise County jail that summer of 1882 was Doc Taylor. Talt Hall was a murderer plain and simple, if those descriptive terms can be applied to such cold-bloodied business. Doc Taylor, on the other hand, was a more complicated individual. No less cold blooded, and no less a killer, Doc Taylor was by far the more interesting of the two.

His real name was Marshall Benton Taylor, although most just called him "Doc," or later, when his reputation as a killer among killers was made, he was sometimes referred to as the "Red Fox." It was this appellation that Fox gave Taylor as one of his characters in *The Trail of the Lonesome Pine*, but mostly, he was just known as Doc Taylor. One source says that he studied medicine at the Louisville Medical College, hence the "Doc," but this may or may not be true. [22] His roamings around Wise and Lee counties Virginia, sometimes in the employ of the local law, were often as mysterious as the man himself, but he quickly became a legend, a boogeyman for unruly children, and the dread of those whom the law wanted for one reason or another.

They said he could appear and disappear at will, and he did not discourage their fears and beliefs. They said that he would be talking to you on a high mountain trail one minute, and the next minute would find you talking to yourself. They said he wore moccasins because they were quieter in the woods, and he wore them backwards so no one could tell in which direction he was really traveling. Soon he became a man of mythical proportions, and a kind of superstitutious dread surrounded even the mention of his name. He carried a heavy Winchester rifle, and when Talt Hall heard that Doc was on

Marshall "Doc" Taylor was something of a Renaissance man in and around Big Stone Gap in the 1890s. Sometimes known as the "Red Fox," among his many talents were doctor, lawman, preacher and killer. Convicted of killing Ira Mullins and most of Mullins' family near Pound Gap, Virginia, "Doc" Taylor was hanged in 1893 (courtesy Wise County Historical Society).

his trail in his capacity of sometime U.S. marshal, he knew he was a doomed man.

Doc Taylor dabbled in just about every occupation that he knew even the slightest bit of information about. Sometimes he took on the duties of doctor and preacher, and sometimes he hunted men who were wanted by the law for the sheer fun of it. He did this either officially as revenue officer and U.S. marshal or on his own. On these jaunts he was as feared a hunter as any who ever took to an outlaw's trail, but Taylor had a darker side that he kept hid, and up until the time he massacred the Mullins family near Pound Gap, he kept his homicidal secret well.

It is unlikely that the Mullins Massacre was Doc's first foray into killing. As secretive as the man was, most people who knew him only suspected that he might be a little "off" in the head, but nothing serious and certainly nothing heavy enough to suggest that darker side of his character. As far as his personal appearance was concerned, he looked little different than his more law abiding neighbors. "He is a man of medium build," wrote a reporter who visited Taylor in the Gladesville jail, "thin and rather stoop-shouldered. He is fifty-six years of age and looks his age, being quite bald on the crown of his head and with touches of gray appearing in the reddish brown hair that covers the lower part of his head on either side and behind. A fringe of red whiskers of a much deeper color than his hair extends down either cheek and under the chin, his upper lip being clean shaven." The amateur phrenologist went on to say that Taylor "has a good forehead and his countenance indicates a degree of intelligence above the average." [23]

One early morning in May 1892, about five months before Talt Hall would swing from the gallows in Gladesville for killing Enos Hilton, Doc Taylor and two brothers, Calvin and Henan Fleming, lay in ambush near Pound Gap, just across the mountain from Jenkins, Kentucky, on the Virginia side of the Kentucky border. Somehow they had learned that Ira Mullins, sworn enemy of Doc Taylor, was leaving the country, fed up, it was said, with feuding and fighting and everything that went with it. He was determined to leave for the West, and take his family with him. Doc and the Fleming brothers lay in the hot sun with their rifles, watching and waiting. For Doc, the opportunity to even a score with an old enemy was the chance of a lifetime, and he was determined to make the best of it that he could.

Some sources say that Mullins was carrying a large sum of money that Doc wanted, which is unlikely since money was somewhat scarce, although killings had been done for only a handful of dollars, or for no gain at all. Others say that Mullins, hating Taylor, reasons unknown, had offered a reward to the person who killed the Red Fox, but for whatever reason, or maybe none at all, when the wagon finally hove into view and range, a volley of shots from the rocks and underbrush mowed down the entire family, excepting two. The Mullins family, although caught off guard, apparently did not go down willingly, nor without a fight. "Hundreds of shots were fired from both sides until well-loaded, high-powered guns, revolvers and the like had been emptied," said one account written somewhat luridly forty years after the fact. [24] The same account lists the dead as Ira Mullins, his wife, two daughters and two sons, with a third son, John Harrison Mullins, escaping through the brush, only to be killed many years later in the "far west." Other accounts insist that a second child, a daughter, also escaped, along with the aforementioned son, and it was their testimony that put Doc Taylor on the gallows.

They had shot the horses pulling the wagon first, and then methodically killed all who would stand still long enough for a decent shot. Years later, the bleached bones of the animals still lay as a stark reminder of the murders alongside the road to Kentucky, and the rocks from which Taylor and the Fleming brothers fired their fatal shots became known as the "killing rocks."

"Horror was upon the face of the Cumberlands. The perpetrators of the deed were unknown and people talked only in whispers of the crime. Suspicion was everywhere." [25] The day after the killings, Doc Taylor showed up at the courthouse in Gladesville, as he had been wont to do, watching, without expression, the continuing trial of the man he had so recently apprehended for murder: Talton Hall. However, it was not long before suspicion fell upon Doc Taylor and the Fleming brothers. Doc took to the brush during that hot summer of 1892, and remained a fugitive for several months. In another twist of irony, it would be one of Taylor's own friends who did to him, what Doc had done to Talton Hall.

"Devil" John Wright was a well known in the mountains as any man, and in many respects, he and Doc Taylor shared many of the same characteristics. Both were well versed in trailing of fugitives, both were good at it, and both enjoyed the challenge of hunting the "most dangerous game" of all. Wright was hesitant, at first, to take up the trail of his old friend, or so the stories go, but when he found out that the capture of Taylor was inevitable, Wright began to lay his snares for his erstwhile friend. For awhile, Wright was unsuccessful, but in time, the Red Fox grew careless and was cornered in a cabin near the Kentucky border, and captured.

Then, a funny thing happened on the way to the jail house. Doc Taylor, in the hands of "Devil" John Wright, escaped. "Somewhere and somehow ... the wily Red Fox evaded the clutches of his captor. Just how, both history and legend are silent." [26] But he did not stay free for long. A man by the name of Ed Hall, with the appropriate nickname of "Bad" Ed Hall, tracked Doc into Norton, Virginia. Legend has it that Doc's son had nailed his father into a shipping crate, and sent it off by rail to West Virginia, but "Bad" Ed boarded the same train, and upon its arrival in West Virginia, Hall arrested both crate and contents.

The murders had been committed by Doc Taylor and the Fleming brothers, although in later years, there was some evidence that the case against Doc was shaky at best. At his trial, detailed by Fox in *Lonesome Pine*, and further told in *Bluegrass and Rhododendron*, Fox painted Taylor as a fiendish criminal and murderer with few if any redeeming characteristics. The case against him, as Fox presented it at any rate, was strong, and conviction an almost foregone conclusion, but the Red Fox was not yet without his surprises. Found guilty, and awaiting his sentence from the judge, Taylor was asked one final question by the trial judge: "'Have you anything to say whereby sentence of death should not be pronounced on you?' The Red Fox rose: 'No,' he said in a shaking voice; 'but I have a friend here who I would like to speak for me.' The judge bent his head a moment over his bench and lifted it: 'It is unusual,' he said; 'but under the circumstances I will grant your request. Who is your friend?' And the Red Fox made the souls of his listeners leap. 'Jesus Christ,' he said." [27]

According to Fox, the judge patiently and solemnly allowed Taylor to read passages from his Bible that insured that his enemies would be pulled asunder. But even a judge's patience has its limits, and he finally called a halt. Another account, certainly more humorous, though likely suspect in regards to its truthfulness, contends that the judge was a little hard of hearing, and when Doc Taylor said he would have Jesus Christ speak for him, the judge "turned around to the clerk and ... just as straight and sober as a judge is supposed to be, and said, 'Well, bring him around Charley, and have him sworn in.'" [28]

There would be little humor in the sentence, however, and death by hanging would be handed down in the little mountain town that had never seen a "legal" hanging before. Now they would have, in the space of a few months' time, two hangings—both legal. Now here they were, both of them, sharing the same jail, sharing the same sentence of death.

They were placed in adjoining cells, side by side, and according to Fox, Doc Taylor's fear of Talton Hall was pitiful to see. "The Red Fox's terror of Hall was pathetic," he wrote, and although Taylor begged to be moved, "the two stayed together — the one waiting for trial, the other for his scaffold, which was building. The sound of saw and hammer could be plainly heard throughout the jail, but Hall said never a word about it." [29]

If Taylor heard the construction outside the jail window, he never mentioned it either. True to his Swedenborgian religious [30] beliefs, he "sang hymns by day, and had visions by night." [31] "I am fully prepared to die if the court decides against me," Taylor told a reporter from Lynchburg, Virginia, while his trial was still going on. "I aim to keep myself constantly prepared. All men must die some time, and my death upon the gallows would only cut me out of a few years of life anyway." [32] But, somewhat surprisingly, he was worried about the appearance of the hanging itself; not in the actual mechanics, but in how it would seem for a man like himself to be hanged. "I have thought about it a good deal," he said, "and I don't think men who are hung suffer much. It is a death that is looked down on, I know, and considered a disgrace, but better men than I am have died on the gallows. Christ himself died an ignominious death, but I don't pretend to compare myself to him." [33]

One day, a note was thrust through the bars to Horace, John's younger brother, when Hall was not looking. "There never was such a helish [sic] on earth as Hall," read Horace, "is [sic] he got mad has done something to the inside of the jail. Look and see what it is. It has tore the pipe loose inside of my cell. Don't say that I said any thing for I am afraid of him." [34] The note has been attributed to Doc, but one must assume that someone of Doc's intelligence and education, was a better scribe than the note implies. In any event, Hall was a bad man and a reluctant prisoner, though not for long, and Doc Taylor was scared to death of him.

Taylor was afraid of Hall, but his fear of his own approaching execution and subsequent death waxed and waned, as did his outward confidence. His visions had told him he would "come out all right" in his upcoming trial; his religious visions told him that, but at night, lying awake listening for Talt Hall to take his revenge, he was not so sure that things would come out all right after all. He was right to be worried.

The day of Hall's hanging was at hand. September had come, and in the mountains, September is the month of most beauty. The trees, poplars first, begin to change their colors in the high places, and soon the whole countryside is startled with the combined colors of autumn. The changing leaves did nothing to soften the angular corners of the rough jail in Gladesville, and the gallows of newly sawn timbers, built inside a tall wooden box to hide the execution from the onlookers, looked innocent enough, and hardly the instrument for punishment that it was built to be. It was the first gallows many of the hundreds of onlookers had ever seen, and before the day was over, many would carry that image back to their mountain cabins and city homes, and their children and grandchildren yet to be born would hear, time and again, how they had seen Talt Hall hanged on a day in September that was made for anything but death, legal or otherwise.

The day for Hall's hanging had come at last. "With the day," wrote John Fox, "through mountain and valley, came in converging lines mountain humanity—men and women, boys and girls, children and babes in arms; all in their Sunday best—the men in jeans, slouched hats, and high boots, the women in gay ribbons and brilliant home-spun; in wagons, on foot and on horses and mules, carrying man and man, man and boy, lover and sweetheart, or husband and wife and child — all moving through the crisp autumn air, past woods of russet and crimson and along brown dirt roads, to the straggling little mountain town." [35]

They had come from miles around. From across Pound Gap from Kentucky they came, and from every county and surrounding territory. Buildings with two stories looked ready to explode with people hanging from the windows. Trees bent under the unaccustomed load of young boys and men who hung precariously from the limbs. Roofs of nearby houses and places of business were covered with spectators that made them look like so many oversized birds, roosting, squirming, and fidgeting for a better advantage. The grounds around the jail were covered with men, women, children, horses, mules, wagons and dogs. It was the biggest thing since the Fourth of July, and no one wanted to miss any part of the spectacle.

There were picnic lunches, talking, laughing, knife swapping and horse trading, and

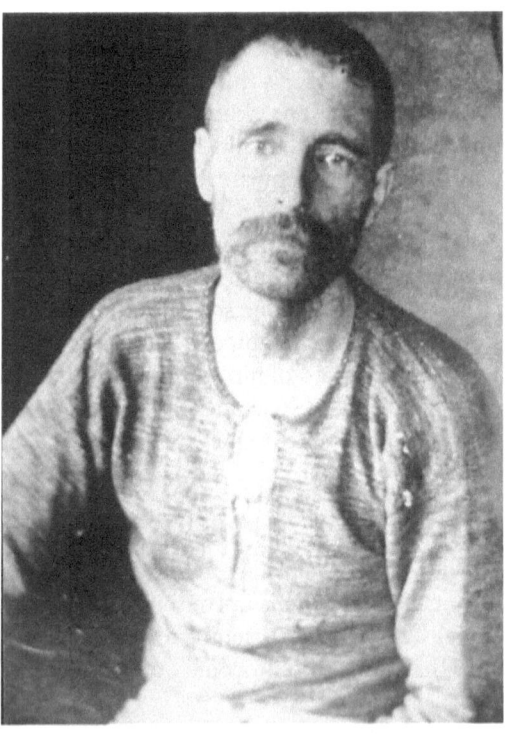

Talton Hall in sadder times. This photograph was taken while Hall was being held by the "Guard" in the Wise County jail. John was among those who guarded Hall to prevent his rescue by any friends he might have. It did not happen, and Hall was hanged a few days afer this photograph was taken (courtesy Wise County Historical Society).

fathers holding their children on their shoulders so that they might have a better view. Too many times death in the mountains was sudden and without warning. It was over before anyone was ready for it, and easily missed if you were not careful, but today would be different. Entertainment, the likes of which few if any had ever seen before, had come to Gladesville, Virginia, and the crowd, excepting one or two in particular, was enthusiastic about it.

There was little to be excited about inside the jail. Both the guards and prisoners were silent and reflective. The sheriff who would carry out the actual execution was nervous, and wished he was anywhere, doing anything else, but what his official duties bade him to do. It was his fervent prayer that nothing would happen to make a mess of things.

Once outside the jail, the crowd waited. Some had come early, long before noon when the scheduled hanging was to take place, and now they, along with those who had come later in the morning, waited. "The neighboring hills were black with people waiting; the housetops were black with men and boys waiting; the trees in the streets were bending under the weight of human bodies; and the jail-yard fence was three feet deep with people hanging to it and hanging about one another's necks—all waiting." [36]

Confession, it has been often said, is good for the soul, and it is very likely that no one is more concerned about his soul than a condemned man. It was the custom in those times for the condemned to confess his crimes, both for what he was being hanged for and others he may have committed that still remained mysteries. There were, of course, those who refused to confess even those crimes that were sending them to the gallows, but most, wishing for a few more minutes of life, went into great detail, prolonging the inevitable as long as possible. Usually, of course, they were apologetic and sorry for the life they had led. Most blamed whiskey, and sometimes bad or loose women for their plight. Sometimes, they blamed both. Most found religion all of a sudden, and were baptized, sometimes on the very day of the hanging. "Bad whiskey," avowed "Bad" Tom Smith on the scaffold in Jackson, Kentucky, in 1895, "and bad women have brought me where I am.... I will give this warning: Don't drink whisky and don't do as I have done.... Again I ask you, take warning from my fate and live better lives than I have lived. I die with no hard feelings toward anybody. Farewell." [37] A few hours before his hanging, "Bad" Tom had been baptized in the nearby Kentucky River.

"At 11:55," wrote Fox for the New York *Herald*, "Hall was allowed to go to the front window of the jail so that his friends might get a last look at him. He was pale, worn and looked sick in body and mind." [38] If they were expecting a confession, the crowd was disappointed, at least for the time being. After about ten minutes of staring out the window at the crowd, Hall simply got up without speaking a word, and turned back to his cell. Finally, it could be put off no longer. Hall passed the cell of Taylor, surprisingly kissed him, forgave him, shook hands with his guards, and walked out to the scaffold. He looked at the rope, and said, "It looks to me as if that rope would break." [39] The sheriff assured him that he had tested it himself (he did not say how he knew, and Hall did not ask), and he was sure it would hold. Hall's legs were bound around his ankles, and his hands tied behind his back. "'I want to see my sister,' he said. A thin woman in black came in from outside and mounted the scaffold. "Don't take anything to heart," he said. "I don't want any more good men killed about me." [40] It was completely contrary to mountain law that his death could go unpunished by his family, and Hall was insistent, but his sister was not persuaded, and did not promise. A black velvet hood was snugged down over his face at two minutes before 12:30, and if one had been closer, they could have seen the cloth move in and out with Talton Hall's last breaths.

It was not the best arrangement that could be had, but not knowing much about hangings, and the building of a gallows, the citizens did the best they knew how. The trap-door would be held closed by a single rope beneath it, stretched tight. As an added insurance, two upright beams were wedged against the bottom of the door. They would have to be knocked loose, and then the rope cut, dropping the door, and the condemned man above it. "As each of these was knocked out," wrote Fox in his description of the hanging, "the door sank an inch, and the suspense was horrible. The poor wretch must have thought that each stroke was the one that was to send him to eternity, but not a muscle moved. All was ready, at last, and the sheriff cried, in a loud voice: 'May God have mercy on this poor man's soul!' and struck the rope with a common hatchet. The black-capped apparition shot down, and the sheriff ran, weeping, out of the door of the box." [41]

Those closest to the gallows heard the blow of a hatchet that cut the rope holding the trap-door closed, and then, "T-h-o-o-m-p! The dangling rope had tightened with a snap and the wind swayed it no more." [42] Less that ten minutes after 12:30, the doctors in attendance announced that "Hall's pulse had ceased to beat. Twenty minutes later he was cut down and place in the coffin. His neck was not broken." [43]

Talton Hall had strangled to death, which is not that uncommon in executions by hanging, and his death could not have been the painless one that Doc Taylor was hoping for. It was Wise County's first hanging, and they knew little or nothing about the necessity of weights and measures to insure the instant death of the condemned.

Inside Hall's cell, Doc Taylor, according to Fox's characterization of him in *The Trail of the Lonesome Pine*, "stood with his watch in his hand and his eyes glued to the second-hand. When it had gone three times around its circuit, he snapped the lid with a sigh of relief and turned to his hammock and his Bible." [44] The crowd, ever curious, gathered around, trying to see the body, but his sister, friends and relatives quickly placed the lid on the rough coffin, and, under guard, quickly left Gladesville for Kentucky. "He told her he wanted to be buried in his underclothes, and anywhere else on earth except in this county. He said he did not want to lie among thieves." And so, Talton Hall, desperado, was taken back to Kentucky and buried, and no doubt the "twittering birds" from time to time sing over his grave. [45]

The tension had been high while they guarded Hall. It was expected that at any minute his friends from Kentucky would cross the border, armed to the teeth, and take the condemned man out of the jail and away from the gallows. Twenty years later, an old man remembered how some dogs chased a cat into the jail. "The guards thought the Kaintuckians was comin' an' they was skeered to death. I seed the Red Fox go plumb under a bed." [46] John did not remember anything like that happening, but he did remember how tight everyone was wound in those days.

Talton Hall was dead, and the guards from the Gap went home, little concerned that anyone would attempt to break Doc Taylor out of the small solid jail. Ironically, it seems that Hall had plenty of friends who, if they had the leadership, would have been willing to break their friend out and set him free, but Taylor apparently had none so adventurous, and so the guards were not worried too much. "The guard down at 'the Gap' had done its duty with Talton Hall," wrote Fox of the event later, "but it was the policy of the guard to let the natives uphold the law whenever they would and could, and so the guard went home to the Gap while the Gladesville natives policed the jail and kept old Doc safe. To be sure, little care was necessary, for the Red Fox did not have the friends who would have flocked to the rescue of Talton Hall." [47]

When the time for Doc Taylor's hanging neared, John did not keep himself away, and went from the Gap to Gladesville to see the end of one of the more notorious of men ever to run the hills and mountains of southwest Virginia. On Sunday, the 22nd day of October 1893, Doc Taylor was led from his jail cell, not to the gallows, but to the court house where he administered his last communion. He asked the curious onlookers to join him, but the crowd remained silent, and none moved forward, indeed, there was hardly any movement at all. "Taylor was dressed in white from top to bottom. He looked very thin and emaciated, as a result of his long confinement.... He said that the white clothes he wore were emblematic of what he would wear in Heaven." [48]

Being the sometime preacher he was, he promised to preach his own funeral sermon, beginning at 9 o'clock on the day of his execution, and continue preaching until the trap was sprung. He asked the crowd to come and "break bread" with him as the disciples had done with Jesus at the Last Supper, but he got no response. When none of the crowd came forward, he reached a small portion of the bread of his communion towards his wife, a small quiet woman, dressed in black, who silently took the bread he held out to her. He began his sermon to the crowd, but it was hard to follow. It was "rambling, denunciatory, and unforgiving. Never did he admit guilt." [49] He paused often to borrow his wife's handkerchief to wipe the tears that streamed from his eyes. Finally, on October 27, 1893, Doc's own time had come, and as promised, he began his final sermon from the upstairs window of the court house. At 2 o'clock in the afternoon, he was led to the gallows, enclosed to hide his final moments from the crowd outside. "He was attired in snow white linen from neck to foot and wore a brown derby hat and neat slippers." [50]

His wife had made him that suit of snow white, sewn from a linen tablecloth, and the hood that would hide his face at the last moment, but many wondered where he had gotten that brown derby hat to top off his unorthodox costume. He had given her instructions that he was not to be buried for three days, and on the third, he would, he promised, rise from the dead.

"On the last day, the Red Fox appeared in his white suit of tablecloth.... He walked firmly to the scaffold-steps and stood there for one moment blinking in the sunlight, his head just visible over the rude box, some twenty feet square, that surrounded him — a rude contrivance to shield the scene of his death. For one moment he looked at the sky and the trees with a face that was white and absolutely expressionless; then he sang one hymn of two verses. His hands were tied with a white handkerchief and a white cap instead of a black one pulled over his face." [51]

Cruelly, but inadvertently, the guards, in removing the props beneath the trap-door, caused it to shift, and Doc, thinking his time had certainly come, fell to his knees "an awful spectacle to witness. He had to be assisted to his feet and stood tottering there only for an instant for the door swung from under him," [52] and the strange old man was launched on the way to that other world in which he believed so firmly. "The body turned slowly round and round for several minutes, then hung perfectly still, not a muscle moving. At the end of nineteen minutes he was cut down and placed in his coffin and delivered to his family." [53]

As she had promised, Doc Taylor's wife did not bury him for three days, keeping vigil over the still, cold form that was her husband. At the end of three days, he did not stir, and on the fourth, the second man hanged in Wise County, Virginia, was buried just like any other. [54]

The Fleming brothers, who had helped the Red Fox murder the Mullins family, would fair little better in the end. Posses scoured the countryside for the two brothers, and would eventually run them to ground in West Virginia. In the meantime, however, the brothers were trying to get out of Virginia, and when John heard that the brothers had been cornered near Pound Gap, he hurried to get to the action. "Ten minutes later I was astride a gray mule, and with an absurd little 32 Smith & Wesson popgun on my hip — the only weapon I could find in town — was on my way to the Pound." [55]

He must have cut quite a figure. One needs little imagination to visualize him bouncing around on a long-eared mule like some latter day Ichabod Crane with a pop-gun flapping at his hip, and topped off with a ridiculous yellow straw hat perched atop his head, making him an easy target for anyone of even mediocre marksmanship. He was, however, unconscious of any danger, and it was just his luck to stop for directions at the very cabin that belonged to the men who were fleeing from the posse that John was trying to join. Embarrassed and not a little ill at ease when he discovered his situation, he tried to bluff his way through.

The old woman there, mother of the Fleming brothers, gave him buttermilk to drink, and refused payment for it. She asked him, mountain like, where he was from and where he was going, and he evaded her questions, thinking to hide his real purpose, but she was too crafty for him, and saw through all that he said. There had been a fight, she told him, and her boys were wounded, maybe dead already. "'I reckon you know my boys is hurt — mebbe they're dead in the woods somewhar now'." [56] She followed him out to his mule, and looked up at him as he mounted again, anxious to be away, and asked him not to "harm 'em," if he caught up with them. "I promised her that if they were caught unharmed, no further harm should come to them; and I rode away, the group sitting motionless and watching me." [57] The posse was almost as difficult to catch up with as the Fleming brothers, but after two hours of steady riding and climbing, he overtook them, and was not altogether pleased with their laughter. "They were much amused," he wrote, "when they saw the Christmas toy with which I was armed." [58] One of the possemen drew out his own pistol that John swore was all of two feet long, and gave it to him to replace his "Christmas toy." The hat would not do either, and they gave him another, more inconspicuous one, which they said was the captured hat of Henan Fleming from the battle earlier that day.

They had surprised each other — the Flemings and the posse — and for a few minutes, they exchanged pleasantries rather warmly. One member of the posse got so excited he shot several times into the ground, causing another to laugh so hard he could not shoot at all. Another, hiding behind a tree hardly big enough to support a bird's nest, looked around and found "seven full-grown men" hiding behind that one tree. Finally, one of them sent a load of buck-shot in the directions of the desperadoes, and by luck or happenstance, wounded Henan Fleming in the shoulder, knocking his hat off in the process. The Flemings fled the scene, and the posse, surprised at their good luck, sat down to cook breakfast, pick up souvenirs, arrested Henan's hat, and talked over what they should do next.

They did not know it, but Cal Fleming had himself been wounded as he turned to run. Both were hurt badly, Cal's wound being the more serious of the two. It was about this time that John came upon the group, still eating, and still milling around.

Among those souvenirs that were gathered from the battleground, was the aforementioned hat. Everyone had a decent hat except John, and no self-respecting member of any posse anywhere would wear a bright yellow straw hat, and nothing would do except that he take the fugitive's abandoned chapeau. "And so," he said, "I wore Heenan's hat — looking for Heenan." [59]

The next morning, they renewed their search. It was dangerous work, for nothing is as dangerous as a scared man, and these men they were hunting were wounded and armed as well. Their search was fruitless. The two had escaped, some said, dressed as women, over the border into Kentucky and beyond their reach, but like Talton Hall and Doc Taylor before them, the Flemings' time was running out quickly.

In West Virginia, three months later, they were caught off guard by three members of that ill-fated Virginia posse. John was not among the three, having retired from man-hunting for the time being. The two brothers were cornered in a country store near Cowden, West Virginia. "Their surrender was demanded," said the Big Stone Gap *Post*, but they refused, going for their guns. The gunfire came from inside the small confines of the store; the crash of gunfire and crockery was heard, along with the thump of bodies hitting the plank floor. "The battle lasted thirty minutes. Calvert Fleming was instantly killed, and Henry Fleming [sic] and Deputy Sheriff Brannon mortally wounded. Deputy Sheriff Seannel was shot through the throat and," said the *Post*, "his recovery is doubtful." [60] The battle, or so the paper said, lasted 30 minutes, between five or six armed men inside the walls of a small building, resulting in one man killed outright, and three others wounded, not counting a bystander who was shot accidently.

Two weeks later, the *Post* had more particulars, though the results were mostly the same. John Branham, Ed Hall and E.J. "Doc" Swindell cornered the Flemings in a country store, but the "battle" did not last 30 minutes. Hall entered the store first, and walked quickly up to Cal, grabbing him by the arm, and throwing his pistol on him at the same time. "Instantly Henan Fleming leveled his pistol and the ball grazed the back of Hall's head, causing him to loose his hold on Cal, who then drew his pistol and began firing. Hall, recovering himself, drove a 44 into Cal Fleming's brain, felling him to the floor." In the meantime Henan Fleming's unerring aim had been doing deadly work. Swindell, before he could fire, either from slowness or fright, received a ball through the neck and made for the door. John Branham, in spite of his bravery, was lying on the floor with one hole just above the left lung, and one through his right lung. He would die nine days later. Hall now turned his attention to Henan, and a bloody duel followed. Standing face to face at a distance of ten paces they fought unflinchingly until, with one bullet through his chin and his hand rendered useless, Henan Fleming ran into a back room to make his escape. Finding no outlet, he ran back to the door and threw up his bloody and still bleeding hands, with the fingers of one hand shot almost off. 'I can't do any more, Ed,' he said; 'I surrender.'" [61]

Cal was killed outright in the gun battle, and Henan (referred to as Henry by the *Post*), was wounded seriously. He survived his wounds, however, and was tried for murder in Wise County on July 24, 1895. After six days of testimony, Henan was acquitted, one of the prosecution's key witnesses, the surviving daughter of Ira Mullins, having been shot and killed from ambush in the meantime, some have said, by another of the Fleming brothers. Apparently, Henan had confessed his part in the Mullins Massacre, but the prosecution was not allowed to use that confession, and Henan walked out of the Wise County court house a free man, and John owed him a new hat.

John's version of the fate of Henan Fleming appeared in his "Man Hunting in the Pound." In that story, Henan is brought back to trial, and, John, thinking Henan might go free, and not wanting any hard feelings to linger over the ownership of an abandoned slouch hat, went to Henan, and asked him if he had ever gotten his hat back. He had not, and John promised that should Henan go free, he would buy him the hat of his choice to replace the borrowed and lost one. According to the story also included in *Bluegrass and Rhododendron*, the last time John saw Henan was when the former fugitive rode out of town, a free man, with a new derby hat on his head. [62]

Henan Fleming returned to West Virginia, apparently wearing that new derby hat, and recovered completely from his wounds, and as far as known, saw no more trouble from the law for the rest of his life, even taking up, or so family tradition says, the life of a protector of the law himself. Ed Hall, however, who was also wounded in the battle to capture the two brothers, did not fair as well. Several months later, recovered from his wound, he was shot from ambush from the window of a store building in Pound, Virginia. No one was ever arrested for the killing.

The times were lively, as John had said of the Gap, but it was not all fun and games. It was not all hangings, shootings and carousing. Big Stone Gap was never without lawlessness, but it was blessed with a majority of people who wanted change from the old ways, and with the help of the Guard of which John Fox was a member, that majority finally won over the minority, and peace, along with law and order, came to the Gap.

"There is not much danger," wrote the Big Stone Gap *Post* early in 1894, soon after the capture of the Fleming brothers, "of a new outbreak of lawlessness, as that class have been taught a lesson that has been indelibly stamped upon their memory that when they commit murder, sooner or later, their punishment will overtake them." [63]

The hangings that fall of 1892 and 1893 did prove to be something of a watershed for Big Stone Gap and Wise County, Virginia. In the paraphrased words of Jack Hale, "Law had come to the Gap, and had come to stay." And it was the volunteer Guard that played an important and indispensable part in that coming of the law; such an organization might not have worked at any other place, at any other time, but for that short, sometimes violent and bloody period of time, the light of the law shone on the Guard and peace reigned in the Gap.

In the excitement of those last few months, the appearance, finally, of "A Mountain Europa" in the September-October issue of the *Century* magazine was almost anticlimatic. It had been so long since the manuscript had left John's hands that it seemed new to him when he read it again that winter. That first story was important to him. It would be printed over and over again by various publishers and added to other collections of stories and novelettes. Some have said that his "Hell-fer-Sartin" was the story that broke down literary doors for John Fox, Jr., but the initial success of "Europa" cannot be downplayed. The success was immediate and gratifying to him. At last, at long last, he was a published author, and although he did not fool himself into thinking it was a great piece of literature like those he had studied at Harvard, it was a good and decently written story. He was pleased and proud. Now, almost 30 years old, still a bachelor with few prospects, and no employment to speak of, the publication of "Europa" was a godsend to his feelings about himself. At last, he could think of laying aside any aspirations of being a "Napoleon of finance." The boom was dead or dying anyway, and anyone could see that, and John himself was relieved to think that part of his life was behind him.

But it was not entirely over. The year 1893 posed new problems, especially financial ones, for John, and the boom would come back to haunt him and his brothers for months and years to come. Like a headless snake that still moved and refused to lie still, the boom at the Gap would continue to twist and squirm, trying to trap the unwary in its coils, and tighten its grip on those already entangled. That year proved to be a rewarding one for John Fox, Jr., as his literary career began to take off, and as he devoted more of his time to writing, but just when he thought he was free of the financial troubles the boom had caused him, an incident would occur that would shame and embarrass him almost to the point of killing his soul.

CHAPTER 9

Feuds and Romance

> This year of 1893 will long be remembered for it's [*sic*] financial distress...
>
> John Fox, Sr.

"A MOUNTAIN EUROPA" WAS PUBLISHED in two parts in *Century* magazine for September and October of 1892, and there was no money to be had from that late publication, and he did not expect any. He had been paid long ago for that story, and as he said so often, the money was not important; he would have paid them twice the $262 to have had it published. The satisfaction he gained was, however, enormous, and few could blame him if his step was a little livelier, his head held a little higher that fall and winter.

It was and is a story as old as time: boy meets girl, boy almost loses girl, boy gets girl again, and girl dies and solves all the problems the author has created for them both. It was his first good story, arguably his favorite, and was doing better than anyone, including John, had hoped for. Readers everywhere seemed to like the story and it was just the kind of encouragement that John needed to do more.

"Europa" had promise, and thousands of readers were ecstatic in their praise of it. His brother Richard wrote him from Harvard in December that "one fellow told me that he had dressed for a foot ball game and sat down to look over the CENTURY. He started your story and got so interested in it that when he arose he had finished the story and had forgotten the game." [1]

By the time "Europa" was published in 1892, John was on the low end of 29 years old. He would be 30 in December, which might seem like something of a late start for someone who had once said that he knew that he wanted to be a writer since the age of 12, [2] and he could not help but wonder at that strangeness himself, or of how events had turned for him. He had gravitated towards all things literary from the earliest age, but somehow, a career in writing had escaped his notice except for his brief stints at newspaper reporting and writing. He had joined a literary society while at the old Kentucky University, "and was even orator at the annual open session of that society, but I wrote neither the editorials of the paper nor even my flowery speech," he wrote years later after his literary reputation was assured, "but I imagine I got a literary reputation that I no more deserved than my younger brother who was at home." [3]

He thought it more than passing strange that all those years before, when he could have devoted himself more to writing, he never thought to do so. "Looking backward," he wrote in 1908, "it strikes me as strange that I did not try paper, magazine, or poetry. Until I went to Harvard I had never seen nor did I know that I should ever see a man who had painted a picture, carved a statue, or written a book." [4] Whether or not he wanted to be, and believed he would become, a writer for all of his adult life, he could not now ignore the stories and people around him in southwestern Virginia. The stories were there, and in his memories of Jellico and Kentucky, and there were plenty of them, but through it all, no matter how good the stories, it was the people of the mountains that fascinated him the most, and they always would.

In the meantime, he had finished a second story—"A Cumberland Vendetta"—and had sent it off to *Century* magazine, and with the initial success of "Europa," they were anxious to use the story as quickly as they could. Name recognition is a fleeting thing, and they wanted to capitalize on his initial success. They sent an artist to look over the locale of the story and get some ideas for illustrations to go with "Vendetta." John met him at the railway station in Pineville, Kentucky, and took him home to the Gap, showing him along the way the mountains and places that inspired that story of feuds and romance.

When John Fox, Jr., wrote "A Cumberland Vendetta," a story of a fictitious Stetson and Lewallen feud, memories of past "troubles" in the mountains and current murders were fresh in the minds of Kentuckians. Newspapers and magazines, with their renditions of murder and mayhem that may or may not have followed strict rules of truth, had made certain that everyone knew the most lurid tales of murders and assassinations that were happening all too frequently in the mountains of Kentucky and neighboring states. The facts of those feuds were sensational enough, but most publications did not hesitate to make them more fantastic, if possible.

Murder and violence of all kinds have always been popular topics for writers to explore, and it is the rare volume that concerns itself with such mischief that does not prove successful. Newspapers of the time were fascinated by the killings they were hearing about in the mountains of one of the country's oldest states, and their readers were hungry, starved for details. Eventually, it got so bad that every fistfight, every quarrel, every incident, no matter how trivial that resulted in a shooting or knifing or killing, was attributed to one Kentucky feud or another.

9. Feuds and Romance

John had seen feuds and feudists firsthand at the Gap, and had seen the bloody results of such feuding, and wondered, with the rest of the country, how those things got started, and what kept them going. Fox related that one such feud started when "two boys were playing marbles in the road along the Cumberland River, and one had a patch on the seat of his trousers. The other boy made fun of it and the boy with the patch went home and told his father. As a result there had already been thirty years of local war." [5] And he would say, by way of explanation of those killings, in *The Trail of the Lonesome Pine*, that "this feud business is a matter of clan-loyalty that goes back to Scotland. They argue this way: You are my friend or my kinsman, your quarrel is my quarrel, and whoever hits you hits me. If you are in trouble, I must not testify against you. If you are an officer, you must not arrest me; you must send me a kindly request to come into court. If I'm innocent and it's perfectly convenient — why, maybe I'll come." [6] Later, in the same book, he would have June Tolliver forgive and excuse her people for what they did. "They were not to blame — her people," thought June Tolliver, "they but did as their fathers had done before them. They had their own code and they lived up to it as best they could, and they had had no chance to learn another." [7] Others, of course, held a less forgiving opinion of the killings.

John knew these people who lived by the feud, and he turned to that segment of Appalachian history several times in his stories and novels. These people treasured family above all else and would not tolerate a wrong, real or imagined. "Oh," he would write, "they were a strange people, these mountaineers — proud, hospitable, good-hearted, and murderous! Religious, too: they talked chiefly of homicide and the Bible." [8]

John Fox, Jr., was fascinated as anyone by the feuds that rocked the Kentucky hills at the close of the nineteenth century, and the Big Stone Gap *Post* and other newspapers, especially those from Kentucky, were full of tales of murder and mayhem. As far as fiction was concerned, he could have done a lot worse in picking those "troubles" to showcase his writing talent for the first time. Fox's first foray into the world of the Kentucky feudist was, of course, with his "A Cumberland Vendetta," soon followed by "The Last Stetson." Neither story is very long, with "The Last Stetson" being the shorter of the two. Both borrow heavily from incidents in the Rowan County, Kentucky, feud between the Martins and the Tollivers of the 1880s.

John had sent "Vendetta" to the *Century* while all the excitement was going on in Gladesville with the hangings of Talt Hall and Doc Taylor, and they had returned it for revisions in much the same manner as they had done "Europa," but by April of 1893, *Century* wrote that they felt the story was just about ready. A few more minor changes, they said, and he could expect to see the story in print. Everything was going so smoothly for him now. The "good time," he assured himself, was here at last. He should have been more pessimistic. An old sore lay festering beneath the financial waters of the Gap, out of sight, out of mind. In July it broke to the surface, and the "headless snake" of the boom reached out to trip him one last time.

He could not bring himself to believe it was happening. He knew things were not good with their real estate business and other ventures, but he tried to not let himself think of how bad they actually were. The papers he held before him brought reality crashing down like falling plaster in a condemned building. He noticed that the papers were shaking, and saw that it was his own hands that trembled.

It was mostly legal mumbo-jumbo having several references to parties of the first part, second part and third parts, but the bottom line of it all was he and Horace were being sued. The failure of the boom had affected plenty of people, not just the Fox family, and now those people whose money they had used for investing wanted it back, but there was no money to be had. The remaining lots were taken from them to satisfy those debts, but it still had not been enough.

He recognized all of the names on the suit, most were all citizens of the Gap and people he met on the street everyday, but the one that hurt him the most was that of James Lane Allen, his friend. He had dedicated "Europa" to Allen, knew him as his friend, teacher, confidant, and now he was shamed to the bottom of his very being. His friends had to sue him for the money he could not repay. He had his doubts, only a few years ago, about borrowing money from his friends, and had often told his brother James that it was bad business to do so. Now, too late, he wished he had listened more to his inner conscience. He filed the papers away, not wanting to look at them any longer, not wanting to admit they even existed. He would let the courts have their way. There was little anyway, he thought ruefully, that he could do about it no matter how hard he wished for it.

He was not the only one humbled by the events of 1893. His father, who perhaps as much as any of them had hoped for success in the Gap, was bruised in spirit as well. Their father was now 62 years old, soon to be 63, and by any generous accounting, could not have many more years left to him. They — James, John and the rest — had promised him success and comfort in his declining years. He would have peace and quiet and rest and everything he needed. The Gap was to be their paradise. They had promised, and he had believed. Now, everything was in shambles. Nothing was panning out as they planned, and coupled with those events was also the realization that it was too late for him to start over anywhere else.

No one felt the pain and remorse of those broken promises more that John Jr. He looked at his father and saw himself, and although the old man said little, John felt he knew his father's thoughts, for they were the same as his own. They were too close as a family, however, and there were no recriminations. They would go on.

By Christmas, things had gotten back to normal. He bore no hard feelings against Lane Allen. He tried to convince himself that he might have done the same had his and Allen's roles been reversed. Richard and Rector were both back in school. Polly, Sidney's widow, had gone back to New York after a lengthy visit, and James was gone off somewhere still trying to corner a fortune. There had been storms and cyclones and disasters of the natural kind, thankfully in other places besides the Gap, and at least, thought John Sr., all the family was well, but what a year it had been! He wrote in his diary on Christmas Day, "This year of 1893 will long be remembered for it's [sic] financial distress," [9] and as he laid his pen aside, he wondered what the next year would bring, and was not sure at all that he cared to know.

By January of 1894, John had published one story, albeit a rather long one, and one that had received praise from nearly all quarters, including one future president: Theodore Roosevelt. With his usual, and sometimes chronic, optimism, he thought of leaving the Gap and joining the speaking circuits that were so popular then. His friend Thomas Nelson Page had recommended it to him some time ago, but John had not thought too seriously about it, mainly because most of those who took to the reading platform read their own writings, and John had little enough of those at that time. Now with the publication of "Europa" and the soon to be published "A Mountain Vendetta" and "The Last Stetson," he was reconsidering Page's proposal.

James Lane Allen was against it. Their friendship had survived the suit of 1893, and although he would always value Allen's friendship and advice, he turned now more and more to his newer friends, especially Thomas Nelson Page, for advice on how to proceed. [10] This, in part, may have accounted for Allen's pessimism concerning the speaking circuit, and though there is little foundation to support it, Allen may have been a little jealous of his one time protégé's success, and perhaps not a little resentful that John searched elsewhere for advice now that he was more or less successful as a writer.

Page was guardedly enthusiastic, but was still more than willing to offer advice and introductions to the right people. Letters of recommendation would not be out of the realm of possibility either, he assured John. "You can say to Coldwell [the president of the Southern Lyceum Bureau in Louisville, Kentucky] that I have heard you," wrote Page in January, "and think you would make a hit.... There is another agent, Major J.B. Pond ... to whom you might write and with whom my name and warm endorsement may help you." [11] Despite Page's assertion of confidence in his friend and fellow Virginian, he could not help but suggest that John try a few engagements for charity first, to see how things would go. John was not interested in charity. Page was a successful writer and lawyer with plenty of money and a literary standing that meant he could sell practically anything he wrote and could afford to be charitable. John was a long way from living on "beef tea and crackers," but not that far.

Once he saw the way things were going, Allen relented, although he could not help but warn John of taking Page's advice on such an important matter. Everything being equal, he thought it might be a good idea and one that John would be successful with, but he disliked that Page had anything to do with it. It may have been genuine concern on Allen's part, or it just might have been the aforementioned unconscious jealousy, seeing his one-time student and who had actively sought his authorial advice seeking counsel of another. "There is only one disagreeable thing about your letter," wrote Allen late in January, "that you should take the step you are about to take at Page's suggestion. Why don't you look inward for counsel." [12]

Allen, however, was nothing if not realistic, and after having told John not to take Page's advice, offered his own. "I believe a lecture on the Cumberland mountaineers would fetch you in a good sum of money, if you lectured so far away that no one knew anything about them ... and you will find my article on the subject ready for you also, and myself ready to collect a royalty for your use of it." [13] John could always count on his friends to be outgoing and generous, especially where money was concerned. It is not known whether or not Allen collected his royalty for the use of any material John used on the speaking circuit. Allen was not a stranger to the mountains and mountaineers of southeastern Kentucky, and had written about them over the years, and his friend was welcome to use any material he wanted, for a price.

Taken around 1894, this photograph appeared in many of the promotional brochures for the Lyceum Bureau (courtesy University of Kentucky).

Once he began to give advice, Allen had trouble restraining himself. He suggested that John might have "Europa" bound into copies that could be sold at the door "together with your photograph," he wrote, and he and John shared the same opinion about "charity entertainments." Allen was not too impressed with "A Mountain Europa," and did not hesitate to say so, more than once. He had told John early on, while "Europa" was still being written, that he did not think too much of it, although he encouraged him to finish writing it while he was still in Jellico, and he still did not like it after *Century* had published it. He neglected to say why.

He was critical of John's voice. "Whenever you read," he said, "you must get your voice out from behind your jaws." [14] He complained too that John had adopted an affected manner of speaking. "You know how delighted I used to be with your reading; but afterwards it seemed to me that you fell under the influence of some one [Page?] who had a peculiar underground croak that sounded like a freak.... When do you want the personal endorsement?" [15]

The announcement that James Lane Allen penned for John to use on his tours was unabashed in its praise of the young author from some practically unknown place called Big Stone Gap. "Now in no species of American short story has there been greater need of an interpreter for the dialogue than in that of the Cumberland Mountains; and this interpretation Mr. Fox is admirably prepared to give," Allen wrote. "For he has lived several years among the native folk, has talked with them, studied them, and become himself their literary interpreter through his splendid work in the *Century* magazine." [16] Allen neglected to mention that John had only published one story of the mountain people, but no matter, for that story did show a great understanding of the ways of the mountains and her people that the public would soon be clamoring for.

Thomas Nelson Page wrote his own flattery for his friend that same month in an introduction to the Southern Lyceum Bureau. "I know Mr. Fox well," he wrote, "and he seems to possess the very qualities which would recommend him to a cultured and refined audience ... he understands the dialect of the mountains of Virginia and Kentucky as few people in this country do, and he is perfectly natural." Page went on to say that he felt John would "prove one of the most popular readers [they could] find in this country." [17]

With introductions in hand, John went to work for the Southern Lyceum Bureau almost at once, and by the middle of February 1894, he wrote his brother James that he was reading in "Georgetown [Kentucky] to-morrow ... night ... Shelbyville [Kentucky] Saturday night next," and Danville, Kentucky, on the following Monday. Page's predictions proved to be right on the money, and John was a success from the very first. His audiences could not get enough, and although the money was slight, and he often had to pay his own expenses going to and from the various cities, it was a steady income, and he found he could work as much as he liked.

Publication of "A Cumberland Vendetta" was being held up by *Century* magazine, and probably would not be published until the middle of the summer, but he could not wait, and no one wanted him to. He had already read much of the novelette to eager audiences, and would read more before *Century* would get around to publishing it in June, July and August of that year. He had even given a reading in his old hometown of Paris, Kentucky, and one supposes it was a charitable reading, and the *Kentuckian-Citizen* was impressed. "To say that the story is well told," they wrote of "Vendetta," "admirably well told; that it catches and holds the fancy throughout, increasing in interest from page to page, without ever flagging for a moment, is certainly the mildest praise that can be spoken of it." [18]

In June, the first of three installments of "A Cumberland Vendetta" appeared in *Century* magazine, but by that time, John was already making a name for himself. He had read and spoke in as far divergent places as Fayetteville, Arkansas, where he delivered before an audience of 800 people, and in a dozen Kentucky towns. For the next four years, he would never stop. He grew farther and farther from Big Stone Gap, and finally, wanting to be closer to the more metropolitan cities, especially New York, his stays in southwestern Virginia grew less and less frequent. He found that he did not miss his home in the Gap as much as he would have thought he would.

Soon after the *Century* began publication in its magazine of "A Cumberland Vendetta," an official in the Civil Service Commission in Washington, D.C., wrote John just after the first installment hit the stands congratulating him on the way "Vendetta" was beginning, and hoping for more of the same in the later installments. "If the rest of the story goes as well as these first chapters you will have made a lasting and real addition to American literature," he wrote. "We have had excellent work done in short stories, but it seems to me a pity that most of our novels just fall short; and so I was peculiarly glad to see one begin with no promise of failure." [19] The letter was signed: Theodore Roosevelt. That letter from a future president of the United States began a long correspondence and friendship that lasted well into the twentieth century. Once the last installment appeared in August, Roosevelt wrote John once more. "Now that it is done," he wrote on August 11, "I feel that I can congratulate you without reservation. It is the best American story I have seen for years, and I firmly believe that it has in it the element of permanence, and that it will last as a fixed addition to our literature. It is excellent." [20]

With the end of "Vendetta," Fox had enough unresolved themes left over to justify a sequel that appeared in *Harper's Weekly* for June 29, 1895, entitled "The Last Stetson," which, for the most part, only tied up the loose ends of "Vendetta." It was successful, however, again drawing praise from Theodore Roosevelt, who was something of a writer himself. "Yesterday coming back from Harvard," he wrote John on June 28, "I read your *Harper's* story; I like it greatly, for though it was a sequel to the feud story it was on an entirely different line; but do keep in mind that novel; I earnestly hope you can write it." [21] The "novel" Roosevelt was speaking of was *The Kentuckians*, and John had already begun writing what would be his longest effort thus far.

Between the publication of "A Cumberland Vendetta" and "The Last Stetson," John wrote in two days, a story that he sent to *Harper's Weekly* that is arguably one of his most famous pieces. Its title was "On Hell-fer-Sartin Creek." "[A]s I was coming home on a train from Cumberland Gap one morning at daybreak," he wrote of the genesis of the story, "I overheard one drummer telling another in a seat in front of me how he had been over in the

Kentucky mountains and had seen two mountaineers take a drink together and then walk out and shoot each other down. This was the genesis of 'Hell-fer-Sartin.'" He had intended to make it a comic piece, "But," he said, "when I started to write it, the story got serious, and so did I." [22]

The story was sent to *Harper's*, and they published it in November of 1894, but, surprisingly, without John's name attached to it in any way. They did, however, send him a check for $6. The story was printed in the back of the magazine among the advertisements. Soon, the humorist Bill Nye and Indiana poet James Whitcomb Riley, with whom John would soon join on the speaking circuit, were entertaining their audiences with the short piece, without knowing who the author was. The mistake was soon rectified. Six weeks after its publication, John went to New York, and visited the offices of *Harper's Weekly*. Explaining who he was, and drawing attention to the absence of his name in connection to "Hell-fer-Sartin," he caused the editor there to blush with embarrassment, but it resulted in another check for $6 to add to the previous one of the same amount. "I had sweated my life's blood on 'A Mountain Europa' and 'A Cumberland Vendetta,'" remembered Fox in 1908, "but I found myself known now for two days' work on 'Hell-fer-Sartin.'" [23] He drew attention to the shortness of the story by commenting that he had seen a criticism of the story that amounted to 2,000 words—the story itself numbers only around 750.

Fox never hesitated to give plenty of credit to this little story's part in his success as a writer. "Europa" and "Vendetta," though successful, had not brought him anything like the national attention he desired as a writer, but "Hell-fer-Sartin" changed all that overnight. He was not unaware of the irony, but neither was he one to dwell upon it too much. Nor was he too much affected by the magazine's lapse in neglecting his name in connection with the story, and the paltry sum he received for it. "A twelve-dollar honorarium and anonymity was my reward," he wrote, "but it started me on a short career of short stories." [24]

Now there blew up a wind around the Fox family, especially John. Like so many times before, this new thing, this literary vocation, became all-encompassing to him, and he devoted himself to it totally. His speaking engagements almost immediately began to take up more and more of his time, keeping him away from home. He was always on the move; he was everywhere, it seemed, at once, speaking here, running to catch a train to another city, speaking there, and then off again to who knew where. He did not slow down, and never paused for a moment. He found that he had mounted a whirlwind and could not let go.

CHAPTER 10

The Whirl-Wind

Everybody is kind to me out here.
                                        John Fox, Jr.

Why that feller don't know how to spell.
                                        Anonymous

WITH THE ACCEPTANCE AND PUBLICATION of "A Cumberland Vendetta," John's literary dam burst at the seams, and in quick succession, stories poured from his pen in such a manner that one wonders where he had been keeping them hidden all those years. He would seldom again have such a burst of literary effort in such a short time, and never again would his stories again be as fresh as those first stories of the mountains.

"A Mountain Europa," after its initial publication in magazine format in 1892, would be brought out as a book in 1899, though its length would hardly qualify it for that format. Charles Scribner's Sons would publish the story in book form again in 1904 and 1914, and it would be combined with "A Cumberland Vendetta" and "The Last Stetson" in book form and published by *Harper's* the same year that "Stetson" appeared (1895), and again by the same company in 1900. Scribner's, not to be outdone, and who became Fox's publishers after 1900, brought out the same combination in 1904 and in 1918. "Few writers," says Warren Titus, "have had as successful an initial effort as Fox." [1]

Reviewers were beginning to like him too. Lawrence Hutton, writing for *Harper's* for November 1895 was lurid in his testimony for "A Cumberland Vendetta." "Primitive passion, white-heat or relentless hate," he wrote, "a feud sacredly transmitted from sire to son, make the romantic tissues, of which ... the ... stories by Mr. John Fox, Jun., are woven." [2] *The Bookman*, equally effusive in its praise, said of the collection put out by *Harper's* that "this volume does not read like fiction. It seems to have been cut out of the Cumberland Mountains by a bold, firm hand.... The atmosphere of the scenery, the purple seas of mountains that wave over and between Virginia and Kentucky, the wreathing veils of mist, the green and bronze of tree and moss-covered slopes, the cool, green shadows, the sharp, massive, grey boulders, the deep sweeps of valley, the odour of the earth, the dripping, sparkling dew, the notes of birds, and the hints of laurel, rhododendron, and violets could not have been given by any save a son of the soil." [3]

Neither was "Hell-fer-Sartin" neglected. In quick succession, Fox had produced such stories as "Courtin' on Cutshin," "Through the Gap," "The Senator's Last Trade," "Grayson's Baby," "The Passing of Abraham Shivers," "Message in the Sand," and "Preachin' on Kingdom Come" and all were gathered together after their magazine publication, and along with the aforementioned "Hell-fer-Sartin" would be published by *Harper's* in 1897. It was an instant hit, and *The Critic* was quick to praise it all, even to the shortness of the pieces, and especially the title piece. They said that "although written entirely in dialect, it combines, perhaps more literary excellences than any other single sketch." [4]

The Bookman was a little more reserved, being suspicious of Fox's use of dialect in his stories, suggesting that he was using "a language we have never heard and ... throwing into disuse our noble English" [5] to create an effect. In other words, they surmised that the story could have been easier told, and read, if done in "plain" English, and they were not the only ones to think so. John would receive a letter from his brother Rector, in college at Harvard, that "our instructor in English ... jumped on the dialect stories in one of his lectures recently and mentioned the *Century*, but no particular story. He said that as the stories were written for the country at large they should be written in a language which the whole country, and not a section of it only, could understand. The instructor," continued Rector somewhat apologetically, "is a young man, however." [6]

With all he was doing—the speaking tours, visits with his publishers and the writing he was doing—he was not spending much time at home in the Gap, but neither was anyone else, and it would prove to be one of the loneliest times for his mother and father. Except for short visits, all, with the exception of Horace and the girls, were off to various places, John traveling most of all. Their father was left to face the creditors and the tax assessors and collectors, and all the subsequent problems left from the final financial crash of 1893. In December of 1894, the senior Fox wrote to James that back taxes and "summonses," of which he could make neither "heads nor tails," were arriving at their home almost daily. The exasperation fairly bleeds from the pages of the letter. "There are now posted at the door of the P. Office," John Sr. wrote, "a number of sheets, say 40 or 50, enough to make a good sized pamphlet, containing farms, mineral lands, town lots, etc. to be sold at the court house door at Wise on Tuesday after the fourth Monday in this month ... to pay back taxes for 1888, 89, 90, 91, 92, 93. I never saw or dreamed of the like." None of the postings were for James, but he "saw several against Johnnie and Horace." [7]

Now, and for some years to come, their father would shoulder the unwelcome burden of the family's finances. Despite the money John would soon be sending home regularly, it would not be enough to settle their debts for more than short periods at a time. John Sr. was left to raise hogs and cattle and corn, and attempt to sell them at enough of a profit to his also cash strapped neighbors, to pay their taxes and debts. "At Christmas," he wrote James assuredly, "or as soon thereafter as the market will justify it, I shall have eight more fat hogs to sell, which are smaller than the others but will bring $50 cash, then I shall have 2 sows and 22 shoats left to sell hereafter as they get older and fat. They are worth to day $75 at a reasonable price. Then I have about 300 bushels of corn ... which by spring will bring I think 75 [cents] per bushel or $225, besides 75 or 100 bushels ... to feed my hogs. Then ... I have 128 doz. oats worth $25 in the stack. So," he concluded, "I hope to be able to pull through." [8]

It would seem an ignominious step down for the once respected school master, but he did not think so himself, or if he did, he was careful not to reveal it to his sons. He knew they were trying as hard as they could, and times being what they were, he would not expect from them more than they could give. As for himself, he would not allow himself to shirk nor complain, nor did he allow himself to stop his own efforts until nearly twenty years later when his heart, taxed also to the limit, finally gave out, and left a family void that nothing could fill. But in the meantime the struggle for financial security went on in the Fox family, no matter where the various pieces of that family were, and the whirlwind that John was now riding in far away places could not help but affect them all, both positively in the money he could now send home and otherwise in the number and duration of his absences from home.

The only thing that John Jr. had done that had resulted in any money at all had been writing and the reading of his and others' works from the stage. Nothing else had brought in anything worthy of mention, and that, with the exception of his newspaper work, had only amounted to less than $300 by the end of 1894. He had tried newspaper writing and had been forced to live, at times, on beef broth and crackers, and 15 cent lunches. He had tried mining in Jellico with much the same results. His "Napoleon of Finance" days in the Gap were best left unspoken of, and his venture into law enforcement was a volunteer effort. He was 30 years old, and in those 30 years had not made enough money to fill one of his hollow teeth, which were now beginning to cause him a lot of pain. No one was more keenly aware of their monetary problems than John Fox, Jr., himself.

The amounts he had received for his writing had been modest, but on the other hand, he had been able to sell nearly everything he wrote, including the newspaper articles about the hanging of Hall and Taylor. That was so much more than he could say about his other ventures, voluntary or not.

Other writers had supplemented their literary incomes by taking up the speaking circuit and going from town to town, reading from their own works or from the works of other. A good speaker, willing to travel and live out of a suitcase, could possibly make more money from reading a story or parts of a novel than he could from writing that same material. "A Mountain Europa" brought John Fox only $262 from its first publisher. Other writings would bring as little as $6. Speaking engagements could, and often did, bring in $100 per night. It was too good a deal to pass up, and late in 1893 and early 1894, he had investigated the possibilities and had found a new and fertile ground for his talents. The people he spoke to liked him, and that was something new to him. He had his friends in the past, to be sure, and most of those who met him thought well of him, but he had never had 800 people stand and applaud him. It was an intoxicating experience, and he was grateful to them, and they were just as grateful that he was bringing to them the culture they could not find anywhere else, even if it was in mountain dialect. They were entranced, and so was he.

He soon found himself rubbing shoulders and sharing the stage with authors and artists, writers and poets, and men from every important profession, and more importantly to him, they accepted him without reservation. He was introduced to Richard Harding Davis, the boy-wonder from Marion, Massachusetts, and began a friendship that lasted until Davis's death in 1916. He knew Charles Dana Gibson of the Gibson Girl fame and nearly anyone who had anything to do with publishing in New York.

Theodore Roosevelt enjoyed his readings and stories so much that he seldom failed to write and tell him so, and would remember John years later and request John's presence at the White House once he was elected president, and John seldom failed to go. By 1895, Fox was moving in ever widening circles of high society and was gaining the respect of those who were vastly more famous and better known than he. He was given farewell dinners by the likes of Gibson, and dined on different occasions with Owen Wister and Frederic Remington, and with Richard Harding Davis and Thomas Nelson Page. He loved it all, and could hardly restrain his joy when in their company. It was the life he had always dreamed of for himself.

He was just about the best reader of his own works that anyone had ever heard, in small town or large. The ladies who heard him wanted to mother him because of his smallish size and shy manner, and men who accompanied those ladies wanted to be his friend. "I remember Fox reading ... one night at the Women's Club in Lexington — my first sight of him," wrote John Wilson Townsend in 1932, "a little fellow with enormous gold-rimmed glasses, entirely covering his fine eyes, the side arms wandering off back of his ears, becoming lost and tangled in his long brown hair, which was past due at the barber's.

He wasn't much of a man, physically about the size of Madison Cawein and Emerson, each weighing something like 130 pounds. But when he began to read and talk one forgot about his unprepossessing appearance. His clothes didn't come from Bond Street; but his brain did. He was," concluded Townsend, "without any question...the best reader of his own stuff I have ever heard." [9]

His clothes didn't come from Bond Street, as Townsend said, and even the most casual observer would have noticed that. His letters home regularly ask for the odd item of clothing, Sid's old hat, and the like. He was forgetful of his clothing as he was of almost everything else, but perhaps not as forgetful as one of his stage partners.

He was paired with James Whitcomb Riley from time to time, and on one occasion, John had remembered everything including his suit. Riley was not so fortunate nor lucky. He had neglected to pack his clothes, and found himself about to go on stage without his customary suit. But on the spur of the moment, he and John worked out a plan. "Fox would speak first and disappear back stage," wrote W.B. Ardery for the *Lexington Herald* in 1973. "Riley then would face the audience in Fox's suit." This went on all night, and no one in the audience the wiser, or if they suspected, they were too polite to say anything. [10]

The people John was speaking to were culturally starved, and cared little or nothing about the clothes he wore on stage, or what he wore under them for that matter. They were interested in him, and the stories he could tell. That is what made the whole experience so heady and intoxicating. Not that there was not ample culture in the nation after the Civil War; far from it: Most of America's most famous and capable authors and artists and poets were working at their peak during the decades following the 1860s. The problem was that many were located in larger cities, principally New York, Chicago and Philadelphia. Mark Twain was doing just fine, and so were authors and poets like Emerson, Whitman, and Thoreau. Even foreign born authors did not neglect America. Dickens and Conrad and Kipling all visited and gave readings from their works to enthusiastic American crowds. And so did a great many others, both famous and not so famous. "As a means of enlightenment," wrote David Kesterson in his biography of Bill Nye, another veteran of the stage, "moral instruction, and entertainment, lecturing had flourished in America since the early 1830's when the American Lyceum was established. Lecture halls ... drew crowds to hear the likes of eloquent Ralph Waldo Emerson and Henry Ward Beecher ... Artemus Ward ... and Mark Twain." [11]

Soon problems developed with the system, or rather, the lack of a system. In the beginning, local "boosters" would write an author or lecturer and ask them to visit their town for a speaking engagement. Once the engagement was performed, the pay might not be forthcoming from tight-fisted city officials, or in a currency that could seem more punishment than pay. Once, somewhere in the South, John was paid in silver dollars, and walked back to his hotel with his pockets bulging, and his trousers threatening to fall around his ankles from the weight. "When he got back to his hotel," said his brother-in-law William Cabell Moore, "he emptied them out on the counter much to the surprise of the clerk who" assumed he had been gambling, and "said enviously, 'Well, you did have a good game, didn't you?'" [12]

The speaker might be expected to pay his own transportation, room and board, and in the end receive only a few dollars for his efforts. It was a situation that was about as organized as a chicken house, and the circumstances begged for management.

Another problem was the distances involved and the modes of travel that were available that made bringing culture to the smaller cities and towns extremely difficult. But they would pay money to see and hear the famous and even the not so famous, and so they came, and spoke, and read, and smiled, and through the auspices of Chautauqua societies and lyceum bureaus, culture was brought to the masses. This was, of course, years before radio and television, and even phonographs were yet to be invented. So, if culture was to be had beyond the printed page, it had to be live, and authors reading their works and the works of others was de rigueur in literary presentations for local Chautauqua societies and the more national lyceums. Once the value of such a network was established, cities and towns of any size at all were quick to capitalize upon the phenomenon and establish societies of their own, and culture penetrated the Midwest, the South, the East and North.

Naturally, authors were usually more than glad to supplement their coffers with money, modest though it was, between the publishing and writing of the next book. Books were not very expensive, at least by today's standards, and there were few who were able to support themselves and their families strictly through their writing efforts. Most held other jobs to keep the wolves away from the door. Some, like the Kentucky poet Madison Cawein, worked in pool rooms or just about anywhere a dollar could be made so they could continue to write. Because they often thought of themselves as famous people and "artists," some required a little more pampering than others. It became the job of the managers, like James Pond, to see that their "stars" were happy.

But the lyceums were outlets for those who were blessed with at least an average speaking voice. The people did not care. Here in their little town was the poet from Indiana, James Whitcomb Riley, reading his poems, and with him might be the humorist Bill Nye, or John Fox, Jr., or any number of other people they had only read about. They were fascinated, and they ate it up with a spoon, and clamored for more.

Fox was in his element on the stage, and the people he spoke to were loud and consistent in their praise of him. James Lane Allen, Fox's fellow author and friend, had said of Fox's reading talent early on that "not only is he a very beautiful reader, but he is the first public reader of the dialect of the Tennessee and Kentucky mountaineers that has yet appeared." [13]

He was able to mimic his characters as effortlessly as a professional actor who had studied for years, and he had sympathy and admiration for his characters, and it showed. He made them alive, and they seemed so real to the people who heard him because they were real, sometimes uncomfortably so to his mountain born listeners. He could do what few writers and speakers hardly dared to do: attempt to mimic perfectly the idiom of the mountain people he wrote of. This he could do effortlessly, but such rendering was not without its dangers. Once, when speaking at Berea College in Kentucky, a school established expressly for mountain youth, Fox found himself in something of a tight spot. He had given a reading from some of his works, in dialect, and "the mountain boys were ready to mob him. They had no comprehension of the nature of fiction. Mr. Fox's stories were either true or false. If they were true, then he was "no gentleman" for telling all the family affairs of people who had befriended him. If they were not true, then, of course, they were libelous upon mountain people." [14]

Those mountain boys in Berea were not overly sensitive; they were, perhaps, more intelligent and shrewd in literary matters than most might have thought. Once, Horace Kephart, author of *Our Southern Highlanders*, gave a copy of one of John's books to a typical mountaineer, to see what he thought of the writing, especially the manner in which Fox rendered the mountain speech. According to Kephart, the man "scanned a few lines of dialogue, then suddenly stared at [Kephart] in amazement." Kephart asked what was wrong, and the mountaineer replied in a shocked voice, "Why, that feller don't know how to spell!" When Kephart explained to him that the dialect style of writing was done that way on purpose, the man was "outraged." "That tale-teller," he said, "is jest makin' fun of the mountain people by misspellin' our talk. You educated folks don't spell your own words the way you say them." It was something of an eye-opening experience for Kephart, and, he said, "gave me a new point of view." [15]

In finding larger audiences for their works through speaking engagements, authors were almost assured that those same people who paid a dollar, sometimes as little as 50 cents, to hear them read from their stories and novels would go out and buy a copy of the book or story for themselves. Assuming, of course, that they were suitably impressed with the speaker, and most, unfamiliar with literary stars, *were* impressed, and *did* buy their books. It was hard work, to be sure, both physically and mentally exhausting, but it could be profitable, and many of them perfected their speaking abilities and materials accordingly, fully realizing that their literary reputation and livelihood might depend upon their successes on the circuit.

However, thinking that their audiences wanted culture and to be enlightened by their speakers could get a speaker in trouble. They soon discovered that most wanted only to be entertained, and nearly every speaker who remained for any length of time on the stage had to quickly change their repository of material to include comic and dialect pieces. It was a valuable lesson for a writer to learn, and many transferred their speaking lessons to their writing. They would give the public — both listening public and reading public — what they wanted. James Whitcomb Riley, friend and fellow speaker, said, "In my readings I had an opportunity to study and find out for myself what the public wants and afterward I would endeavor to use the knowledge gained in my writing." [16]

Fox stayed with the speaking circuit longer than most, years in fact, where most "retired" after only a few months or only a year or so. Few could stand the endless travel and cities that soon grew to resemble each other too much. Not many could keep up the pace required, and the constant reading and speaking the same material over and over again. It would be several years, however, before John tired enough of the seemingly endless travel and speaking engagements that never seemed to end. He was popular, but that popularity, and the satisfaction he gained personally from being on the same stage with the likes of Riley, and rubbing shoulders with the likes of Mark Twain among others, could not sustain him, and his health began to suffer. His writing did as well, for involvement with the lyceum mandated a major investment of time on the part of the speaker. Bill Nye, the nineteenth century humorist, and himself a popular circuit speaker, was not immune from the problems of speaking tours either. "The history of the lecture industry of America," he wrote, "is one of alternate elation and depression. It is a history of alternate failure and success, written with the heart's blood of the lecturer." Later, he said, "The lecturer has two or three great obstacles to overcome which the actor has not ... he has no scenery, he has to occupy the entire evening alone, and there is no division into three or four acts with a chance for the audience to rest and run down the show. And yet the lecturer often starts out fearlessly without training, or with training that is far worse than none, and on the reputation he has made in some totally different art he fearlessly rushes in where angels would naturally hang back." [17] What every speaker or prospective speaker needed was an agent, whether they knew it or not.

It had been Thomas Nelson Page who had first mentioned John's talents to Page's own lecture agent, a man by the name of Major J.B. Pond. He had written the Major, who had guided Page's own speaking engagements, introducing Fox, and warned Pond that if he did not sign Fox to a contract soon, someone else would. The Major did not hesitate for long, and was never sorry of his decision. When he wrote of the men and women he had been an agent for in his book *Eccentricities of Genius*, Pond said of Fox, "Mr. Fox is surely one of the most popular Southern authors of the time, and is very much appreciated in the South...because of the high character of his literary work." Of Fox's presence and ability on the platform, he said that "he has a capital presence, a magnetic force and manner, and a most telling voice at his command. On the platform he is pretty much what he is off it, except that he is sensitively watchful of doing his work well.... He has no mannerisms and gives no evidence of effort. He simply tells and lives in the telling. What he gives is truly his own work. His dialect is perfect, but it is human and actual, not a mere caricature. The figures he gives are wholesome and clean, as is the man who presents them." [18]

James Pond was impressed with his new speaker, and said so in expansive terms to Thomas Nelson Page who had introduced the two men early in 1894. Pond was cautious at first, no doubt having had any number of would-be speakers take up his time with less than adequate talent, but he found out soon enough that Fox was the real thing. "It was one of the most delightful entertainments I have heard," he wrote Page early in 1895 after having John read for himself and his guests. "It is absolute novelty [and]," he continued, "I am certain if he has got the energy, he has the ability, which is extraordinary." [19]

Everything he did now revolved around his new career, even his trips home. He would miss Christmas at the Gap in 1894, and it would not be the last time he would not be there for that holiday. He wrote his mother two days before that Christmas that it was unlikely that he would be home before February of next year. He hoped for January, but with his readings, he thought February more likely. The money was pouring in, he wrote, but "My expenses are rather heavy and I haven't yet been able to get proper clothes [but] everyone is kind to me out here." [20] He was making the inevitable friends and receiving invitations to parties, but did not accept any, he wrote, because of his lack of proper clothing.

The money he was making was steady and ignited his enthusiasm, and he was more hopeful of success than he had been in a long while. James Lane Allen was near him in New York, and brother James was there also, still hopeful that their business interests would soon change for the better. "Oh! things are bound to come in time," John wrote to his mother. "Don't let papa work too hard; and don't work too hard yourself. Worry [?] through the winter the best you can — all of you — for it is the last of the kind, I believe, that you will ever have to go through." [21] He was wrong, of course, but perhaps it was mostly wishful thinking on his part. Then again, maybe he thought if he wished hard enough on Christmas, that wish was bound to come true. It did not.

Between the publication of "A Cumberland Vendetta" in 1894 and the onset of the Spanish-American War in 1898 he wrote and published at least 13 stories and articles, one novelette, and his first full length novel; managed to speak in dozens of cities and towns all over the East and Midwest — and still managed to keep his sanity and heath; and met nearly every famous and near-famous personage in the United States, including one future president, which by any measure is pretty good doins' for an old Kentucky boy from a little wide place in the road called Stony Point.

10. The Whirl-Wind

On one of his rare trips home, he went over into Kentucky and attended his third hanging. In June of 1895 in Breathitt County, Kentucky, "Bad" Tom Smith fell through the trap-door in Jackson, the county seat, in front of a crowd reminiscent of the ones who watched Talton Hall and Doc Taylor swing a couple of years before. Very likely, many of the same ones who saw Hall and Taylor hang were on hand to see Tom Smith swing from the gallows.

John had not seen such a circus atmosphere since Talton Hall and Doc Taylor was hanged three years before, and in some ways, he thought it was worse. The jail was full of people — reporters, deputies and guards, and the just plain morbidly curious who outnumbered everyone else. Tom's wife, whom he had so recently abandoned in Perry County, came to see him, dragging their three children along with her.

"Bad" Tom was not anything like John expected him to be. He did not really look like a killer; few murderers do. "He was a good-looking fellow," he wrote just after the hanging, "just over thirty, with a pallid face, a black mustache ... of the mountain dandy — black hair, and black upper and under lashes that literally lay out on his cheeks. They eye under them was blue, languid, and bold only when it looked into a woman's. He played the banjo and sang, and, as he himself said, women would leave their husbands to follow him." [22]

On the morning of the hanging, "Bad" Tom — no longer, at least outwardly, as bad as he once was— walked out of the jail under heavy guard and was led to the muddy river that flows hard against Jackson and was baptized. Soon enough he mounted the raw scaffold where the now obligatory scaffold confession and prayer began. It was scorching hot, even for July in Kentucky, but Tom preached and confessed for nearly an hour, wringing out every minute he could, with his sister at his side. Confessions make people nervous, especially where murder is concerned, and some in the crowd uneasily wondered if their own names might be mentioned in connection with some of the dastardly deeds Tom was wringing out one by one.

John, on the scaffold himself to hear Tom's last words, stayed as long as he could, swinging down into the crowd, close to the last minute, "near [a] little woman in black. She insisted that [her] children should be held where they could see. A mountaineer spoke to her, holding out his hand, 'How air ye?' 'I'm feeling mighty good now', she said. She was the widow of Bad Tom's last victim.'" [23]

At last, Tom kissed his sister, "knelt down on the trap door and prayed a long and hysterical prayer, his legs were then pinioned, the black cap adjusted and in the presence of 4,000 or 5,000 people his soul went out into eternity.... Just as the trap was sprung he was heard to cry out in an agonizing voice: 'Save me, O God, save me.'" [24]

John returned home to Big Stone Gap, thoughtful and not a little depressed by what he had seen. "Bad" Tom was dead, but the feud still lived.

"Bad" Tom was dead, but his many victims did not arise from their graves. It ought to have been different, somehow, but it was not. It was too harsh, too cold, and too unforgiving in these mountains.

He wrote of the execution for *Harper's Weekly*, and received the munificent sum of $30 for his effort. It seemed to him that he could not write enough; everything was grist for his pen, and he was surprised at how little effort it seemed to take for him to write these early days. All that would change in a matter of a few years, and the effort would come harder and harder as he grew older, and new ideas and new inspiration would become more and more rare. But despite the quantity and quality of the stories he was writing now, his best work was still years ahead, when he would, at times, have to force himself to sit down and simply write.

He spent most of the winter of 1895-96 away from home, reading in one place or another, and working on his novel *The Kentuckians*, and the coming of spring did not improve things. He was constantly somewhere besides the Gap — Shelbyville, Kentucky; New York City; Lexington and Louisville, Kentucky; and, of course, Frankfort, Kentucky — and in November he made his reading at Berea, Kentucky, where he was almost mobbed by irate mountain students who thought he was misrepresenting their people and their ways.

By late fall of 1896, he had finished his first full-length novel, although it too is barely long enough — some 20 pages or so longer than "A Cumberland Vendetta"— to deserve that title of novel. The advent of his "novel," *The Kentuckians*, in July of 1897, set in Frankfort, Kentucky, caused Charles Wingate, an old school chum, to write in *The Critic* that "the name of John Fox, Jr. ... will be well known in this country before many years, or I miss my prophecy." [27] He was right; it would be.

John had chosen Frankfort, Kentucky, as the setting for this first novel, and between speaking engagements, he made his second home there during 1895 and 1896. While there, he shared rooms with the poet Robert Burns Wilson, whom he openly admired. The house where they lived, at the corner of Ann and Clinton Street in Frankfort, still stands, and it is hardly a stone's throw from the state capitol that is the centerpiece of the story.

Fox was already living up to Windgate's prophecy, at least in Frankfort, where he was seldom without a dinner invitation, and it was a rare ball that was held while he was in town that did not include him among the guests. "Hobnobbing with the poet-artist Robert Burns Wilson," wrote Willard Rouse Jillson, "Fox found ready entree as a dinner and party guest to many of the most cultured homes of Frankfort." [25] One evening while walking in a garden with a young lady named Maria Lindsey, he was boldly asked by her, "how he dared to write so frankly and describe intimately the lives, habits and language of the mountain folk of Kentucky" as he had so recently done in "A Cumberland Vendetta." Fox answered by saying to the curious young lady, "Well, when I am in the mountains, I dress and act and live like the people down there," which is true enough even if he elaborated enough to impress Miss Lindsey, and perhaps remembering his close call at Berea, he continued somewhat melodramatically, "but if I was known as the writer of my stories, my life wouldn't be worth much." [26]

He was in no danger in Frankfort, however, unless he should fall victim to the wily charms of the young Southern belles he strolled and danced with. But none of them held a lasting attraction for him. They were interested, and he oftentimes was as interested as they were, but the thoughts of a serious relationship never seemed to occur to him, and he escaped, though in later years, alone, he might have regretted his ability to run so fast from their attentions.

Harper's published the finished novel, more like a novelette, in four parts in the July, August, September and October issues of their magazine, and almost immediately it was brought out in book form by the same company in December.

John (left), Robert Burns Wilson and two unidentified women. Probably taken in Frankfort, Kentucky, around 1895 while John was staying there gathering information for his next novel (courtesy Duncan Tavern).

The theme of the book, once the romantic scenes involving the three main characters are dismissed, is a serious one, and again uses the well-worn themes of Kentucky feuds and feudists, along with a healthy mix of Kentucky politics, which to some my seem a double dose of the same thing. It was a short first novel, but one that was successful for him. Coming on the heels of his success on the speaking circuit and with the publication of his short stories previously, the book sold well, and it did not hurt that the critics liked it. [27]

Before, during and after the publication of *The Kentuckians*, John continued to play the part of speaker and author, and played it as well as he could, but like others, he soon found it too difficult to write and speak of his writing at the same time. He missed his home and family in Virginia and Kentucky. He missed the leisure time, and he missed his mother and father and all the rest of his family most of all. In the end, the decision to go home and stay there more often would have come easier for him if it had not been for the money he was making. The people he was seeing, and the places he traveled to without end, hypnotized him with their attractions, and he could not give them up, at least not yet.

But in a land far away to the south of Big Stone Gap, events were shaping John's future for him and would keep him from home for several months yet. Before the end of 1898, he would see men die violently and would hear the whistle and zing of offensive bullet and shell. He would see the litter of the battlefield, and the awfulness of disease and sickness and the gut-wrenching stenches of war. Before the end of 1898, he would find himself quarantined in the tropical and oppressive heat of Tampa, Florida, finally making his way to Louisville in August, delirious, and wandering the streets, not knowing where or who he was, and five days later, on August 8, he would arrive home in the Gap, still seriously ill, much weakened and unsure in his step, but home at last.

CHAPTER 11

The Spanish-American War

We were all a little mad.

       Richard Harding Davis

It is all like a big picnic.

       John Fox, Jr.

HE HAD SPENT A GOOD DEAL OF TIME on *The Kentuckians*, dedicated it to his father, and *Harper's* had paid him $1,000 for the serial rights. He had, in the meantime, sent them another story called "An Intersectional Episode" which they had declined to publish, citing lack of space in the magazine. When he received the manuscript back, he filed it away and forgot about it. It was not that great a story in the first place, and as far as is known, he never revised it, nor made any other attempt to have it published at *Harper's* nor anywhere else.

  The first months of 1898 continued to be active ones for John. He found that, tired as he was of the constant pressure of deadlines and train schedules that went along with traveling the speaking circuits, he still enjoyed being away from home enough to pick up the circuit again after Christmas. The cities he spoke in were endless that winter of '97 and spring of '98. He told the same stories over and over until he thought he might sneak a nap in the middle of one, keep talking, and no one would be the wiser. However, he liked it, despite the drawbacks, but mostly he liked the people. He slummed with James Whitcomb Riley, Henry Clay, Jr., and Robert Burns Wilson. He dined with Kentucky governors, ate lunches in the company of the likes of Mark Twain, Thomas Nelson Page, Theodore Roosevelt, and Richard Harding Davis, of whom he was growing more and more fond. He was becoming known by and beginning to know a myriad of other people, famous and not so famous.

He became one of them, not as a "hanger-on," but as an accepted equal, a writer who showed great promise, and they genuinely liked him and counted him as one of their best friends. He in turn returned their friendship and missed them and their company when he was away from them.

Every trip that he now made back to Big Stone Gap reminded him of how small and confining the town was, and the memories of his financial troubles there did nothing to make his visits home any more pleasant. Every trip to New York, Boston, Chicago and Richmond, Virginia, only emphasized the differences between large towns and home town. He would always be a small-town boy, however, and the homesickness he felt when away from the Gap never left him for long, and he often found himself torn between what he had to do outside the Gap and the isolation he felt when he was away from his home and his family.

The publication of his stories, novelettes, and now a novel over the last five or six years was fast making him a marketable author with a name well recognized by the American reading public. His stories were gaining in popularity in Europe as well. *Harper's*, his publisher, was still being a little conservative, or so John thought, in their estimation of his worth, although they were hesitant to say so. John made the mistake of thinking everything he wrote was good enough to publish, but it was not. *Harper's* had the unenviable task of telling him so in a way that only the best of editors can do. By the early spring of 1898, he was a little put out with *Harper's*, but the Spanish-American War would intervene before he could make any hasty decision about them and their association with him.

He had begun to think of changing publishers, although he had to admit that *Harper's* had not really mistreated him, and he would not change for another couple of years. They were giving him good advice, but he was impatient with the progress of his success. However, a seed of discontent had been planted, and it was only a matter of time before he would move to Scribner's, and stay there the rest of his writing career. [1]

By the early spring of 1898, the thing that would send him to a foreign land for the first time, and nearly kill him in the process, was already rolling inevitably toward an unknown end, and even the president of the United States, arguably already one of the most powerful offices in the world, seemed powerless to stop it, or even come up with a reasonable plan should the worst happen. On February 15, the U.S. battleship *Maine* was blown up in Havana harbor, and the snowball that began in Thomas Jefferson's administration had finally gained enough mass to cause an avalanche. But John was oblivious to it all. He had fallen in love.

Her name was Currie Duke, and she was not just any girl, but was the daughter of General Basil Duke, who was not just any general. Duke had been the right hand and the brother-in-law of General John Hunt Morgan, immortal figure of the Confederacy and beloved Kentuckian from the late Civil War.

Morgan had been killed before the war's end, but Duke had survived. Upon his second-in-command, was bestowed all the laurels that had been reserved for Morgan. The high society circles of Lexington and Louisville and everywhere in between knew no higher society figure than Basil Duke, and by extension, his daughter.

It was not a new acquaintance. They had met before, and had sparred a little, and flirted a little as so often happens, but they had parted without any real feelings other that a casual friendship between them. Now, they had reacquainted themselves and their relationship took a decidedly more serious turn. During that spring of 1898, John courted her as often as his speaking engagements would allow. She was flirtatious with him, allowing him to kiss her one day, and refusing her favors the next. He was more adamant than usual, recognizing in her everything he desired and did not have, but she was cautious and it is not unlikely that the hidden American class system, even in Kentucky, demanded that she, the daughter of the famous and respected Basil Duke, should be more circumspect in her romantic entanglements. She was his one true love — everyone, then and now, seems to agree on that point — and the one he would have been most happy with, but in the end, in the real world, John's position in society and his prospects for advancement — though moving steadily upward — were not enough. The Spanish-American War intervened at the worst possible time, and Currie would marry another. But he always loved her. [2]

Despite the personal concerns that were occupying most of his thoughts, he could not completely neglect the coming of the Cuban War; no one could. It was not long before he began to entertain thoughts of going there as a correspondent for anyone who would have him. *Harper's* was his natural first choice, but even they fell into second place while he was trying to decide whether or not to serve in the army, preferably with Theodore Roosevelt, whom he knew and who was already thinking of forming his own command for the upcoming war. John knew that being a soldier would almost certainly bring more adventure, and he was giving serious thought to that if Roosevelt would take him along, but he also knew that journalism was what he was best suited for. He was determined to go, however, when and if the war actually began, regardless of position or capacity.

John was not the only one with misgivings and hesitations about what to do. His compadre, Richard Harding Davis was, if anything, more distressed about much the same decision as was John. Davis wrote to his brother Charles after making his decision to remain a correspondent that "the man that enlists or volunteers ... and the man who doesn't enlist at all ... is much better off than I will be writing about what other men do and not doing it myself.... I'll always feel I lost in character.... It was a great opportunity missed but the sad thing is that if it came again I doubt if I'd take it. I just don't want to make the sacrifice and I am sorry to find I am not of the stuff that can make it." [3] Davis should not have been so hard on himself, and his self-depreciating matter did not fit the picture of his usual self, nor was that the image that he presented while he was in Cuba where he gained the respect and admiration of nearly everyone who knew and met him there. The decision to remain true to his calling of journalist, however, was a difficult one for him to make, and it was a difficult one for others who, like Davis and John Fox, Jr., remained journalists and did not become soldiers. John, perhaps, had more of a chance at being a soldier than Davis. He had already talked with Roosevelt who had no illusions as to what role he should play in the upcoming war, and Roosevelt had made it plain—at least to John's way of thinking—that there would be a place for him in any outfit that Roosevelt commanded. Through their correspondence about John's writings, they had become friends—not as close as they would become later—but friends nonetheless, and when Roosevelt mentioned that he would have a place for John, he was taken at his word. However, John was a journalist first, but the attraction of being a soldier in this great war to come was strong and undeniable.

Eventually, both Davis and Fox would make the choice to remain true to their aptitudes and report the war. Volunteer and part-time soldiers were a dime a dozen that spring of 1898, and literally everyone wanted to fight the Spanish, and the decision that both of them finally made was not an easy choice for either of them, but then choices of that nature seldom are.

War itself, has seasons, much the same as the weather. The revolutions in Cuba in the nineteenth century that culminated in the Spanish-American War of 1898 had waxed and waned from season to season all through that century without any resolution whatsoever. There were periods of peace followed by periods of insurrection, put down by the Spanish authorities in Cuba in varying degrees of cruelty and harshness. One should not forget, however, that as far as the cruelties of war, especially revolutions, there are usually enough recriminations on both sides enough to go around, especially in regards to cruelty. Blame, and those who deserve it during wartime, is not a rarity. Americans, however, tended to side more often than not with the rebels especially when the newspapers of the day could be biased in their reporting of what was going on in our backyard. Americans soon came to visualize the Spanish in Cuba as being the oppressors, and would seldom, if ever, entertain any serious thoughts that the situation there might not all be the fault of those Spanish rulers.

After a period of relative peace, the final insurrection broke out in February 1895. Everything went downhill from there. Ninety miles away in the United States, there was concern about what was going on in Cuba. It was an island that was just too close to be ignored, but the difference now was that the United States had time to act on that concern should the need arise. "To our innocent and altruistic publics the Cuban insurrection of 1895 seemed a simple case of an oppressed people rising spontaneously to free itself of an alien tyranny. Actually, however, it was a far more complex phenomenon than that; behind it there lay a long and sometimes rather dubious history, of which the American people remembered very little and understood even less." [4] Before 1895, the United States had been tied up in various wars and expansions, including the Civil War (the end of which removed slavery, and one of the principal reasons the South had supported annexation of Cuba), [5] and various other national concerns, but by the time the century was drawing to a close, much of what occupied the national consciousness had been settled, or at the very least, moved to the back burner of the national oven.

Perhaps most importantly, the frontier had at last been settled. The frontier was no more, and Americans felt a new sense of growth — of stature that made them individually, and as a country, want to stretch as far as they could, both figuratively and literally — but with the frontiers closed, they found themselves with nowhere to turn to exercise that growth. Perhaps the pistol shot that sounded the charge in the Oklahoma Territory for the last, vast free land in 1889 signaled the beginning of the end of that frontier that served as a comfortable buffer, a border and a challenge for generations of Americans, but it may have also signaled a beginning of American expansion abroad, even if it was less than a hundred miles away. But the challenge was on the horizon, and in some ways, a more dangerous horizon than the late, lamented frontier. "The nation had completed its appointed task in the romantic West and found itself looking beyond the borders with a new energy, a new sense of union, a new self-consciousness." [6] In retrospect, it seemed that the nation was looking for somewhere else to go, and in 1898 Cuba looked like a good place to start.

After the *Maine* went down, everyone was shouting for war, even poets. Robert Burns Wilson, John's old Frankfort pal, quit writing about flowers and ferns and trees long enough to write a battle song entitled "Remember the Maine." John thought it was just possibly the best thing he had ever read, and in his enthusiasm called it "a ringing corker." [7] It was soon published by the New York *Herald* near the middle of April, but by that time John had other matters more pressing.

But not everyone was as enthusiastic as John and Burns Wilson over the war. John's friend, Thomas Nelson Page, was totally against the idea, and just possibly might have been, along with Mark Twain who also disapproved of the coming war, the only sane person in America. "Your belligerent mountaineers," Page wrote John early in April, "have a fine opportunity now to direct their surplus bloodthirstiness against the enemies of their country unless they prefer to conserve it for their neighbors." [8]

John was ahead of the pack, and despite the disapproval of his friend Thomas Nelson Page, he was, by the first of April, trying to get to Cuba. He wrote to Theodore Roosevelt, but Roosevelt was not encouraging, saying that it seemed "unlikely" that he would be raising a company of Kentuckians or Virginians, and that he did not see any "immediate" chance of John going with him. John was adamant, however, but it did him little good with Roosevelt until the first part of May, when the future president wrote, "If you want to go along as correspondent you shall always have a place with our regiment." [9] It was not exactly what John wanted, nor what he thought Roosevelt had previously promised, but he soon gave up any real notions of becoming a soldier, and began to set his sights more and more upon convincing *Harper's* to use him as a correspondent, just as Roosevelt had suggested that he might do.

The United States was woefully unprepared for war of any kind. The navy was bad enough, but the army was even worse, if possible. It was limited in size to only 25,000 men, and had not seen any action against a foreign enemy since the Mexican war 50 years before. Worse still, the United States had not fought any war at all, excepting the Indian Wars, since the Civil War nearly 35 years before. President McKinley was chagrined to discover that it would take two months by conservative estimates to gather and train a force of sufficient size to invade the island of Cuba. It was, to say the least, embarrassing.

The American soldier was shamefully unready, and not especially through any fault of his own. Roosevelt in his autobiography supposed that "the United States will always be unready for war," and surmised that we continued to reinvent the wheel whenever war actually came to us. "Americans learn only from catastrophes," he wrote in the chapter of his autobiography entitled "The War of America the Unready," and by default, it is the common soldier who bears the brunt on his superior's ignorance. [10]

The government set about trying to play catch-up for more than 30 years of neglect. Training camps were established in different parts of the country. San Antonio, Texas, hosted the "Rough Riders," and the citizens there were not too sorry to see the hard-fighting, fun-loving volunteers leave for Tampa, Florida, in May. They had been quartered outside of San Antonio proper, but apparently not far enough away for some of the citizens of the town, and so it was with mixed feelings that the more sedate inhabitants of San Antonio saw the rowdy soldiers leave.

The old Civil War battlefield of Chickamauga in Georgia was another training camp set up to teach every sort of volunteer the rudiments of military life. [11] In some ways, it was nothing more than organized chaos at those camps, but the American public was impatient for any news, and *Harper's* decided to send John to see what was going on down in Georgia. "Will you go to Chickamauga," they wired him on April 21, "and send three thousand words," [12] and there would be more. They asked him to consider going to other camps in New Orleans, Mobile and eventually, Tampa, but they wisely did not promise anything past Chickamauga. It was not Cuba, where the real story was supposed to be, but Tampa was very close, and who knew what could happen when and if he arrived there?

He wasted no time in accepting *Harper's* offer. At Chickamauga, not far from Chattanooga, Tennessee, he did not find "much to exhilarate" him he wrote his mother, but managed to write a "letter" for *Harper's* that he did not think much of. There simply was not enough excitement in Chickamauga to fire up anyone's imagination. Camp life, as any soldier can relate, is dull, more often than not, and writing about it does not improve its flavor.

Spring was going full blast when he arrived in Chattanooga, and without a doubt, spring is the best time to be either in Kentucky or Tennessee. He found a flurry of activity, and not a little confusion in the hastily constructed army camp situated on nearly "six thousand acres of creek-bottom and wooded hills." [13] They were not using all of the old battlefield, of course, and were lumped together, row upon row of tents and equipage, and parade grounds had been hurriedly constructed to try and train volunteers with two left feet, the highly desirable trait of marching in order. Baron Von Steuben of Revolutionary War fame would have felt right at home with these soldiers.

But they were so eager. Eager to fight, and few could blame them if they thought marching in a straight line, and learning close-order drill would be of little use to them on the battlefields, and so they mingled, unorganized and unmindful. "On the streets gray slouched hats everywhere," John wrote for *Harper's*, "hats with little brass crosses that were not symbols of religion, crossed rifles, crossed sabers, tiny crossed cannon; men with big pistols and belts of big gleaming cartridges; soldiers white and black; soldiers everywhere, some swaggering and ogling and loud of voice, but nearly all quiet, orderly, well-behaved." [14]

The civilians in and around Chattanooga and Chickamauga were all agog at the influx, and most had never seen anything like it in their lives. In a scene reminiscent of the first battle of Manassas in the Civil War, the citizens came out on Sundays to see the spectacle. It "was a vast local holiday for a swarm

John Jr. in his Spanish-American War "uniform." He went as a correspondent for *Harper's*, and vowed never to go again under those circumstances. This photograph may have been made for Frederic Remington who intended to do a portrait of John, but for one reason or another, it apparently was never done (courtesy Duncan Tavern).

of curious civilians to the same spot; for hundreds of miles around: on train, farm-wagon, spring-wagon, buggy, horse-back, foot; on bicycle, in open landau, carriage cart; in express-wagons, baggage-wagons, omnibuses; in barges with projecting additions and other land craft beyond classification or description." [15]

The weather had remained exceptional, but by Wednesday after his arrival, the rains began and was something of a portent of things to come. John noted in his first "letter" to *Harper's* that on that day, "everybody tried to keep dry." [16]

In the way only spring can change from good to bad, the weather in Camp Thomas at Chickamauga went from good to bad and back to good again. Through it all, John remained stoic. His tent, pitched with the privates of the army, drew dampness from the heavy dew that came every night and left slowly in the mornings, and only one who has been there can truly appreciate how welcome the rising sun is, and how long it takes sometimes for it to rise. But even that early in the war, he knew that the real story would be with the men of the army: the ordinary soldier who was anything but ordinary. The generals and colonels might get the glory, as usually happens, but he chose to stay where he felt more at home, and despite the hardships that decision imposed upon him, he did not waver; he stayed put alongside the common soldier.

He ate hardtack fried in bacon grease with sergeants and privates, drank their coffee, and listened to their stories and complaints, of which there were few, and was impressed with what he saw. He was proud too. Among the men there were a lot of Kentuckians, and he felt a friendship and camaraderie there.

But he was right about one thing: there was not much going on, at least not much exciting. In his second missive to *Harper's* from Camp Thomas, he mentioned that there was a daily sick call for the sick to come forward and receive quinine, the only known accepted remedy for malaria. He did not devote too much space to the sickness that was starting to become a serious problem in the camps, however, and remarkably, in light of his own fragile health, he did not seem too sympathetic with those who claimed sickness in a military camp. He would assume that no one in the field would report to sick call without actually being sick, and those who did at Camp Thomas only did so only because they were tired. That is what he wrote *Harper's*, but the fact was, things were quite a bit darker than he or any of the other correspondents let on.

There was an unwritten understanding among the tenuous relationship between the military and the journalists that the journalists would turn a blind eye and a deaf ear to mismanagement, and most of them did. To do otherwise could, and did, bring down the wrath of the army upon the unfortunate correspondent who wrote anything contrary to what the military wanted told, and any credentials the journalist might have could be quickly withdrawn. In short, the offending newspaperman would be quickly out of a job. Most saw the wisdom of going along to get along, leaving the truth to be written at a later, more convenient time, but all knew what was going on: the sickness, the lack of provisions and the poor quality of uniforms and supplies. They could not help but notice, and it galled them not to be able to do anything about it, but most of them wisely kept their mouths shut and wrote little of the real situation in the army camps for their newspapers or magazines.

It was a time when "good" reporters did not report what was then termed "unacceptable" stories, especially when they put the government or the army in a bad light, as the lack of sanitation and scarcity of medical supplies was sure to do. Mothers all over the country would rise up in arms if they learned their sons were not being taken well care of, and although women's suffrage was a long way off, when motivated, those same women could become a formidable force, whether they voted or not. Some reporters were even accused of treason when they went a little too far. It was far safer to stay away from controversial issues, but as in most things, the truth will come out sooner or later. The trouble was, men and boys were sick and dying, and few back home knew anything about the gravity of the situation. The situation would finally come to a head for John in the person of a correspondent for the New York *Herald* named Poultney Bigelow and Bigelow's nemesis Richard Harding Davis. Their falling out over the truth and whether it should be printed or not would pull Caspar Whitney, correspondent for *Harper's*, into the controversy and would unnecessarily exacerbate John's already shaky relationship with *Harper's*. In the end, John would be the only loser in the whole journalistic mess.

Clara Barton, because of her national and international reputation, was somewhat immune from the dangers of reporting the army's mismanagement, and wrote that same year that "here in Chickamauga, men fell from the ranks day after day," and even the most hardy were not spared. "When the plague descended on the camp, and a full realization of present and impending horrors was forced upon all intelligent minds," she continued, "frantic efforts were made to stay the progress of the destroyer, but the seeds had been sown, and the epidemic was fated to run its course." [17] The epidemic was malaria, and not unlikely, yellow fever, both of which often reached epidemic proportions throughout the war. It has been said that the only good thing that came from the Spanish-American War was the discovery of the cause of both yellow fever and malaria, and their subsequent successful treatments. [18] Be that as it may, for the time being, both diseases were a scourge that doctors and hospital workers, especially in the crowded and generally unsanitary military camps, were helpless to alleviate, and most correspondents were hesitant to report.

There would be worse to come. Once the army actually arrived in Cuba, yellow fever and malaria only intensified on the diseases' home court. To make matters even worse, medical supplies were either in short supply or, in some cases, nonexistent. Doctors themselves were terrified by the disease and sometimes chose to isolate those who showed signs of the diseases, especially yellow fever, and hope for the best. "Medicines of all descriptions were in short supply, and quinine tablets were so scarce that they were being sold on the black market for several dollars apiece." [19] The sicknesses and the silence about those diseases went on unabated, and the efforts of those concerned — like Clara Barton — were woefully inadequate. Throughout the rest of the training and into the War proper, it would be these diseases — malaria and yellow fever — that would cause more trouble for the American soldier than did any of the shells and bullets that the Spanish enemy would fire in their direction.

It was also at Chickamauga that John got his first close look at the Negro troops, and he surprised himself by being impressed. The times were certainly different then, worse, racially, than at any other time since the Civil War. Racism was an accepted, if not respected — at least in most quarters — institution, and even John Fox, Jr., was not immune. But now, former slaves and the children of slaves had become soldiers, and although invariably under white command, few would deny that they were fighters.

The black troopers "were big fellows," he remembered, "and some of them splendidly made, and the invariable testimony of the officers is that they make good soldiers." [20] Their commanding officer, however, got tired of repeatedly answering some local military genius as to whether or not his Negro troops would fight. Racism being what it was in those days, it was supposed and generally agreed that colored troops would be too lazy to fight, or perhaps too unintelligent. Their commanding officer had no doubts, however, about their fighting abilities. "The trouble," he told John, "is going to be to restrain them when their blood is up." [21]

John thought this just might be the case when one of them, a private, fairly made his skin crawl. "'I want to see a heart beat,'" said one small, black, savage private. "'I've always wanted to see a heart beat, and the fust Spaniard I ketch I'm going to cut him open and look at his heart beat.'" [22] He was somewhat amazed that the Negro troops thought they were the only ones in the army capable of driving the Spaniards from Cuba, and from what he had seen, he thought they just might be able to pull it off if they were left alone.

In his third and final correspondence to *Harper's* from Chickamauga, John related more of what his feelings were, and incidentally, the feelings of a vast majority of the country at that time. He and others had begun to think that the war with Spain would serve a more nobler purpose than freeing Cuba from Spanish tyranny. This war might, for once and for all, bring the still divided North and South back together again. In the camps of Chickamauga and other training camps all over the country, a mixture of Northerner and Southerner was the rule rather than the exception. It had been 35 years since the Civil War officially ended, but this war, small though it may be, might be the thing that could finally bring closure to that late, terrible, tragic struggle.

Soon enough, he was done at Chickamauga. The army was breaking camp with their eyes cast in the direction of Tampa, and beyond that Cuba. Not all were going, of course, and he felt it keenly himself that he might be left behind in this one great adventure of his lifetime. He still retained high hopes, however, and was waiting in nearby Chattanooga to hear from *Harper's*, hoping they would send him on to New Orleans, Mobile, and finally, Tampa where there was sure to be more going on. They had insinuated in their previous wire to him at Big Stone Gap that they might want him, and so he cooled his heels in Chattanooga waiting for instructions.

When the news came from *Harper's*, wired to him the same day he was writing to his mother from Chattanooga, it was not good. After all the hullabaloo and war fever, it began to look like the army was not going to move after all anytime soon. It might be weeks, and many thought it would be August or later before anything of interest happened. Some even suggested the war would be averted or settled by the Navy in a few weeks' time. As it was, *Harper's* already had Caspar Whitney on the payroll, doing the same things that John had wanted them to hire him to do. At the present state of things, that was one correspondent too many. "I am very sorry," wrote Henry Loomis Nelson for *Harper's*, "to be obliged to tell you that it will be impossible for us to arrange for a series of articles from you at present.... As nearly as we can ascertain from Washington, there is no disposition to move the troops for some time to come and therefore ... we do not see our way clear to ask you to represent us at present." [23]

They left their options open, however, and dangled a little bait in front of him that they might use him if things changed. It was not enough to keep him waiting in Chattanooga at his own expense. By the fourth of May he was back in Kentucky, making arrangements for readings and lectures in Lexington, but before he left Chattanooga, *Harper's* had wired him to go to Atlanta. He had already committed himself to readings in Kentucky, however, and now he chafed under the delay and finished up his speaking engagements as quickly as he could. As soon as possible, he left Lexington and headed south to the army's camp in Atlanta.

He arrived there just in time to witness the same lack of activity that he had seen and reported on in Chickamauga, but now there was a new rumor among the camps, and one that dampened everyone's spirits. All those men, in all those camps, after all their training and waiting and marching, might not go anywhere until late in the summer. Those more pessimistic kept saying that they might not go at all. The navy, some continued to say, might end the war before it ever got off to a good start. That, the soldiers and volunteers, would not, could not accept, and they worried about it.

Under other circumstances, at another time, he would have been glad to return to Kentucky, and that's where he found himself near the first of June, leaving Atlanta a few days before. He went to Lexington where the home of Henry Clay had been usurped as a local training facility, admittedly a smaller one than Chickamauga or Atlanta. He saw, and reported, the riding of a coward out of camp on a rail, commented upon the rejection of mountain boys because they were too tall, and the resulting dejection they felt when those mountain boys were told they must go home. The rejected soldiers, he said, took turns weeping and then cursing their luck.

His humor began to show through his dispatches from Lexington to *Harper's* when he told of soldiers, hungry and no food in sight, asked their commanding officer if he had a calendar so that they might eat the "dates." John also related that when the word got out that food was scarce in camp, "it touched the hearts of the mothers in bear-grass and blue-grass, and now these boys have the fat of the land in plenty." [24]

After Lexington, feeling blue and discouraged, he went home to the Gap, home to mother and father, sisters and brothers. He realized that he had no prospects of going anywhere now, and could not see any way out of his predicament.

He could not be happy with being idle when he was so sure that something must be happening somewhere and that he should be covering whatever it was, and he implored *Harper's* to send him somewhere, anywhere. Action was what he wanted to see. There was a war going on, or at the very least, rumors of a war, and if somebody did not do something soon, he would miss what he thought was a chance of a lifetime. *Harper's* could not relent. "We have so many correspondents with the Army at Tampa, appointed before you communicated with us, that they are falling over each other and complaining of mutual interference. Your letters," they assured him, "have been admirable and everyone has enjoyed them to the utmost. If we can find a place for you, you may be sure we'll utilize you, but just at present we do not see a place." [25]

Harper's was right about one thing: Tampa, Florida, *was* crowded, and mostly with correspondents. "Tampa was the scene of the wildest confusion," wrote Roosevelt, who was already there, "There were miles of tracks loaded with cars of the contents of which nobody seemed to have any definite knowledge," and, he added, "there was no semblance of order." [26] By the middle of May, when *Harper's* was wiring John that there was no room for him, there was already nearly 130 correspondents at Tampa, and coupled with the number of soldiers arriving nearly every day, the little Florida town went through something of a population explosion.

Tampa was a good size town in 1898, numbering around 10,000 souls, not counting newspaper reporters and the military that invaded the town that summer. "There were three banks in Tampa, one theater, gravel roads, wooden sidewalks, a trolley line to Lakeland, and a railroad to the north and south with a single track to Port Tampa. Also there were some general stores, electric lights, a few telephones that seldom gave good service, a telegraph office, and always and everywhere sand. But above all there was the Tampa Bay Hotel." [27]

The Tampa Bay Hotel was an oasis in the middle of nowhere, and the officers and journalists quickly invaded and made it their headquarters. The hotel itself "could be believed only when actually seen ... its red-brick fireproof walls contained close to five hundred rooms on five floors. The building covered six acres to include a gaming casino and a swimming pool; and the whole was topped off with a great silver dome and thirteen silver minarets bearing crescent moons, one for each month of the Moslem year." [28] It was fantastic, and opulence of the finest sort, but even rich surroundings and warm, sunny weather can become old in time.

"This was the rocking-chair period of the war," wrote Richard Harding Davis, who was there under the auspices of the New York *Herald* and *Scribner's* magazine among others. "It was an army of occupation but it occupied the piazza of a hotel." [29] While it lasted, everyone enjoyed themselves completely. Casper Whitney, who had the job with *Harper's* that John wanted, was already ensconced at Tampa. "Here at Tampa," wrote Whitney, "correspondents continue to arrive and the tradesman to prosper. If we do not start shortly, additional transports will be required to convey the correspondents and their purchases." Like idle tourists, which they were, the journalists were buying all kinds of useless materials and souvenirs. "Some startling outfits are resulting from the delay and the growing crop of newspaper men," continued Whitney. "I saw one a few days ago buying an armory of knives that would put to shame the heaviest villain that ever trod the melodramatic stage." [30]

Perhaps no one enjoyed themselves as much as the old soldiers from the Civil War, many of them brought out of retirement to fight a new war, who gathered together on that same porch and fought the Civil War all over again. "Those were the best days of the time of waiting," continued Davis. "Officers who had not met in years, men who had been classmates at West Point, men who had fought together and against each other in the last war, who had parted at army posts all over the West ... were gathered together apparently for an instant onslaught on a common enemy, and were left to dangle and dawdle under electric lights and silver minarets." [31]

They were not, however, blind to the situation they were in, nor at the comic character of their sojourn at the hotel. "One imaginative young officer compared it to the ball at Brussels on the night before Waterloo; another, less imaginative, with a long iced drink at his elbow and a cigar between his teeth, gazed at the colored electric lights, the palm-trees, the whirling figures in the ball-room, and remarked sententiously: 'Gentlemen, as General Sherman truly said, "war is hell."'" [32]

The idle time played on the nerves of the correspondents, and some began to feel guilty for getting paid by their respective newspapers and magazines for doing relatively nothing to earn those dollars, meager though they were for most of them. A few gave up in disgust, believing that they were never going to move from Tampa and go to where any real action was, or was to be, and Caspar Whitney was ready to resign, but not over boredom.

Whitney was a veteran journalist and author and was not used to "rocking chairs" of any description. He had traveled all over the world, exploring all parts of North and South America, hunting as he went along in Mexico, the East and West Indies, Malay, and India, and writing about it. He had written books, magazine articles, newspaper articles, everything. The constant delays and vacillation of the army at Tampa chafed him raw, but he tendered his resignation to *Harper's* over an article in that same magazine by Poultney Bigelow that condemned the army in Tampa for the mismanagement of nearly everything. Everyone knew it was the truth, but no one thought it a good idea to say anything about it just now; no one, that is, except Bigelow. That *Harper's* would print the article, and furthermore uphold it in an editorial, was more than Whitney thought he could honorably accept from his employer. As far as Whitney was concerned, both *Harper's* and Bigelow had broken the unwritten code — ridiculous as it might seem, especially in light of future journalistic codes. Bigelow had scooped them all, had written a story about disease and dying in the army camps, and he blamed the army — from top to bottom — for it. Caspar Whitney was not the only one who was mad about it. It was a great story, and one that almost by common consensus the correspondents had agreed not to write until a later date, and most felt that Bigelow had betrayed them by violating that unspoken agreement. It was Whitney, however, who was working for the magazine who published the article, and it was he who felt that he could do no less than offer his resignation to *Harper's*.

Whitney's resignation letter may have only been a gesture on his part, but *Harper's* took it seriously, and immediately contacted John to take Whitney's place. In the meantime, Whitney offered his services to *Century* magazine, and they readily accepted. By circumstance and accident and being in the right place at the right time, John would go to Cuba, but so would Caspar Whitney. But the reason for Whitney's resignation in the first place would simmer in the hot Cuban sun until August, and then break forth anew. At the time, John knew nothing about it, but that knowledge would soon be his to simmer over, also in Cuba.

Finally, *Harper's* had asked him to come on to Tampa, and they did not have to ask him twice. He was fairly packed before their cable to him hit the floor in Big Stone Gap. On June 2, 1898, he left for Florida. *Harper's* decision to send him was purely a business one. On the one hand, one more correspondent could not hurt, and the amount of money they would be out if his assignment turned out to be a total loss was not substantial. Neither were they blind to his growing dissatisfaction with them as his publishers, and it is not unlikely that they sought to dampen his wandering spirit a bit by giving in to his wishes to go to Cuba. After all, they reasoned that the war just might be over before he got there anyway. They could not see how they could lose. They told him nothing of Whitney's resignation, nor the reasons for it, and he did not find out about it until he arrived in Florida.

11. The Spanish-American War

When John arrived in Tampa, he had a hard time believing all that he saw there. He thought he had seen a lot of things, but he had seen nothing that could prepare him for Tampa that June. He found that *Harper's* had not exaggerated about the number of correspondents—if anything they had underestimated the number—and coupled with nearly 25,000 soldiers running everywhere in a town of 10,000 "normal" inhabitants, it was simply chaos. He was adaptable, however, and soon began to make himself at home. "I am in fine condition and am most cheerful," he wrote to the "folks" at home on June 10, but persistent rumors that peace might break out before the war really got off the ground dampened his spirits a little, but not much. He need not have worried. And even if everything came to naught, he had another reason for doing what he was doing. [33]

He was gathering information by the carload for a Civil War novel he had already begun writing, and everything he saw and heard in the camps and at Tampa was just what he wanted. Like Stephen Crane he was attempting to write a war novel without actually having seen war, and found it was not as easy as he first thought. Already, however, he was learning that all wars are much the same, and a description of one nicely describes another. The Spanish-American War would, of course, be different from the Civil War, but when it was all boiled down to its basic components, men killing other men as fast as they could, one would hardly be able to tell the difference. Literary license would take care of the rest.

Right now he had other things to worry about, which was how to get to Cuba with the army, when and if that army should go. Again, he should not have worried. Things were moving faster now, and in a matter of a few days he would indeed find himself on the way to Cuba and to the war he so much wanted to see for himself firsthand.

While at Tampa, John mingled with the officers and especially the regular soldier, and although he was impressed with the spirit of the men, he had reservations about the way the government was handling the whole thing. He was not the only one. Bigelow's article in *Harper's* had opened the eyes of many, and possibly removed the blinders of others. Richard Harding Davis was there, and though he would stick close to the Tampa Hotel, he too was disturbed by what he saw in the camps. "Davis was appalled," wrote Arthur Lubow "to see the men sleeping on blankets on the ground." [34] Anyone knowing the sensibilities of Davis might shrug that off with little concern, but there was more. "No one had bothered to build rain trenches around the tents or gutters along the company streets. In one regiment's camp, the open latrines had been placed fifty feet from the first row of tents, on the windward side. The wind had been blowing from that direction for a week. Garbage also was burned windward, so smoke and latrine odors swept across the camp." [35]

It would only be after the war was over that any real charges of mismanagement by the army could be made by any of the journalists or anyone else who saw what was going on in the camps. To do so now could invite immediate charges of treason from the army. The reporters — Bigelow being the most notable exception — held their counsel, and wrote nothing, although privately they were "appalled" at some of the things they saw.

Davis was upset at the conditions in the camps, but he was almost as troubled by the way some of the Rough Riders were making free use of his hotel room. "They use my bathroom continuously," he complained to his brother Charles, "and I never open the door without finding a heap of dirty canvas on the floor and a cheery voice splashing about in the tub and ... then an utterly strange and utterly nude giant will appear from the bath room." [36] They were Rough Riders — Roosevelt's men — however, and they were almost by definition adventurous and completely without affectation of any kind; just the kind that Davis admired and openly loved to be around, despite their somewhat barbaric and casual manner.

The volunteers, sent to the training camps by their native states — usually under less than adequate command — were as rough looking, casually dressed and, for the most part, undisciplined as could possibly be imagined. They arrived dressed in a variety of uniforms, or no uniforms at all, and a great many without weapons of any kind. The state governments assumed, perhaps craftily, that the federal government would supply all the equipage that the volunteers would need. Some had even raided their state arsenals of every old-fashioned firearm they could lay their hands on with the expectation that the government in Washington would replace those with more modern rifles, which the volunteer would naturally bring back home with him after the war was over. "Several of the volunteer regiments came here without uniforms," said General Nelson Miles, "several without arms, and some without blankets, tents or camp equipage. The 32nd Michigan, which is among the best, came without arms." [37] All in all, the states could not lose. At a time when state militias played an important part in keeping the peace in their respective states, the prospect — intentional or not — of having those volunteer militias rearmed and reequipped at no cost to their native states was just too good of a deal to pass up. It was a good deal for the states, but it caused havoc in Tampa and the other training camps.

John almost missed the whole thing. On the last day of May, General William Shafter received his orders from Washington to leave Tampa as soon as possible and, under the protection of a naval convoy, proceed to Cuba. Had it not been for the incompetence of the army coupled with the availability of only one railroad out of Tampa, the army would have been long gone by the time John left Big Stone Gap on the second day of June. As it was, the delays kept the army on the mainland until June 14. There had been several false starts, and when the final pushoff had been accomplished, it was somewhat anticlimactic.

Two days before, John, safely in Tampa and mindful of the delays and rumors of sailing dates that never materialized, wrote that "it is after midnight, and I cannot get a row boat out to the transport *Iroquois*, that is to carry me to Cuba — and we sail to-morrow. We have been sailing to-morrow," he observed dryly, "for several days." [38]

He was close. The convoy did not sail on the 13th, but the next day, cheered out of harbor by two or three black women, and about the same number of soldiers and a handful of others, 32 troop ships carrying 16,000 men steamed out of Tampa Bay with every intention of not stopping until the shore of Cuba was in sight.

Richard Harding Davis, on board the *Seguranca*, was overjoyed but at the same time "suspicious and wary." Previous delays had made him and the other correspondents pessimistic. There was no singing or cheering when they left on the 14th, the crowds "had done that on the morning of June 8th, and [the ships] had been ingloriously towed back to the dock; they had done it again on the morning of June 19th, and had immediately dropped anchor a few hundred yards off shore." [39]

This time, however, it was the real thing, but it was no picnic on either the *Iroquois* or the *Seguranca*, nor any of the other transports. The ships were crowded and smelled. John had never been to sea before, and in that he was not alone. "Probably half of the men forming the expedition had never been to sea before," wrote Davis, and added: "They probably will desire never to again.... They will not wish to go again, because their first experience was more full of discomfort than any other trip they are likely to take could possibly be." [40] They moved at a maddening pace, slower than anyone thought possible without coming to a complete standstill, which they sometimes did. It was a hesitant flotilla, and the pace did nothing to assuage the tempers of the men on board. "We traveled at the rate of seven miles an hour, with long pauses for thought and consultation. Sometimes we moved at the rate of four miles an hour, and frequently we did not move at all." [41]

By this time, John was being a little less timid in reporting the conditions, especially bad ones, that he found all around. He certainly thought that if those in command were not directly responsible for the horrible conditions on board the transports, then they were indirectly responsible. "I make out, with an active use of inference and common-sense, merely that most men in public life do no more than they have to do," and that included not providing the food and support necessary for an army to fight. [42]

Even some time after the war ended, Davis had no trouble remembering. "The water on board the ship was so bad that it could not be used for purposes of shaving," he wrote. "It smelled like a frog-pond or a stable-yard, and it tasted as it smelt." The doctors at Tampa who did not have to drink it declared that, despite its smell, it was quite safe to drink. The food was not much better. [43]

The men, however, were in reasonably good spirits between bouts of seasickness and meals of bad food and water. Even the horses in the bottom of the transports were seasick, and that was something no one had seen before. Naturally, they were unprepared for any such circumstance. Just what do you do with a seasick animal that weighs nearly half a ton? Surprisingly, there were little or no complaints from the soldiers. "No one complained and no one grumbled. The soldiers turned over to sleep on the bare decks [and] bombarded each other with jokes on the cheerful fact that they were hungry and thirsty and sore for sleep. But for all that," continued Davis, "our army's greatest invasion of a foreign land was completely successful, but chiefly so, one cannot help thinking, because the Lord looks after his own." [44]

John's own transport, the *Iroquois*, had an inauspicious beginning, grounding herself on a sand-bar within sight of the dock at St. Petersburg, Florida. They loosed themselves without a lot of trouble and dropped anchor for the night. The next morning, they got under way in earnest. Six days later, John saw the coast of Cuba for the first time. He wrote home that morning — June 19 — from the *Iroquois* just off the far western tip of Cuba, Cape Maisi. "This is Sunday morning," he wrote, "and we are near the point of Cuba and expect to land about 40 miles from Santiago. The trip has been glorious — cool, a strong wind night and day and I cannot get over the idea that it is simply a big picnic — a trip to Coney Island to get cool. At night with the lights of the transports all around it looks merely as though we were sailing ... in a harbor. So far, the terrors of the tropics have been a bugbear and if this keeps up I consider myself lucky to get such a trip, even if I weren't being paid $10 a day just for expenses." [45]

He did not think he would be in much danger and was mesmerized by his surroundings: the lights, the ships, the peace and quiet of the ocean. He wrote, "I don't believe we are going to have much fighting — if any." Two days later, on the 22nd, the army landed at Daiquiri. [46]

The night before the landing things were tense on board the *Iroquois*, and John rightly assumed that things were a little tight on all the ships standing off shore. The word had come down, finally. "Orders came at dusk," he wrote of the 21st of June, "that we were to force a landing at daybreak ... day broke and showed the big emerald-green mountains rosy with light, the coast foothills still in shadow, the spray breaking high against the low rock wall that makes the coast-line, and the ship floating alone in a little bay that curved like a horseshoe shore wards. And we kept on waiting. Other transports gathered; the lifting mist showed a long pier stretching into the bay and loaded with tram-cars, and the frame houses of a little mining-town — Baiquiri [sic]. Flames shot up from the town — the Spaniards were burning up the property that might be of value to us. The war-ships steamed lazily in." [47]

The day before, the navy had bombarded the coastline, but there was no response from the shore. "The *New Orleans* gave a deep, hoarse cough as a signal, and a puff of smoke from her starboard side was quickly answered by a cloud of smoke high on the hill where her shell burst. At once the *Castine* and the *Annapolis* began to clear the throats of their guns. The little *Vixen* barked a note of sharp inquiry; the tug *Wompatuck*, with all the pride of a recruit at his first target practice, barked too, and all were fused into an angry chorus. At first they searched that hill, smoke belching from the ships like the spray spouting along the coast, and the smoke forming into the clouds and rising on the hill as mist forms and rises there after a rain." [48]

The Spaniards, it was discovered, had abandoned the site, and likely lost the best and only chance of doing irreparable damage to the invading Americans, and just perhaps ending the war before it really got started. From on board the *Iroquois*, John saw it all, and thought it was "a beautiful bombardment of a big rocky hill to stir up masked batteries ... but there was not a response and we landed without firing a shot." [49] He thought the landing was too easy to be believed. Only two men were killed, and they drowned tragically in the offloading. The horses for the cavalry were pushed overboard since there was no place they could be unloaded otherwise. Confused, some of the horses began swimming out to sea before an enterprising soldier began sounding assembly on his bugle. The horses, trained cavalry mounts that they were, immediately turned toward shore and safety. All in all, only a handful of horses were lost, but as things turned out, they were not needed anyway.

Again, John wondered if there would be a fight at all. "We are not apt to have any more fighting until we besiege Santiago — and then, perhaps, very little," he wrote. [50] But there was a darker side to this war, this rebellion, and it was not long before they were all introduced to it.

Once on shore, the army had captured a Spaniard, and the insurgents, who were with the Americans, and whose bravery John did not think too much of, had executed the prisoner with their machetes. It was a cruel footnote to the years of bloody warfare that had been going on in Cuba long before the Americans landed at Daiquiri. It would not be the last time that the Americans questioned their motives for fighting for the freedom of a people who were now beginning to seem a little more bloodthirsty than the Americans had first supposed.

The island of Cuba can be envisioned as a long, narrow island somewhat reminiscent of a shrimp with its head to the east and its tail fairly brushing Key West, Florida. Just around the eastern tip, or head, and just underneath that head, lies Santiago de Cuba. The war, with the exception of Guantanamo, was fought in an area approximately ten square miles in size. Along the coast to the west lies Daiquiri, and then a short distance farther on the coast lies Siboney. Just north of Siboney is Las Guásimas, and just northwest of Las Guásimas, and behind Santiago, lies El Caney, where the war in Cuba, for all intents and purposes, started and ended.

In the meantime, John was not the only one reminded of Coney Island. A few miles down the coast, the army made a second landing at Siboney. Davis was lulled by the unreality of the scene. "It was one of the most weird and remarkable scenes of the war," he wrote, "probably of any war.... An army was being landed on an enemy's coast at the dead of night, but with somewhat more of cheers and shrieks and laughter than rise from the bathers in the surf at Coney Island on a hot Sunday. It was a pandemonium of noises. The men still to be landed from the 'prison hulks', as they called the transports, were singing in chorus, the men already on shore were dancing naked around campfires on the beach, or shouting with delight as they plunged into the first bath that had offered in seven days." [51] They found Siboney abandoned by the Spaniards as well.

Major-General Joseph Wheeler, a general in the Confederacy during the Civil War when he was a lot younger, took command, since Shafter had elected to stay on board the *Segruanca*, and ordered a general advance as soon as it was light enough to march. Wheeler, well advanced in years and so thin that he looked anemic, had brought his son along to show him a "real" war, and unlike his commander, Wheeler proved to be exceptionally capable during the Cuban campaign, despite his age and stature. By the end of the war in Cuba, Wheeler had gained for himself additional laurels, while Shafter would become, arguably, the most hated of those in command there.

That morning, the Rough Riders had entered Siboney only to find the Spaniards had abandoned that post the same as Daiquiri. By now, all were beginning to wonder if they would ever catch up with the enemy. Turning north, about three miles inland, the Rough Riders came to a place without a name, but it was soon to have one. It was called Las Guásimas for the thick, tangled, low-growing trees and the nuts that they bore. It "is a tree that bears nuts and has low wide-spreading boughs," wrote John, "from which it is easy to swing a noose. Therefore the Cubans use it for the fattening of pigs and the hanging of criminals." [52] Roosevelt, with his Rough Riders, remembered it as "a mountainous country covered with thick jungle, a most confusing country." [53]

A man could get lost in such a place. On either side of the road that narrowly made its way towards El Caney were stands of low-growing shrubs and stunted trees higher than a man's head. No one could see more than a yard or so into the dense growth, and once the army entered that maze, more than one wondered what was hidden behind the greenery. It was blisteringly hot. No wind or breeze could get through to the men who crowded the two trails that roughly paralleled each other before meeting just in front of a hidden Spanish position. The enemy that had eluded them thus far was just around the corner, at the end of that claustrophobic maze, waiting for the American army. Things were about to get lively.

CHAPTER 12

The Horizontal Hail-Storm

> They tried to catch us with their hands.
> Spanish soldier

 THEY HAD HEARD OF jungle and perhaps read of what the tropical foliage was like in popular magazines of the day, but nothing they had read or heard prepared them for what they found at Las Guásimas. "The ground was covered with high grass and cactus and vines so that you could not see twenty feet ahead," said Richard Harding Davis. "The men had to beat the vines with their carbines to get through them. We had not run fifty yards through the jungle before [the Spaniards] opened on us with a quick firing gun at a hundred yards." [1]

At first no one did anything; they were not even aware that they were being shot at, only that someone was shooting nearby, and a few leaves and twigs were falling among the American soldiers. But they *were* being shot at, and it only took a few seconds for even the most green of recruits to realize it. At that point, everyone knew instantly what to do, and the ground and what little cover it offered was suddenly covered with soldiers protecting themselves as best they could.

The American soldiers' first hostile encounter with the Spanish soldiers was also their first encounter with smokeless powder on the battlefield. They never saw where the bullets that cut through the underbrush at Las Guásimas came from, only heard the zip and zing of bullets passing close by, or sometimes heard it hit flesh, sometimes their own. There was nothing to shoot at in return. No tell-tell smoke that most of them were used to that would reveal the enemy's position. "There was a good deal of cracking of rifles way off in front of us," wrote Roosevelt some years later, "but as they used smokeless powder we had no idea as to exactly where they were, or who they were shooting at. Then it dawned on us that we were the target. The bullets began to come overhead, making a sound like the ripping of a silk dress, with sometimes a kind of pop; a few of my men fell, and I deployed the rest, making them lie down and get behind trees." [2]

The thickets on either side were almost solid green, remembered Stephen Crane, who was there covering his first "real" war for the New York *World*, "out of which these swift steel things were born supernaturally. It was a volunteer regiment ... going into its first action against an enemy of unknown force, in a country where the vegetation was thicker than fur on a cat." [3]

That first engagement of Roosevelt's Rough Riders was later described as an ambush by many of the correspondents there, and as much the same by those far behind the lines who got their information from others. Others, mostly the officers in charge, including Roosevelt denied this. [4]

Ambushed or not — and if it was not, it was the next thing to it — the volunteers were pinned down and being shot to pieces. "The enemy's fire was exceedingly heavy and their aim was low," wrote Davis, who was in the thick of things, even to the point of taking up a rifle himself, endeared himself forever with Roosevelt. "Whether the Spaniards saw us or not we could not tell; we certainly saw nothing of the Spaniards, except a few on the ridge across the valley. The fire against us was not more than fifty to eighty yards away, and so hot that our men could only lie flat in the grass and fire in that direction." [5]

Up ahead, Major William Beach, chief engineer of the Cavalry Division, was pinned down with the rest of the Americans and saw the casualties begin to pile up. "Time crawled," he said, "it did not fly; the sun seemed to stand still as in the days of Joshua of old, and it was too awfully hot and oppressive for words." [6] Finally, reinforcements arrived, the very sight of which put the Spaniards into flight. "As the firing grew fainter, General Wheeler, sprang to his feet and shouted, 'We've got the Yankees on the run.'" [7] For a moment, the old Confederate general had forgotten what enemy he was fighting, but it lifted the spirits of all in hearing, and they responded with a charge with Roosevelt in the lead, put the Spanish into permanent flight, and the Battle of Las Guásimas was over.

The Spaniards were dismayed at the American methods of fighting. "When we fired a volley," one of the prisoners said later, "instead of falling back they came forward. That is not the way to fight, to come closer at every volley." [8]

The dismayed enemy retreated upon Santiago, where they announced they had been attacked by the entire American army. One of the residents of Santiago asked one of the soldiers if those Americans fought well. "'Well!', he replied disgustedly, 'they tried to catch us with their hands.'" [9]

Miles behind the battlefield and hours too late, John Fox was trying to catch up with the action once more. He had been kept on his transport with the other correspondents and had been the last to be released, angry at the delay, but resigned and determined to catch up with the rest as soon as he could. The road to the battlefield was easy enough to follow. "I was following the trail of an army by the signs in its wake," he wrote, "the debris of the last night's camp: cans, bits of hardtack crackers, bad odors, by-and-by odds and ends that the soldiers discarded as the sun got warm and the pack heavy — drawers, under-shirts, coats, blankets, knapsacks, an occasional gauntlet or leggin, bits of fat bacon, canned meats, hardtack — and a swarm of buzzards in the path, in the trees, and wheeling in the air — and smiling Cubans picking up everything they could eat or wear." [10]

He was surprised at how much trash was on a battlefield. He had thought to see the bodies of dead and wounded, had expected it really, but the reality was litter almost beyond description. In battle, he was learning, men do not care very much for equipment other than their rifles, and nearly everything else is expendable. Davis was finding much the same thing. He had returned to the battlefield later in the day to find "the rocks on either side ... spattered with blood and the rank grass matted with it. Blanket-rolls, haversacks, carbines, and canteens had been abandoned all along its length, so that the trail looked as though a retreating army had fled along it, rather than that one company had fought its way through it to the front." [11]

John finally reached the front, or what there was of it. He made his way through a group of Cuban rebels and curious natives, still trying to find some Americans. "There was a little creek next," he wrote later, "and climbing the bank of the other side, I stopped short, with a start, in the road. To the right and on a sloping bank lay eight gray shapes muffled from head to foot, and I though of the men I had seen asleep on the deck of the transport at dawn. Only these were rigid, and I should have known that all of them were in their last sleep but one, who lay with his left knee bent and upright, his left elbow thrust from his blanket, and his hand on his heart. He slept like a child." [12] He had found the front. Those were not the first dead men he had ever seen, but those lifeless shapes whose faces he could not see for the blankets they were wrapped in affected him deeply.

He reached the front after everything had settled down and joined up with Frederic Remington, who would be providing *Harper's* with illustrations of the war for their magazine if he ever got close enough to see anything before it was over. Remington and John had crossed on the *Iroquois* together, but had become separated at the landing. Now, both were afraid they had missed the one and only battle of the war by only a few hours, and both felt bad luck all around. Remington felt much the same, and between them, they decided that it would not happen again. They would find the front no matter where it was, and stay there until something happened. "We had not seen this fight," wrote Remington later, "of the cavalry brigade, and this was because we were not at the front. We would not let it happen again. We slung our packs and most industriously plodded up the Via del Rey until we got to within hailing distance of the picket posts, and Fox said: 'Now, Frederic, we will stay here. They will pull off no more fights of which we are not a party of the first part'. And stay we did. If General Lawton moved ahead, we went up and cultivated Lawton; but if General Chaffee got ahead, we were his friends, and gathered at his mess fire. To be popular with us it was necessary for a general to have command of the advance." [13] From that day onward, true to their words, they got to see it all, not always together, but both were always "within hailing distance" of the front.

For a week after Las Guásimas, the army stayed put, with Fox and Remington eating bacon, hardtack and coffee. "By God," complained Remington who was used to getting regular meals and plenty of them, "I haven't had enough to eat since I left Tampa."[14] Then it began to rain as it only can in the tropics, and Davis wrote that it was "very rough living. I have a cot raised off the ground in the Colonel's [Roosevelt's] tent and am very well off." [15] Remington and Fox were, however, excluded from the shelter. "Mr. John Fox and I had no cover to keep the rain out, and our determination to stay up in front hindered us from making friends with any one who had," wrote Remington. "Even the private soldiers had their dog-tents, but we had nothing except our two rubber ponchos. At evening, after we had 'bummed' some crackers and coffee from some good-natured officer, we repaired to our neck of the woods, and stood gazing at our mushy beds. It was good, soft, soggy mud, and on it, or rather in it, we laid one poncho, and over that we spread the other.

"'Say, Frederic, that means my death; I am subject to malaria.'

"'Exactly so, John. This cold of mine will end in congestion of the lungs, or possibly bronchial consumption. Can you suggest any remedy?'

"'The fare to New York,'" said John, as we turned into our wallow.'" [16]

Finally, even Fox had enough. Friendship could be pushed just so far, even for a Southern gentleman, and he "left their soggy camp to find a more sheltered niche with Roosevelt." [17] It could not rain forever, and by the last of June he was writing home that the weather reminded him of Big Stone Gap, and if it were not for the different foliage, he would be fooled completely into thinking he was home. He also noted that he was in good health, despite the weather and absence of anything resembling decent food.

For a week they waited, milling about in the heat and sand and the tropical rain that came now every day, cooling at first, and then, once it stopped, making conditions more miserable. They were plagued with all sorts of strange creatures, most of them of the crawling persuasion, trying to share their beds, wet or dry. They were nearly out of tobacco, and although the supply ships held plenty, it remained safely stored off shore. The men, wrote Davis, suffered greatly from the lack of it, and he "observed the enlisted men smoking dried horse droppings, grass, roots, and tea. If it could be found, a plug of tobacco normally worth eight cents went for two dollars." [18] But it was not just

12. The Horizontal Hail-Storm

tobacco — all rations were in short supply, and what rations they had were woefully unsuited to the tropical weather: canned pork and fat bacon which spoiled almost immediately if it was not spoiled already, and hardtack, a kind of tasteless, thick, unsalted cracker, and coffee.

There were few real complaints however, for even the blindest of them all could see that they were almost within sight of Santiago — the Cuban capitol — and with the exception of one or two minor fortified hills, the way stood clear into the city and victory for the American army.

On July 2, John wrote home from the battlefield. "We had a big battle yesterday," he wrote, "and I'm all right. I have been under fire and have heard the whistle of Mauser bullets, and shell and been potted at by sharp-shooters, so I've had all the experience I want and hereafter, shall take good care not to expose myself again." [19] He had mentioned several times in his letters home since leaving Tampa that he was getting plenty of "experience," presumably for the novel he was writing, but like many before him, the stories of battles, and what battle actually turned out to be, was a little disconcerting. There is little in life that will get a man's attention quicker than the "whistle of Mauser bullets."

They had lain for nearly a week in one spot after Las Guásimas, doing little except that hardest of times for soldiers: waiting. They sweltered in the heat, fumed, dawdled, slept, scratched vermin, and wrinkled their noses at their own smell. To the left of the American position was a low hill called El Pozo, from which one could see most of the valley that stretched out before them all the way to the coast and Santiago. On June 29, John, Richard Harding Davis and Frederic Remington had taken it upon themselves to climb the hill. They could see the Spaniards on nearby San Juan Hill and thought the Spanish could have thrown a few shots their way, but the Spaniards were unco-operative and they neglected to send any shells over for the three correspondents' entertainment. Over in Santiago, they could see the flags of the Red Cross "floating over a long hospital and the place looked as peaceful as though it were Sunday in a New England village." [20] To the right of El Pozo was the village of El Caney of fairly good size with a mixture of large homes and buildings, dirt streets, and other homes of the meanest sort. It should not have been much of a challenge, they thought, for any army that wished to take it. In fact, nothing they saw that day seemed to offer much of a challenge for the American army.

Out from the hill of El Pozo and the village of El Caney, once you had left the woods and thickets behind those two places, stretched an open landscape dotted here and there with bushes and trees and knee to waist-high grass. Beyond the meadow was a range of low hills that were collectively known as the San Juan Hills. It was upon these hills, and two in particular, that the Spaniards would make their last stand to save and protect the city of Santiago. It would be up to the Americans to cross that meadow, in the bright sunshine, and take it away from the entrenched enemy. But first, El Caney would have to be taken. They did not know it then, but it would take them ten hours before they would be able to take possession of what seemed at first to be an almost insignificant Spanish fortification.

Malaria had felled several of the officers by this time, including Wheeler and General Young. General Wood, who commanded the Rough Riders, was given Young's command, and Roosevelt stepped up to command the Rough Riders. The plan was to send 8,000 men including Roosevelt and the Rough Riders toward San Juan, and position themselves around the edge of the clearing facing outward toward the valley and hills beyond. The heat, the humidity, the rain, and the incessant waiting for something to happen, the Americans thought, was worse than anything they could possibly find on those hills in front of them, and it made them all anxious.

At daybreak on the morning of July 1, exactly 35 years since Gettysburg, the pounding boom of artillery reaching El Caney, awakened all who were not awake already. Captain Allyn Capon, whose son had been killed at Las Guásimas, started the festivities from El Pozo with his artillery. John had left headquarters early that morning and joined Capron's battery before dawn. It was a good place to be. Their position was an envious one to anyone wanting to see everything there was to see about the coming battle, but for anyone wanting a safer spot, nearly anywhere would have been preferable. The fire from the Spaniards inside the town had been pouring toward the American positions for several hours with little effect. Most was small arms fire. The Americans returned cannon fire, and the noise was terrific. "The artillery kept up a steady fire on the fort and town and finally demolished the fort. Several times the Spaniards were driven from it. But each time they returned before our infantry could approach it." [21]

The position that Capron took — according to his orders to get as close to El Caney as he thought prudent — was smack in the middle of the meadow, almost exactly between San Juan and El Caney. John could see everything. "The hill had been cleared of bushes," he wrote later that day, "the four guns unlimbered and thrown into position against Caney, the caissons drawn to the rear, the horses gathered into the bushes to one side, and officers, aides, and correspondents walked the length of the hill, or stood in groups watching with field-glasses the red town.... While everybody spoke in a quiet voice and nobody laughed or jested, nobody was as solemn as the occasion seemed to me to demand." [22]

John was nervous and fidgety and at the same time excited about the whole thing, including his position in the line of expected fire from the Spaniards. He thought that anyone not as nervous as he was ignorant of the danger they might possibly be in. He kept quiet, however, looking toward the Spanish positions all around them and wondering what a shot from their battery would bring in return from the Spaniards. "One of those guns was to fire presently, and it was not going to be a salute, nor a sunset gun, nor a Christmas or Fourth of July explosion. It would be loaded with a shell, and was to be aimed at living men, and was meant to kill somebody." [23] And with words so simple as "Give 'em a shell, boys," from Capron, the battle and the killing was begun from the spot where John stood and saw it all.

"There was a cap explosion at the butt of the gun," he said, "a bulging white cloud from the muzzle, the trail bounced from its shallow trench, and the wheels whirled back twice on the rebound, and the shell was hissing through the air as iron hisses when a blacksmith thrusts it red-hot into cold water. You could hear that awful hiss so plainly that you seemed to be following the shell with your naked eye; you could hear it above the reverberating roar of the gun up and down the coast mountain; hear it until six seconds later a puff of smoke answered beyond the Spanish column where the shell burst." [24]

The first shot went wide, and did no damage to either man or real estate, but it started something. "The first shot started the ripping of cloth, the far-away rumble of wagons over cobblestones, or softened stage-hail and stage-thunder all around block-house, stone fort and town," John wrote. "At first it was a desultory fire, like the popping of a bunch of fire-crackers that have to be relighted several times, and we could hear the hiss of the bullets even that far away; but at times the fire was as steady as the sputter of a Gatling gun that I heard later in the day. But the powder was smokeless, and we could see nothing other than the straw hats of the little devils in blue, who blazed away from their trenches around the fort, and minded the shells bursting over and around them as little as though they had been bursting snowballs." [25]

Soon enough, the blockhouse at Caney was silenced by the artillery, and most of the rest of the battle was fought with small arms fire. John left Capron's battery and moved toward the San Juan Hills on the left of Capron's position. The Spanish had begun to return a little artillery fire of their own, and their first shot toward El Pozo and Grime's battery managed to kill a horse and a man standing beside it. In quick succession, however, their fire became more intense and more accurate, and men began to fall and be tossed into the air by the exploding shells. One shell fell in among a group of Cubans rebels huddled together, killing several and wounding nearly all the rest.

Now came what John had feared with the first shot of Capron's toward Caney. Dead men were one thing, and horrible enough, but the wounded, somehow, seemed to strike him the worst. "At this point," he wrote after the shells began to explode on El Pozo, "began the central lane of death, and the terrible procession to the rear was on its way. Men with arms in slings; men with trousers torn away at the knee and bandaged legs; men with brow, face, mouth, or throat swathed; men with no shirts, but a broad swathe around the chest or stomach — each bandage grotesquely pictured with human figures printed to show how it should be bound on whatever part of the body the bullet entered. Men staggering along unaided, or between two comrades, or borne on litters, some white and quiet, some groaning and blood-stained, some conscious, some dying, some using a rifle for support, or a stick thrust through the side of a tomato-can, and not a crutch to be seen." [26]

He met a wounded soldier coming down the road with his bloody arm in a sling, heading toward the rear. "Take this road," said the wounded man, "I don't know where that one goes, but I know this one. I went up this one and I brought back a souvenir," he added cheerily. [27] Another wounded man John met on the road more than once during his comings and goings throughout the day kept getting a little farther toward the rear each time he saw him, refusing any assistance from anyone. Finally, the last time John saw him was at Siboney, where he still waited unattended in the hot sun. John asked him if there was anything he could do for him. "Well," replied the wounded and exhausted soldier, if it was not too much trouble, "I'd like a canned peach." John left in search of such a small thing for a wounded man; found it, but when he returned, he found that the soldier had boarded a transport and was gone. "He had been shot lying down, and the bullet went through his hat, scraped through the skin along his temple, passed through the palm of his hand, and lodged in his chest. His name was Cosby," wrote John, "I hope he will pull through, and I'd like him to know that I got his canned peach." [28]

Caspar Whitney was himself shaken by what he saw of the battlefield that day. All along the road to and from San Juan were dead men and men dying in agony and wounded being carried by their fellows or limping, barely moving on their own toward the rear and Siboney, "all strangely silent, bandaged and bloody. Beyond the second crossing the road was strewn with parts of clothes, blanket-rolls, cups, pieces of bacon, empty cans, cartridges; at the forks the marks of bullets were everywhere — the trees shot through and through. There were plenty of live bullets coming over the ridge that morning too." [29]

John had gotten behind the front lines, but safety was not to be found even there. Spanish guerrillas were hiding in the tops of palm trees firing at everything American. They shot the wounded, even those on litters, shot at doctors and doctors' aides — nothing that moved or fell into their sights was safe. John and the rest sought refuge in the thickets filled with cactus, but it was a relative safety at best. "Bullets spattered around us [and] perhaps they were only spent bullets, as some still assert, but they whistled from both sides of the road, and were lively in number, and came near enough getting both of us to make us care very little about their previous history. And it was this way until we got under the trenches, where there was a horizontal hail-storm sweeping steadily over our heads." [30] The noise was tremendous.

The Americans, advancing to the edge of the clearing, had been pushed from behind by others coming up, and eventually pushed out into the open from the narrow road. In the open meadow it was breathlessly hot. The tropical sun bore down incessantly. The grass was high enough to keep any breeze that blew from reaching the already sweat-soaked shirts and woolen trousers of the soldiers. Once pushed out into the open end of the road, the soldiers had started fanning out on both sides, trying not to advance any farther than they had to, and looking for cover where there was none. It was then that the Spaniards had decided enough was enough, and opened fire even more so than they had done previously.

No one could go anywhere. Those already exposed were being crowded farther still out into the open meadow by those still coming up from the rear. Those close enough to the action to see what was going on were helpless to do anything about it. Amazingly, the Americans had, for some reason, been ordered not to return the Spanish fire, and so, for what seemed like an eternity, they had flattened themselves as low as they could get while the bullets of the Spanish cut the grass waving over their heads. It was almost like Las Guásimas all over again, except this time the enemy could be seen, but few were foolish enough to risk more than a fleeting look.

Men continued to fall wounded and dead to the ground. The Spaniards did not have to be good shots, only mechanical ones. The Americans were grouped so tightly together at the mouth of the road from Las Guásimas that even a mediocre shot would hit something every now and then. To make matters worse, someone, probably in the War Department back in Washington, thought it would be a good idea to take along an observation balloon to spy on the enemy. Then someone else thought the best place to raise the balloon was directly over the Americans. The Spanish, not overburdened with luck in this war, needed but little urging to take advantage of another's stupidity, and directed their fire in, around and below the balloon with devastating effect. Observers in the balloon dutifully noted that the Spanish were indeed entrenched upon the hills where the firing was coming from. Finally, someone — and one may only assume it was the Spanish — shot a little too high, and brought the balloon down to earth, where it remained unmourned, at least by the Americans underneath it.

While all this was going on, the battle for El Caney had struck a snag. What was thought to need only a couple of hours at most to complete was now close to almost four times that. The Spaniards were fighting this time, and not running away. It was a turn of events that left many on the American side bewildered. The Spaniards were supposed to run, and when they did not, the Americans did not have much of a secondary plan to fall back on. El Caney was well positioned so that any attack on the San Juan heights would put the American flank and rear in danger if it was not secured. El Caney would need to be silenced before the attack, or at the very least, kept occupied by a separate American force while the primary attack on the heights above and behind Santiago took place. It would be nearly dark before El Caney finally fell, and the battle for San Juan would already be over by then.

Sporadic firing had begun to come from the American ranks as they milled around trying to follow conflicting or impossible orders and at the same time stay under cover — any cover. This was the fire that sounded to John like the "popping of firecrackers" from his position with Capron's battery. "From time to time," wrote Roosevelt, "some of our men would fall, and I sent repeated word to the rear to try to get authority to attack the hills in front. Finally General Sumner ... sent word to advance." [31] Roosevelt's orders had been to use his volunteer Rough Riders in support of the regular troops. Unfortunately, the regular troops did not receive the order to advance, and when Roosevelt attempted to make his way through their ranks with his Rough Riders, he found the regulars still hugging the ground, having received no orders to the contrary. "Where I struck the regulars," he wrote, "there was no one of superior rank to mine, and after asking why they did not charge, and being answered that they had no orders, I said I would give the order. There was naturally a little reluctance shown by the elderly officer in command to accept my order, so I said, 'Then let my men through, sir'." [32] This incident would at first give the Seventy-First New York a reputation for cowardice, which they did not in any way deserve. Roosevelt was brash, and in a hurry to get to the war. The men of the Seventy-First and their commanders were only following orders — foolish and contradictory though they might have been — something that Roosevelt neglected to do from time to time himself.

Not to be left behind as Roosevelt and his men broke through, most of the younger officers and enlisted men jumped to their feet, said Roosevelt, and followed. "I have seen many illustrations and pictures of this charge on the San Juan hills," wrote Richard Harding Davis, "but none of them seem to show it just as I remember it.... Instead of which I think the thing which impressed one the most, when our men started from cover, was that they were so few. It seemed as if someone had made an awful and terrible mistake." [33] But charge they did. Dismounted and on foot, the Rough Riders waded through the tall grass with dubious assistance from Grimes Battery on El Pozo. The artillery, using black powder, virtually obscured their targets for a full minute before they could witness the effect of the shell just fired. Usually the shell hit before the smoke cleared, and the effect of the explosion, if any, was in doubt, and a lot can happen on a battlefield in 60 seconds.

The Spanish were not napping. "The crash of the Spanish fire became uproarious," says Stephen Crane, "and the air simply whistled. I heard a quavering voice near my shoulder, and turning, I beheld Jimmie [Jimmie Hare, photographer for *McClure's Magazine*] — with a face bloodless, white as paper. He looked at me with eyes opened extremely wide. 'Say', he said, 'this is pretty hot, ain't it?' I was delighted. I knew exactly what he meant." [34]

Frederic Remington, who had only a few days ago moaned that he might have missed the only battle of the war when he missed Las Guásimas, got plenty this time. "The air was absolutely crowded with Spanish bullets," he reported.

12. The Horizontal Hail-Storm

"The shrapnel came screaming over. A ball struck in front of me and filled my hair and face with sand.... It jolted my glass and my nerves, and I beat a masterly retreat, crawling rapidly backwards." [35]

Once the charge began, everyone else began to move, nearly at the same moment. "General Hawkins, who was left with only the Sixth and Sixteenth Infantry, saw that he must either withdraw his brigade or lead it forward, and but one of these two movements was possible for that brave old general. His trumpets sounded forward, and in the front line with his men the general himself stepped into the tall grass of that little valley of death, and made for the entrenchments on top of the vertical hill. Colonel Wood's brigade was charging at the same time to the right, and later, the 'Rough Riders,' who had worked their way to the front, broke through a skirmish-line of regulars, and with Roosevelt leading them on horseback took a fortified casa to the right — Roosevelt killing a Spaniard with his pistol." [36]

The battle continued to rage and the noise and whoosh of bullet and shell seemed to the soldiers to intensify, even when they thought it not possible. Stephen Crane, who had begun the battle with General Lawton at El Caney, had a superior view of the battlefield. "The two lines of battle were royally whacking away at each other," he wrote, "and there was no rest or peace in all that region. The modern bullet is a fairy-flying bird. It rakes the air with its hot spitting song at distances which, as a usual thing, places the whole landscape in the danger-zone. There was no direction from which they did not come. A chart of their courses over one's head would have resembled a spider's web." [37]

The men could hardly run; the high grass forbade it, and so they had waded, said Davis, "as though they were wading waist high water, moving slowly, carefully, with strenuous effort. It was much more wonderful than any swinging charge could have been. They walked to greet death at every step, many of them, as they advanced, sinking suddenly or pitching forward and disappearing in the high grass, but the others waded on, stubbornly, forming a thin blue line that kept creeping higher and higher up the hill. It was as inevitable as the rising tide. It was a miracle of self-sacrifice, a triumph of bulldog courage, which one watched breathless with wonder. The fire of the Spanish riflemen, who still stuck bravely to their posts, doubled and trebled in fierceness, the crests of the hills crackled and burst in amazed roars, and rippled with waves of tiny flame. But the blue line crept steadily up and on, and then, near the top, the broken fragments gathered together with a sudden burst of speed, the Spaniards appeared for a moment outlined against the sky and poised for instant flight, fired a last volley and fled." [38]

By the time the Americans reached the top, those that were left were breathless, soaking in sweat from the heat and exertion and fear. But it was not over. Roosevelt regrouped his men and another charge, this time down the side of the hill just captured and up San Juan Hill. By the time they got to the top of the second hill, it was almost over. General Hamilton Hawkins, who had gotten there first with his men, should have gotten the credit for taking the most famous hill in American history, but he did not. Roosevelt was more flamboyant and picturesque, and immediately became the darling of the newspapers. Not to lessen his role in any way, but the man who took San Juan Hill was not the same man who later became president of the United States.

But for now, all was jubilation. The soldiers— regulars and Rough Riders— sprawled exhausted wherever they fell, sodden, wounded, dirty and miserably tired. They were so dirty afterwards. Sweat drenched them and made the accumulation of dirt and grime and powder residue stick to their bodies like glue. Even before the battles of the last two days, the men had not had any opportunity to bathe, and even if they had, the Cuban climate would not have allowed their cleanliness to last very long. The dirt and filth of battle only made it worse.

The battles were over, though they did not realize it at the moment. But for a time, they could bask in their triumph and catch their breaths. One soldier, sitting on top of San Juan Hill with the others who had survived that vicious battle, somewhat tersely summarized the whole thing by looking around at where he had come from and where he had gotten to, with nothing short of amazement, and observed, "Well, hell, here we are." [39]

In camp that night, the peace and quiet was almost surrealistic. After the rough-and-tumble, knock-down-and-drag-out fight, the whistling bullet and shell of one of the longest days most had ever known, cheating death and worse every minute, the letdown for them all was immense and total. "I never saw a more peaceful moonlit night than the night that closed over the field," remembered John after the battle and darkness fell. "It was hard to realize that the day had not been a terrible dream." [40] It was light enough still for him to stand and see the battlefield. He could see where the first shot from Capron's battery was fired, and could see much of the resulting battlefield. All those men, who yesterday were still alive, who ate their suppers unknowing of what awaited them — putting it out of their minds should it try to worm its way in with doubts and visions they did not want to see or know. And so, they had laughed softly with their comrades and ignored the reality of what, in the morning, they were about to do and become.

The aftermath of the battle for San Juan Hill was, in its own way, worse than the battle itself. "It was a terrible night after the battle," wrote John to his family, "the stream of wounded, the dead lying by the road but it seems that I have seen it all before. I shall not wish, or need, to see it again." [41] He was tired and on the verge of being sicker than he had ever been in his life. But those final dispatches from Cuba were some of the best he had written in those few short weeks of war. No one wrote anything any better. His first "letters" back to *Harper's* were not much to speak of; there was not much to write about in the training camps at Lexington, Chickamauga and Atlanta, but once

12. The Horizontal Hail-Storm

on the battlefield, he did well, and by the end of the war in Cuba, he had hit his stride, and *Harper's* could do no less than be pleased. [42]

No one who had never seen a battlefield and its aftermath could be prepared for it. Veterans soon grew jaded and learned not to be especially upset at what they saw. "The sight that greeted us on going into the so-called hospital grounds was something indescribable," wrote Clara Barton, founder of the Red Cross. "The land was perfectly level — no drainage whatever, covered with long, tangled grass, skirted by trees, brush and shrubbery — a few little dog tents, not much larger than would have been made of an ordinary tablecloth thrown over a short rail, and under these lay huddled together the men fresh from the field or from the operating tables, with no covering over them save such as had clung to them through their troubles, and in the majority of cases no blanket under them. Those who had come from the tables, having been compelled to leave all the clothing they had, as having been too wet, muddy and bloody to be retained by them, were entirely nude, lying on the stubble grass, the sun fitfully dealing with them, sometimes clouding over, and again streaming out in a blaze above them. As we passed, we drew our hats over our eyes, turning our faces away as much as possible for the delicacy of the poor fellows who lay there with no shelter either from the elements or the eyes of the passers-by." [43]

All of the correspondents, not just John, were maddened by the lack of preparations the government had done for the expected wounded and sick. It was as if they thought they could fight a war, and no one be hurt enough to need attention. "The poor wounded have had a hard time," said John in something of an understatement, but full of sympathy. "Apparently the powers thought there were not going to be any wounded; and, without doubt, few on Cuban soil dreamed there would be so many. The wounded at Las Guásimas were carried three miles to Siboney by hand, for when that fight was over not a wagon for ammunition, supplies, or hospital needs was on shore." [44]

John was not the only one of the correspondents who was disturbed at what he saw during and after the battles, especially where the wounded were concerned. Stephen Crane and Jimmy Hare were brought to a virtual standstill by the scene behind the lines at San Juan. "There they stopped short," wrote Cecil Carnes, Jimmy Hare's biographer, "immobilized by horror at the sight before them. A hundred broken men, the human wreckage brought from a few square yards of the battlefield, were receiving first aid though still under fire." [45] Later, they were just as astounded "by the sight of wounded men forced to drag themselves for miles to the nearest hospital in default of ambulances which should have been there and were not." [46]

"No where," wrote John, "were hospital preparations complete enough in tents, medicine, nurses, or surgeons, on the field, in the rear, or at Siboney." [47] The soldiers, however, suffered in relative silence, and he was amazed at their courage and self-sacrifice. He had never seen or heard of anything like it. It was as if a bullet in the chest, maimed arm, or crippled leg meant nothing to them, and he never got over what he had seen and was never able to tell completely the horrors he saw that day and the days following the battle.

At first it had been the bullets of the Spaniards that the American soldiers needed to fear, but now it was disease. Malaria and yellow fever struck the camps with a vengeance three days after the battle for San Juan. Roosevelt, several years after the fact, noted that "the health of the troops was not good, and speedily became very bad." [48] Once it started, it spread throughout all the camps; no one was safe, not even correspondents from Big Stone Gap, Virginia.

In a few days, Santiago surrendered. The Americans had surrounded the city and had lain in the trenches and "bomb-proofs" they had dug as protection from the Spanish shells that still came from time to time from Santiago. John observed that the wooden planks that served as a roof for the bomb shelters were pretty thin, considering what they were there to protect the soldiers from, and the dirt piled on top of the boards was pretty shallow as well.

On the night of July 13, a fierce and horrific storm struck the battlefield and those Spaniards and Americans camped in trenches and canvas tents suffered through it as best they could. It was the worst storm many had ever seen. The lightning lit up the battlefield with ghostly lights and shadows, and the thunder sounded like cannon-shot of nearly two weeks ago when the heights before Santiago were first taken.

The raindrops on top of the bomb-proof he shared with about 20 others had fallen hard enough to remind him of the bullets that spattered around them. "I can almost liken an action of this sort at the trenches," he wrote, "to an expected rain-storm in which there is a good deal of lightning. It thunders, the drops begin to patter, and you run for shelter, and you lie in comfort and in safety, except for the chance of being struck by lightning, which flashes sometimes uncomfortably close. By-and-by the storm passes, the rain quiets down to random drops, and you come out into the air, look around at the heavens, and stretch yourself." [49]

The Spanish fleet, anchored in Santiago harbor — preferring sinking to dishonorable capture, now that the war was effectively lost for them — steamed out into the ocean and was promptly sunk by the American navy. For John Fox, Jr., the war was over, but its effects would linger a while longer.

John's handwriting, which grew steadily worse over the years, was unusually erratic and irregular when he wrote home to his family on July 5, and it is very possible that he was showing the first signs of typhoid fever. "Don't worry about me when you hear of a battle," he wrote. "I am going to take care of myself but if anything should happen to me by accident you will get a cable to that effect as soon as you read of the fight. So, until you get a cable, you will know that I'm all right." [50]

He was not all right, however, and if he was not sick on the 5th, he was soon after. He went back to Siboney after the surrender to retrieve his belongings that he had thoughtlessly left on board the *Iroquois* in those days when he was excited about getting to Cuba. It seemed years had passed since then. His mind had been on reaching the front and seeing firsthand the glory that was war, and absentminded as he was, he had left nearly everything except what he carried in his pockets. He later told Thomas Nelson Page that he had gone through the whole war with nothing more than an empty tomato can for a mess kit, and, he said, in the company of a Negro who was as lost as he was, and a single shirt that he wore for more than two weeks without changing it. John had arrived back at Tampa on board the *Arkansas*, the first transport to leave Siboney for Tampa Bay. On board had been Caspar Whitney and Jimmy Hare, who was himself already feverish and sick. John, before boarding, had gone back to the transports to see if he could find his belongings without much luck, and managed to lose the rest of what he had, including one of his final dispatches for *Harper's*. At the end of his July 30 article in *Harper's*, the publishers made the notation that the conclusion had been lost, and that John was too sick "from exposure to rewrite it at present." [51] He did rewrite it, however, and sooner that *Harper's* had any right to believe he would, considering how sick he actually was. At the transport that he and Frederic Remington had sailed on from Florida, he had found little of either's belongings, but he gathered up what there was of them and "put them on a wagon — all of them — with my own, gave the driver gold to carry them to Siboney — and I never," he wrote Remington by way of apology, "saw yours or mine ... again." [52] When he arrived back in Tampa, he arrived there with virtually nothing but the clothes, rotten and smelly though they were, on his back. But he was really too sick to care.

His family was worried that he might not be well, or at the very least something was not as it should be, despite his protestations to the contrary. His last letter from Cuba did not sound right nor look right to them, and they knew him better than anyone, and were concerned. Back in Big Stone Gap, John's father had noted in his diary for July 27, that "Johnnie is quarantined at Tampa on his way home from Cuba." [53]

Thomas Nelson Page wrote in 1919 that John had shown up in Kentucky with a temperature of 104 degrees. "He turned up at last in Kentucky," wrote Page, "arriving with a temperature of 104 degrees, which he declared to be nothing, adding that this was General Wheeler's temperature when he went into the battle of Santiago, and that night he [Wheeler] was normal. While in the hospital a fellow patient sent by way of a pretty nurse a card inquiring how he was. The card was sent back with the reply: "'Worse, send [the pretty nurse] to enquire often'." [54] He was getting better. There is not much that a long neglected bath, a change of clothes and a pretty nurse cannot help.

He was stuck once again in Tampa, and no one who has never been homesick can fully appreciate the almost physical pain of not being allowed to simply go home. "Damn Cuby, Cubyans and the red handed war," he moaned to Remington. "I get away from here on Sunday & then Ho for the Heaven of Home!" [55] He need not have worried too much about Remington's gear. Remington was so glad to get back to his home in New York and out of Cuba—and, not incidentally, back where he could get plenty to eat—it is unlikely that he would have cared if he had left stark naked. He wrote John near the first of August not to worry about the loss of their belongings, and John did not.

By the third day of August, John's father was genuinely concerned about his errant son. "We were disappointed very much," he wrote in his diary for August 3, 1898, "that Johnnie did not come home. He is in Louisville, but too sick to come home." [56] He had gotten away from Tampa just in time for typhoid fever to strike him harder than anything had ever affected him in his life. He wound up in Louisville, not knowing how he got there, nor where he was going. Delirious, he wandered the streets until a friend, said Thomas Nelson Page, saw him and took him to a hospital. After a week in a Louisville hospital, he finally left and headed for home, still terribly week, but clear headed enough now to know what direction to take. It was an improvement. [57]

He arrived home in the Gap on Monday morning, August 8, and had to be helped inside the house. He was still too weak to walk any distance by himself. "Johnnie came home from Louisville this morning," wrote John Sr., "much to our surprise as we thought him too sick to travel. He has seen and suffered much since he started to Cuba on June 2." [58]

"I went all to pieces when I struck Tampa," he wrote Remington on August 15, "and had to lie up for a week in the hospital at Louisville, with a temperature playing about the merry height of 106. I am still weak but am getting all right again." [59] He was weaker than he would have liked anyone to know. He could not hold a pen, and he had to have any letters he wanted written, done so by someone else. But he was home.

All in all, he received for 64 days as correspondent, a total of $1,200 for the articles he wrote, and $10 per diem as correspondent. The case of typhoid fever he brought back to Louisville with him was "free of charge." But he never forgot Cuba, and once away, he missed the comrades he had met there, more than he would have thought possible. He missed Frederic Remington, and Davis, and Roosevelt, and Caspar Whitney and Stephen Crane even, whom he barely knew. Now that Cuba was behind him, he missed them all and wondered where they were and what they were doing.

CHAPTER 13

The Road to Recovery

I want to get these things out of my mind and
the fever out of my blood...
Frank Norris

Doc, those oysters were no damned good.
William Goebel

THE JOHN FOX, JR., WHO CAME BACK from Cuba in August of 1898, was markedly different from the man who went there only a little over two months earlier. He had seen war, and he could not help but be affected by it. No one could. Nearly all who went there — like Frank Norris — came back with a fever of one kind or another, and it would be months and years before some recovered their health and sanity. For others, it would take longer to recover, and for some, that recovery never came. But recovery or not, they were all changed.

By the time that Santiago finally surrendered and the Spanish fleet that had been holed up in the harbor there had been sunk, most of the correspondents could think of little else but home and how to get there the quickest. They had all had as much of "Cuby and Cubyans," of heat and disease, and of boredom and dirty shirts and being shot at that they could stand.

Davis went on to Puerto Rico, where the Spanish-American War would take a turn for the worse, and reported from there, eventually writing a book about his experiences in both theaters of war.

Perhaps the most tragic of all was Stephen Crane, who died in 1900 in England. By the war's end in Cuba, nearly all were thoroughly sick of the whole thing. Many, like John, and even Stephen Crane, who wrote so well of war, had never seen a battle before, had never seen what shot and shell can do to living flesh, had never seen the life literally torn out of men. "I want to get these things out of my mind," said Frank Norris after it was all over, "and the fever out of my blood," but he never did. [1]

Not everyone who went to Cuba came back with "fever in their blood," at least not the fever of disease, and of those who did, a great many recovered quickly. Several capitalized upon their experiences, especially politicians— Theodore Roosevelt being the most notable. In a frenzy of political activity after the end of the war in Cuba, Roosevelt returned home, was elected governor of New York in 1898, elected vice-president when William McKinley won, a second term in 1900, and gained the presidency when McKinley was assassinated late in 1901. In little more than three years, he had gone from assistant secretary of the navy to president of the United States. [2]

All who went to Cuba that summer of 1898 recovered in their own way, some better than others, but few could forget where they had been and what they had seen, and most felt a strangeness, almost a longing, that remained unexplained for some time after they made it back to their homes.

Now that John was home in his beloved Gap, he was surprised and troubled by that strangeness he felt. He had only been away for a few months, but now everything seemed changed — different somehow. He was idle, still weak but gaining strength every day, but he almost felt guilty with his idleness somehow. It was hard for him to explain this to his family, and perhaps it was because he did not understand it himself. He was restless, impatient with being sick, and did not know what to do.

There was one thing he did know, however: Never again, he swore, would he go anywhere as a correspondent for anyone. He had his fill of writing about war, or so he thought, and he had conveniently forgotten that it was himself, not his editors, who insisted that he go to Cuba in the first place.

Writing was again his salvation. Before the war with Spain, he had begun writing two novels at the same time. The one that would take on the characteristics of the Spanish-American War would be *Crittenden*, but it would be two years before he would have the book ready for publication, and by that time interest in the war had waned. The second, published in 1903, would be his first bestseller and arguably his most famous book. He had written Thomas Nelson Page near the middle of April of 1898 that he had just finished his first hour's work on what was eventually to become *The Little Shepherd of Kingdom Come.*

Now, back home from Cuba, he found himself wanted by nearly everyone. Florence and Tom Page wrote, asking him to come and recuperate from his Spanish fever at their home. He could have the "chills" there as well as anywhere else, they reminded him. Theodore Roosevelt wanted to see him whenever he felt up to it. The Southern Lyceum Bureau of Louisville was eager to see him well once more, and back on the road. He would be much in demand now, and he knew it. Others were independently asking him to come to their towns and speak of the war and his writing, and he felt confident enough to circumvent the Bureau and set his own terms and handle his own speaking engagements. In reply to one such request from New Hampshire, he told Tom Page, "I telegraphed terms—$150—got an offer of $100, declined less than $150 and have heard nothing further. That's my price for single lectures hereafter; the Lyceum is getting that figure for me in the South, and if I'm not worth it elsewhere, I'll wait until I am." [3]

He was still a little put out by *Harper's*, and their finding themselves in financial straits did nothing to help the situation. Royalty payments from his previously published works were slow in coming, and he was constantly strapped for cash. They intended to use his dispatches from Cuba in a book about the war they were going to publish, and asked him for permission to use them all, but he held them at arm's length. He might want to use the material himself, but finally in desperation they offered him $200 for the use of the dispatches, and he accepted. Later, in *Crittenden*, he would use the material anyway, at times lifting whole sections of his dispatches verbatim and placing them in the manuscript.

Now, he found himself embroiled in an old argument in which he had no business. He was done with *Harper's*, although it would take a few more months and hard feelings on both sides before the final break came. The details and reasons for his leaving *Harper's* would not reflect well upon him, and he knew it, but he found he had soon gone too far to turn back. That the reason for the break was not entirely either *Harper's* nor John's fault only made the divorce all the more painful for them both. In Tampa, and later in Cuba, he had jumped into the middle of an argument: a family squabble, really, and a dangerous one at that.

Poultney Bigelow, in Cuba for the New York *Herald*, was a person that few liked and none admired, except perhaps Frederic Remington, who had too few friends to be choosy. Bigelow had

After the Spanish-American War, a more mature looking John Fox, Jr., appeared. This photograph also appeared in many of the promotional brochures when he rejoined the speaking circuits after the war (courtesy University of Kentucky).

many of the same traits as Richard Harding Davis, but where Davis's friends overlooked Davis's faults and even gloried in his sometimes fanciful demeanor, they could not abide the same things in another man. Bigelow did little or nothing to make things better. Perhaps he knew there was no use in trying. Bigelow had another aggravating trait: he told the truth, regardless of whether or not people wanted to hear it, or whether the time for it was right or not.

In late May of 1898, Bigelow had written an article condemning the shortcomings of the army, and in doing so had broken the unwritten law of correspondents, at least at that time, of writing about things the army did not want the public to be told. All the correspondents in Tampa and Cuba, including Davis, Caspar Whitney and Fox, knew what was going on: the mismanagement, the lack of supplies and rations, the totally useless equipment, woolen uniforms for a tropical war, and the list went on and on. The difference was that Bigelow had written about it, while the others sat on their stories until a more appropriate time — after the war was over. In effect, Bigelow had "scooped" the competition, and they did not like it or him.

To make matters worse, *Harper's* had printed the article in their May 28 edition, along with an editorial that seemed to support Bigelow's claims, and Bigelow did not even work for them — he worked for the *Herald*. Davis was outraged, though one wonders why, except perhaps that Bigelow had beaten him to the journalistic punch. It was Caspar Whitney, however, who had felt more aggrieved than anyone. Whitney felt that Bigelow had usurped his position with *Harper's*. After all, Whitney reasoned, *he* was the *Harper's* correspondent in Tampa, and presumably would hold the same position if he ever got to Cuba, and Bigelow was little more than someone without any credentials who had written an unsolicited article. *Harper's* had accepted it and printed it without checking to see if there was any truth to Bigelow's assertions, or perhaps more likely they failed to check with Whitney to see if it was all right for them to publish the article.

In any event, the controversy, usually known as the Davis-Bigelow Controversy, resulted in Whitney's resignation from *Harper's*, and there is enough evidence to suggest that Whitney had a fondness for resignations, and for immediate withdrawals of the same. This particular resignation brought about *Harper's* sending for John and his subsequent arrival in Tampa to take Whitney's place. By the time John had gotten to Tampa, Whitney had signed on with *Century* magazine.

The anger of all the participants almost seems childish at this date, but they took the matter seriously, far too seriously for what it really was. The reality was that they had the whole Spanish War to stew about it. By the time they got back to Tampa, they were all three wild about it, angry all out of proportion as to what the situation really deserved from them.

It was not just the other correspondents who had been furious. The article had caused a scandal all over the country. The army had been embarrassed and withdrew Bigelow's credentials, which cost him his job with the New York *Herald*. Whitney had written *Harper's* condemning them for publishing the article, but they stuck to their guns. Things were starting to get out of hand: Whitney had resigned and *Harper's* had accepted, and had immediately asked John to take Whitney's place.

Now, back in Tampa and in quarantine, John got into the act. He was friends with both Davis and Whitney, and that friendship had deepened in Cuba. Although he had not many dealings with Bigelow and really knew little about him personally, John sided completely with his friends, and by the time he had arrived back in Tampa, he had worked himself up into quite a furor over *Harper's* real or supposed treatment of Whitney and Davis. Furthermore, *Harper's* was dropping heavy hints that they wanted John to go to Hawaii and send back what dispatches he thought might interest *Harper's* readers, but John had no intention of going to Hawaii for *Harper's* or anyone else. In anger, he dashed off a letter to *Harper's*, asking them for a promise that as correspondent in Whitney's place, he would not be treated the same way that *Harper's* had treated Whitney. *Harper's* was not completely sure what John was talking about, nor why he seemed so mad over something that was to them simply business that went on every day.

He was mad — outraged, really — and it's difficult to understand why he was so upset. None of the other participants in the quarrel seemed to take it as hard as John did, but once he got going, his anger took over and remained long past the time when others would have went on about their business and forgotten the matter. Davis and Whitney did exactly that, leaving John holding the proverbial bag alone. It was a bag, however, that he seemed more than willing to hold by himself.

The letter John wrote to *Harper's* in defense of his friends is almost incoherent in places, and even at this late date the reader can almost feel the anger and intensity. From quarantine in Tampa, he had written to *Harper's* that he had "read Whitney's article and Bigelow's article and the Weekly editorial when I got to Tampa. [4] I sympathized fully with Whitney for resigning and particularly after the editorial the general effect of which, as I gathered it then and remember it now, was certainly in favor, if not in defense of Bigelow — upholding, in other words, a chance contributor against a regular staff correspondent." [5]

Harper's was not to be swayed, however, and understandably sounded a little mystified in their reply to him near the middle of August. They had replaced Whitney with John, wrote H.L. Nelson for *Harper's*, "because Mr. Whitney's resignation gave me an opportunity to help you to the chance of seeing the war which you had asked of Mr. Alden. Of course, it was an experiment on both sides, so far as the war correspondence went, but it has turned out well. You have your experience and your material and we have had some admirable print from you for which I thank you." [6] They would not, however, apologize for something that they, frankly, thought was none of John's business, and it was not. "I do not see any reason," continued Nelson, "for an answer to my letter of June 14th. It is unnecessary to discuss our relations with Mr. Whitney, especially now that he has acknowledged that he was in the wrong. I had hoped that your swift condemnation of us, morally and professionally, would come to be regretted by you, but if you still think that you were right in demanding a promise of good faith, let it go at that. I am very sorry and shall always be sorry, but I know that we did nothing to warrant such a feeling on your part." [7]

He had gotten himself into a mess and was embarrassed and angry at the same time — not at *Harper's* so much any more, but at himself for acting so unprofessionally. It did not make matters any better when Whitney went on the Hawaiian assignment for *Harper's*—the same assignment that John had refused — and by the middle of 1899, Davis would also be publishing again in *Harper's*. Finally, John let the matter drop completely. After things had turned out the way they did, he felt like a pure fool. It was not the first time he had let his anger get the better of his judgement, but he hoped it would be the last. [8]

He had gotten his strength back by the end of August, though he was still a little unsteady on his feet. Christmas found him in Washington, D.C., at the Pages along with Robert Burns Wilson among others, enjoying himself immensely, though he was a little embarrassed by the number of presents he received from Tom and the rest. He wrote home in high spirits that day, surrounded by the friends he loved most in the whole world, that he was sure the good times had come at last. For him that day, Cuba seemed a long way in the past, but he visited there every day in the novel he was writing. In the writing of *Crittenden*, he relived many of those moments of that war, and finally succeeded in getting those horrors out of his mind, and occasionally even the fever out of his blood.

By the first of 1899, he was solidly back on the lecture circuit and spending most of his time in New York. He was also writing furiously on *Crittenden* and had seen Currie Duke again. She was married now to Wilbur Knox Mathews whose business was investing the money left to him by his father who had made his fortune years before in the California gold fields. They were still friends— he and Currie — and although she seemed happy, he could not help but wonder if she really was, and possibly wondered at the chance that he had missed when he went south for *Harper's* nearly a year ago and left Currie to marry someone else. [9]

Back home in Virginia, he received a telegram from Richard Harding Davis, informing him that he and Cecil Clark were to be married. It did not come as any great surprise to Fox. He had known about Davis's courtship for some time. The telegram was followed by a letter asking John to be an usher at the wedding in Marion, Massachusetts. He could not decline and was really

13. The Road to Recovery

looking forward to it. There would be the kind of people there whose company he always enjoyed. There would be the rich and famous from all over the Northeast, and any wedding of Richard Harding Davis would for sure be something to remember. He arrived in New York and left there on the 3rd of May for Marion.

The wedding on the 4th was everything he had expected it to be. The bride was beautiful, dressed in white but strangely unsmiling. Davis looked stiff and ill at ease in the wedding photograph, as did many of the others. It did not seem to be a happy party. Only John strikes a less than serious pose, leaning slightly backwards, cigarette in hand, thoroughly enjoying himself. Neither the bride nor the groom appear half as happy.

As per the groom's instructions, the wedding party vacated the town promptly at 5 o'clock, including John. "I have engaged a squad of rough riders to police the town," wrote an ecstatic Davis to John before the wedding, "and any one of the wedding party who is found on the premises get's shot — If he misses the train to Boston that is no excuse. My guard of honor never miss." [10]

Apparently, everyone got into the theatrics of the thing, though no one seriously gave any thought to getting shot at by rough riders, except possibly Davis, who looked like he wished someone would shoot him quickly and be done with it. It would not be too long before he might have given serious thought to doing it himself. [11]

John returned to the Gap on May 22 after stopping over with the Pages for awhile. Tom Page had read his almost completed novel and had criticized it badly. John was used to criticism, but coming from his dearest and closest friend, he could not hide his hurt feelings.

His bruised ego mended soon — he could not stay mad at Tom Page very long — and by the middle of July he had joined his friend and Tom's family at the Pages' cottage at York Harbor, Maine. He had assumed that *Harper's* would want to bring out *Crittenden* in serial form in their monthly magazine in the same way they had done his other novels, long and short, but *Harper's* was holding back. The manuscript was too much concerned with war, they told him, and most people did not want to read about war anymore. The fervor of the Spanish-American War had died down drastically, and any books written about that war would not sell very well. They sent the manuscript back to John at York Harbor with instructions to edit the war material out of the story as much as he could and send it back. They would skip the usual serialization and bring it out as a book right away.

This did not suit him at all. He had not made any money since the last *Harper's* check for $200 for them to use his Spanish-American War material in their illustrated history of the war. That would soon be a year ago, and he needed the money that a serialization would bring him. On July 28, still at the Pages' revising the novel, he wrote to his brother James that "I've had a row with the Harpers which will result, I think, in my going — lock, stock and barrel — to the Scribners. This, of course," he cautioned his brother, "is confidential. The Harpers think the reaction against war material so strong now that it would harm them and me to publish the novel as a serial." [12] He mentioned that he thought that Scribner's would offer him $1,000 up front for the book by August, and he needed the money.

He was desperate and short-tempered. When James heard that *Harper's* would probably have published the story some months ago, closer to the end of the war when such books were more popular, he did not hesitate to tell John that if he had finished the novel six months sooner, then John would not be in the shape he now found himself. "You irritate me," John replied to James testily, "and widen the breach between us with such remarks as the last line of your letter—'only you sh'd have had the story in 6 months ago' ... I ask you now to drop, hereafter, all expression of that attitude towards me ...you had the right to it once — but you lost it nine years go— utterly — and, in that time, you've done very little to regain it." [13]

Henry Alden of *Harper's* stopped by to see John at York Harbor, to try and smooth things over, but it was already too late. Even Caspar Whitney tried to get John to stay with *Harper's*, and in a letter dated August 2 he urged John to be good "and faithful." [14] It was to no avail. He had already, last month, offered the manuscript to Scribner's, and Tom Page, already at Scribner's, was no little influence upon his decision, having practically offered John's work to Scribner's himself. John had taken the "initiative," however, in offering *Crittenden* to Scribner's, and, he wrote on July 26, "if you don't see your way to using the novel, it will go elsewhere and anywhere but back to the *Harper's* for book publication." [15] In the same letter, he offered them the second of the two novels he had been working on: *The Little Shepherd of Kingdom Come*. The break was complete, and Scribner's would remain his publisher for the rest of his life.

He struggled all that summer and fall, revising and rewriting *Crittenden*, mostly staying away from the Gap, hoping the book would be done by November. But there was always something to keep him from working. He once said that the only place on earth he could really write was at the Gap, and there was likely a lot of truth to that. He always had trouble with deadlines of any sort, however, and his guilt at not completing the book sooner could explain why he had gotten so angry at James. Anyway, there were distractions aplenty in New York, at Oyster Bay with Roosevelt, in Virginia and Maine with the Pages, with Richard Harding Davis, and at parties with Finley Peter Dunne and actress Ethel Barrymore.

It was far too easy for him to drop the dreary revision of *Crittenden* and go to parties, dinners and boating excursions. He lived on the largess of his friends and was welcome to it. He was always the welcome guest, and the Pages even set aside a room that was always ready for him at any time in their home.

"He came to pay the writer a visit," wrote Page in 1919, speaking of himself, "in the winter of 1894, and was so delightful a guest, not only to host and hostess, but to the children of the family, that he was not allowed to leave for five months." [16] But John's conscience bothered him about his procrastinations and the charity he was receiving from his friends.

November came, and the revision was still not finished. Neither did December bring any news to Scribner's of a book ready for the publisher. The Spanish-American War was fast fading from the public consciousness, and John was beginning to worry that *Harper's* may have been right after all. He continued to plug away at the manuscript in bits and pieces of spare time, but it went slowly. It was during this time that New York and the Pages' in Maine became his primary homes, and he stayed away from Big Stone Gap more than usual. Except for railway trips through the state, he had not been back to Kentucky since leaving the hospital in Louisville that fall of 1898. But that neglect of his native state of Kentucky was about to change.

Back home, Kentucky was going through what can only be, for want of a better characterization, referred to as turbulent times. A raucous election campaign for governor had set the whole state on edge. Because of the divisiveness and rancor between the two parties—Democrats and Republicans—and cries of stealing of the election, nearly everyone felt that something was going to happen, but nothing had prepared them for the events of that February, nor for the rest of that summer of 1900, in a little capitol town on the banks of the Kentucky River. In February, John wrote to James that he was on his way from New York to Frankfort. He was going there to attend the funeral of William Goebel, Kentucky's assassinated governor-elect.

By the time the election for governor had rolled around in November of 1899, there were nearly 200,000 people in Kentucky who did not like Democrat William Goebel enough to vote for him, but it is unlikely that very many of those 200,000 thought seriously about killing the man. But someone thought about it, and someone did kill him, but they were not necessarily the same man. Goebel's road to assassination was not a terribly complicated one as politics go, even Kentucky politics. He had managed in his relatively short career as politician to make some very important people and institutions very angry. The Republicans naturally hated the man, but that was their job. Goebel was a special man, and one who drew hatred like a lightning rod draws fire from the sky. The Louisville & Nashville railroad had spent in the neighborhood of $500,000 to see that he was defeated in his bid for governor, and afterwards they admitted that they would have spent twice that if they had thought it necessary. He had enemies, and plenty of them, and one wonders how he survived as long as he did without being shot.

Goebel was from northern Kentucky, the first son of immigrant German parents, and, in time, studied law and became a partner in a law firm in Covington. He soon gained a reputation for ruthlessness, which is of itself not a bad trait for a lawyer. He was successful but not especially satisfied, and by the last month in 1887 had been sworn in as state senator from northern Kentucky and was off to Frankfort, the capitol.

There, his reputation as a cold and unlikable individual, if anything, was enhanced. He was not a pleasant man, seemingly unfriendly and a little uncomfortable around both men and women, especially those he was not acquainted with, but he was smart, both friends and enemies granted him that. His one great enemy was the L&N railroad, perhaps the largest corporation in Kentucky at that time, and growing. Goebel thought they were growing at the expense of the "common people," and sought to rein them in a little. He did this, but did not neglect his other causes: civil rights for both women and blacks, the outlawing of pool halls and lotteries, and more and better rights for the ordinary working man. All gained him supporters across the state, but for every supporter in controversial causes there were at least as many enemies to be gained. If there could be only one word to characterize the man, it was "controversial," and Goebel gloried in it.

The political situation in Kentucky was ripe for violence by the time the election of 1899 rolled around. The first Republican governor ever elected in Kentucky up to that time was William O'Connell Bradley, who had been elected in 1894. Not even the original Republican and native son of Kentucky, Abraham Lincoln, had been able to carry the state some thirty-odd years before. Kentucky was Democratic, and Republicans stood little chance at winning the governorship, and when Bradley won, the Democrats were furious and shocked and vowed not to let it happen again.

"To his supporters," John would write years later in *The Heart of the Hills* of Goebel, "he was the enemy of corporations, the friend of widows and orphans, the champion of the poor — this man; to his enemies, he was the most malign figure that had ever thrust head above the horizon of Kentucky politics [and] to both he was the autocrat, cold, exacting, imperious, and his election bill would make him ... master of the commonwealth." [17]

William Goebel, a Democrat, was instrumental in getting the notorious Goebel Election Law passed, which virtually assured the election of a Democrat, and practically made the election of a Republican impossible in Kentucky, as if it was not hard enough already. The law provided for the Democratically controlled legislature to appoint three election commissioners, who would themselves name local election commissioners, who would naturally be Democrats because of the makeup of the legislature and the politicians who would name them to their posts. The local officials so named would in turn name the officers who would preside at the local polls and decide who could and could not vote and preside over the counting of the ballots. The result was that Democrats would control the elections from top to bottom.

Even so, Goebel narrowly lost the election to Republican William S. Taylor by a little over 2,000 votes. Goebel uncharacteristically conceded the election to the Republicans, and Taylor was sworn in. Then, after everyone thought things were settled, the Democratic State Central Committee decided to get into the act and contest the election, which meant the legislature in Frankfort still controlled by the Democrats would decide who won. The outcome was never in any real doubt.

Goebel went along with the Committee, and the Republicans, crying "foul," sent out the word to the heavily Republican mountain counties that the Democrats were about to steal the election. In practically no time at all, Frankfort was flooded with mountain people, nearly all heavily armed, coming to town to take back, by force if necessary, the governorship should the Democratic Committee give the election to Goebel.

As he walked up to the capitol on January 30, where the Committee was meeting to decide his fate and that of the governorship, Goebel was shot from a window of the state house, next door to the legislature. He was mortally wounded, the bullet passing completely through his body and doing considerable damage. The legislature quickly declared Goebel the winner of the election. He was sworn in the next day and died on February 3. Before his death, Democrats in attendance said Goebel's last words were, "Tell my friends to be brave and fearless and loyal to the great common people." Those words would find their way to the impressive monuments erected over his grave, and in front of the old capitol building in Frankfort. Another version says that sometime before lapsing into a coma, the governor-elect asked for his favorite food— oysters— and after eating, turned to the doctor and said, "Doc, those oysters were no damn good."

At any rate, last words or no, Goebel was dead, and despite the dangers and nervousness still surrounding the assassination and its aftermath, John could not keep himself away from the spectacle of thousands of people, mountaineers and Bluegrass politicians, flooding the streets of Frankfort, while the rarity of political assassination stunned the state, and, in fact, the country.

John was there to see it all. The coffin carrying all that remained of what was arguably the most hated and most revered man in Kentucky at that time was taken to the Capitol Hotel, where it lay in state a stone's throw from where the assassination had taken place. The next day, February 8, a procession marched through the streets of Frankfort, winding its way up one street and down another. Down Wapping Street it went, silently in the rain. From Wapping Street, it made its way over to Wilkinson, and then to Broadway. Up Broadway it turned to Ann Street, passing near where John had roomed with Robert Burns Wilson a few years before. From Ann Street, the procession made its way back to the Capitol Hotel, where the casket was placed in a hearse, and still in the rain, began its trip to the cemetery on top of the hill east of town.

They moved slowly up the steep hill, passing near the spot where, in less than ten years, a cornerstone would be laid by Theodore Roosevelt to begin a new Kentucky capitol. Near the top of the hill, the procession turned right into the cemetery grounds, and still climbing, they made their way toward the highest point on the grounds. The crowd, including John, followed closely, suffering the rain and cold until finally they stopped.

They milled around as mourners are wont to do while everyone found their place. Soon enough, the casket was placed in a vault to await the building of a suitable monument. Music, prayers and speeches kept them all in the rain that refused to let up. John, along with the other mourners and curious, shrugged against the rain, shaking, chilled to the bone, knowing he would be sick, feeling those old familiar warning signs of illness already.

The outpouring of tributes to Goebel flooded the front pages of Kentucky papers. One would not have imagined how hated Goebel was by what the papers printed after the assassination and funeral. One is reminded of the story of a man who, upon hearing an especially flattering funeral oration of a man he knew to be quite the opposite, arose and went to look in the coffin to see who was really there.

John's father, himself no friend of the slain Goebel, was not to be taken in by the newspaper tributes to the slain governor. "Did you ever read such a lot of 'bosh' and 'rot' as is served up by the Courier-J [the Louisville *Courier-Journal*]?" He wrote to James, "We have been taking it all along, but I think I'll stop it, am so much disgusted." [18]

When the funeral was over, John lost little time in leaving Frankfort. He always loved the old town on the river, and still did, but he determined to get away as soon as he could. As close as he was to Virginia, he turned toward the Gap, and home. He arrived back at the Gap on February 12, sick with a cold that steadily got worse.

John lay in bed for several days, trying to recover from the chill and cold he had brought back with him from Kentucky. By the end of February, he was well enough, he thought, to travel to Indianapolis to visit James Whitcomb Riley, with whom he had shared the stage before the Spanish-American War. He did not find the Indiana poet in especially good health, and by the time he had gotten back to Lexington he was sick again himself. A reporter for the *Lexington-Leader* found him "bundled up in bed at the Phoenix Hotel and ... was greeted with that same old rasping laugh and that enormous mouth full of teeth." [19] The reporter found him nursing himself with a variety of pills, potions and powders, and a stomach so weak and queasy that it would not abide anything but hot water which John was taking in huge amounts. But he assured the reporter that he was fine, or would be soon. In the meantime, revision of *Crittenden* languished.

Scribner's was not pressing him about manuscripts, either *Crittenden* or any other stories he was working on, but seemed to be satisfied with just having him on their list of authors, and would patiently wait for whatever he could send them, only asking that he send them along as quickly as he could. *Harper's* apparently was not harboring any hard feelings about their divorce, at least

13. The Road to Recovery

after the fact had been established, and they even considered — at John's request — sending him to South Africa to cover the Boer War. They let the matter drop, however, and there is some indication that Tom Page worked behind the scenes to keep him in the states, thinking, and rightly so, that John's health was too fragile for such a trip.

John's request to go to Africa may have been halfhearted anyway. Richard Harding Davis and others from the Spanish War were going, and it had been long enough that he remembered their comradeship in Cuba and the hardships and privations of war more fondly than he would have thought possible. He missed all of them — Remington, Whitney, Davis and Roosevelt — and the action. It obviously had also been long enough for him to have reconsidered his admonition to himself about never going into the war fields as correspondent again. At any rate, he did not go for *Harper's* nor for anyone else, and if he really regretted not going, he did not reveal it to anyone.

It did not take very much to get John to his wit's end; it never did. Any amount of stress could set him off against whomever was handiest. He had done so in Jellico and had most recently castigated *Harper's* unmercifully over the Bigelow-Davis-Whitney controversy. Now it was his brother James who again fell under his sights, and perhaps, not without some justification.

He was finishing up *Crittenden* and working on *The Little Shepherd* at the same time, trying to decide whether to let another publisher have the serial rights, which Scribner's really did not want, or to forget the serialization all together and try to keep up with his social positions in New York and other northern cities. At the same time, he was still involved in land schemes in the Gap and surrounding areas, principally on behalf of his brother James, who still — perhaps the only family member to do so — held out hope for a bonanza and fortune in the mountains.

Few things are more stressful than those activities that involve money, or rather the potential for making the same, and his striving for investors on behalf of James brought their relationship to the breaking point in the summer of 1900. James was eager and impatient for progress with his schemes. John wanted to do all he could, but felt that James wanted him to work full-time, to the exclusion of all else, including writing, in favor of finding and snaring investors. In short, James wanted things to return to the way they were ten years before, when the boom was going full blast in the Gap and elsewhere, and when he was in charge, and John tacitly worked for him.

John had been places and had done things since those times. He was nearing 40 years of age and was not about to give up his own lifestyle and venture again into those dark waters of real estate and finance. He had drowned there once before and did not care to swim in those waters again.

He thought James was being a little "cocky" in his letters to himself and others, and told him so. James thought John was neglecting their business interests in favor of a leisurely lifestyle that brought in little or no money. "I warn you once more," wrote John to James after a particularly irksome letter. "If you write to me any more about my (supposed) sins of omission — I won't answer your letter at all. I am doing what I think is my duty and if you will judge yourself and myself by results— maybe you'll manage to acquire a little more respect for me and a more becoming humility for yourself. For Heaven's sake let me alone. I'm working hard and I think I'm doing my own work and, perhaps the greater part of yours." [20]

Finally, at last, *Crittenden* was finished, but in the hands of Scribner's, it went nowhere. They thought it a good effort, as much or more than they expected, really. They wanted to go slow, however, "and as this is the first book that we are to publish for you," they wrote, "we have no direct knowledge of its selling power," [21] and they wanted to publish it at the best possible time.

The national election was coming up that year, which, they reasoned, would take attention away from anything not political, including novels. Also, the holidays were on the horizon and they thought the shops and papers would be too "busy" to show off a new novel to its best advantage. They advocated waiting, perhaps as long as sometime next year, when they thought the market would be better, or at least more receptive. Finally, against their better judgement, Scribner's released the novel in November of 1900, and *Harper's* was proved right about the war content of the book.

Crittenden, some might argue, was not the worst thing John had written so far, but it was close, very close. Much of what he had written before was, even by the standards of the time, overly sentimental with a little too much romance of which he was never able to write convincingly. *Crittenden* was the worst of the lot. It told a story that was beginning to be shopworn by now, but it seemed he could not bring himself to edit out those sugar-sweet descriptions of hero and heroine alike.

Crittenden is remarkable more for what it is not than for what it is. It should have, in the natural progression of things, shown Fox as a developing, maturing writer, but it did not. If it is remarkable at all, it is so because of the novel it foreshadowed: *The Little Shepherd of Kingdom Come*.

Crittenden is primarily a love story with the usual pitfalls and blinding sentimentality of such things, but even for John the romance and overly played sentiment is too much here. Surely, no one talked the way these people talk or acted the way they do in this book. People in love do all sorts of foolish things, and say things later that they might regret, and do all manner of strange and unusual things, but not like this. This time, John overplayed his hand, and the book, in places, can be embarrassing to read. Perhaps in his day some lovers talked and acted the way he says they did, but if they did the translation of that love loses something when put to paper.

The book was not the success he had hoped it would be. This was not all his fault, although his waiting so long after the war to finish the book certainly had something to do with its lack of success. War novels that immediately follow on the coattails of a particular war quickly fall out of popular favor, as a rule, and *Crittenden* would likely have been no exception. That it came two years after the end of the war did not help any at all.

Perhaps most important to him was that — somewhat surprisingly — his brother James liked the book. "I have been absorbed by your book," he wrote just after the book appeared on store shelves, "that I finished it on the streetcar and at lunch this afternoon. It is far and away the finest thing you have done — you have given us generously of your best, and it is all to your lasting credit." [22]

John had high ideals while writing *Crittenden*, but for the most part the book failed to gain a respectable response with the American reading public. It is likely that by the time the final draft was in Scribner's hands he had lost interest too.

His interest could now turn fully to his other novel he had been working on at the same time he was giving birth to *Crittenden*. It was different from anything he had written so far, but would still have bits and pieces of his other works. He was excited about it, more than he had been about any other book or story he had written, even more excited than that night so very long ago when he ran through the mud of Big Stone Gap with a check for $262 clenched tightly in his hand. As it turned out, he was justified in being excited. It would be the universal story of a boy and his dog; a story of social outcasts trying, in their plainspoken way, to fit where they might not belong; and of a war that changed people and a country in ways they never imagined they could be changed. He was about to write his first best-selling novel.

CHAPTER 14

The Crook of the Shepherd

> Here for six months, I was a Napoleon of finance, met my Waterloo, and went to the Hell-ena of debt for ten years until I was plucked out by the crook of the "Little Shepherd."
>
> John Fox, Jr.

> Please make it very exciting and don't make the men cry too much.
>
> Margaret
Edith
Eileen
Katharyn

IN FEBRUARY OF 1901, JOHN found himself sitting with his brother James, Richard and wife Louise, Sidney's widow Pollie, and his sister Elizabeth in New York City, watching younger brother Rector Kerr Fox marry Hilda Seccomb. Life was passing him by, and he was beginning to wonder, just a little, if he might be a bachelor for the rest of his life. He was pushing 40 years old with no prospects in sight. It was not so bad when his older brothers married, but when those who were younger than he was— Rector, and only four months ago, Richard — married, he could hear the clock ticking against him. He had not given up marrying someday, and even hoped that he would, but he had to admit to himself, especially in times like this, that things did not look all that bright where matrimony and John Fox, Jr., were concerned.

Months later, after Rector's wedding, his mother would write to him, concerned because she had heard that he had gotten married himself, which came as something of a surprise to her, since he had said nothing to her about it. It was all untrue, but it did not keep her from worrying. He told her it was a "piece of absurdity" and wished she would not worry so much about him, and especially not about something so ridiculous and untrue. "Haven't you enough real worries?" he asked her. [1]

Early in 1898, he had written Tom Page that he had worked his first hour on a Civil War novel. It would be a year before he would write another line after that first chapter of *The Little Shepherd of Kingdom Come* was finished, but after that first hesitant start he would soon find his stride, and the story would begin to come almost effortlessly for him. The Spanish-American War intervened, as did the writing and revisions of *Crittenden*, and his bouts of typhoid and malaria and perhaps yellow fever, but once it was really started, the novel took on a life of its own. [2]

By the first of 1901, he felt free enough from his other obligations to devote nearly all his time to Chad and Melissa and the Major and all the rest of the characters of *The Little Shepherd of Kingdom Come*. His first effort for Scribner's had not been a good one, probably his worst, and he was appreciative of that fact. Scribner's liked him both personally and professionally, and he counted Charles Scribner as one of his best friends. Therefore, he could not help but feel some regret over the failure of *Crittenden* to do better than it had. Then, too, Tom Page had used his considerable influence both with John and Scribner's to affect the marriage, and he felt keenly that the debt he owned Tom Page for effecting his association with Scribner's was yet to be paid in a good and best-selling novel for them. [3]

After his switch to Scribner's he continued to work on shorter tales as well as the new novel, and in fairly quick succession he produced "Down the Kentucky on a Raft," his first for Scribner's, and they published it in their magazine for June 1900. There followed "To the Breaks of Sandy," published in September 1900, and "The Southern Mountaineer," published in April and May of 1901. To make up enough to justify bringing out a book, Scribner's retrieved several other tales previously published elsewhere. "Fox Hunting in Kentucky" and "After Br'er Rabbit in the Blue-grass" had been published by *Century* magazine in August and November of 1896, "Br'er Coon in Old Kentucky" was published in 1898 by *Century* magazine, and "Man-hunting in the Pound" and "The Hanging of Talton Hall" were published by *Outing* magazine in July of 1900 and October of 1901.

It was still not enough, and John produced "The Kentucky Mountaineer," "Civilizing the Cumberland," and "The Red Fox of the Mountains." Gathered all together under the title of *Bluegrass and Rhododendron*, Scribner's brought out the book in October of 1901. It was moderately successful even though it was something of a mirror-image of his first works before the Spanish-American War, and, indeed, three of the tales had been written and published before John had gone off to war in 1898. The collection was fairly well received, though it certainly did not make any huge waves in the publishing industry. [4]

Finley Peter Dunne (left), Richard Harding Davis and John. Probably taken at Davis's home in Mt. Kisco, New York (the Alfred A. Knopf Company).

14. The Crook of the Shepherd

He was busy, and staying occupied with his writing kept him from thinking too much about his personal and money problems, but he could not put those old thoughts of financial inadequacy behind him and out of his thoughts entirely. He was still having recurrences of malaria off and on which gradually weakened him and put him in the hospital from time to time, but through that summer of 1901 he was able to work steadily on the best thing he had ever written so far. In September, he was able to write Scribner's that he had just gotten out of the hospital again. "I've had a tough time for nearly two months," he said, "but I've got over 40,000 words of the 'Little Shepherd' in shape for the editorial eye." [5]

He was exaggerating just a little. In fact, he had only finished the first 56 pages of the novel, but the rest had definitely been roughed out and needed only revising and perhaps dressing up here and there to be finished. He expected it to be a little longer than the 40,000 words, but likely not more than 50,000. Near the last of November 1901, he was able to write Scribner's that he had just "finished up 'The Shepherd' so that it could be published just as it stands now," [6] but he felt it would need going over again before it would be satisfactory to him. Strangely enough, he said his reason for getting it in shape so quickly was in case he should suddenly shuffle off this "mortal coil" and leave Scribner's without a publishable novel. His recurrent sicknesses were beginning to weigh heavily upon his nerves and thoughts, and it seemed to him that he recovered from one bout of illness only to be prostrated by another. He had begun to think — at times — that he might not recover at all. His spirits rose, however, when Scribner's advanced him $1,000 for *The Little Shepherd*, which was half what he had received in royalties from *Crittenden*. [7]

All that summer he was sick, sometimes too sick to work on the novel, but on the days when he could he plugged along, writing and rewriting, adding and discarding. And on top of everything else, his eyes—for some time now, troublesome—began to plague him again, but lack of money kept him from doing anything about them for a while, except to rest them when they grew too tired to focus upon what he was doing. There were numerous times in his life when he would have to suffer needlessly with one ailment or another without the benefit of a doctor, simply because he was too short of money to place himself under a doctor's care.

Finally, the January 1903 edition of *Scribner's* hit the stands, and everyone at Scribner's held their collective breaths. The first installment of *The Little Shepherd of Kingdom Come* was included, and the plan was for future installments to appear monthly until August of that year, when the novel would be concluded. John was hardly breathing at all. He and they may have suspected, but they did not — could not know — that he was about to become the proverbial overnight sensation. The manuscript was not finished when the first installment appeared, but both he and Scribner's were anxious to begin serialization, and so it was rushed into the magazine even while he continued to

John (right) counted among his friends many of the day's literary personalities. Here he is shown with Finley Peter Dunne (left), author of the "Mr. Dooley" stories (courtesy John Fox, Jr., Museum).

revise, rewrite, and even compose the rest of the novel. He knew it was good from the way the writing went, better and easier than anything he had ever written. There was no way it could be bad, he thought, no way it could not succeed where *Crittenden* had not done as well as expected, but he was reserved in his feelings about the future of the new novel. After so many other great expectations in the past that failed to materialize, he could not keep himself from being cautious this time.

Almost immediately after the first installment appeared, he was sick again, probably with the continuing touch of malaria. It could not have come at a worse time, considering the amount of work yet to be done on the manuscript. Cuba would kill him yet, he thought. He had gotten progressively worse over the last several months, losing weight, and for someone who was never of any great size, the loss of a few pounds showed quickly. Finally, he took himself to a doctor in Baltimore, Maryland, and was beginning to see some progress in his condition. He feared a relapse, however, and writing to his mother on February 1, 1903, he said he would likely stay with the Pages, and under the care of the doctor in Baltimore, as long as his doctor thought it necessary. "I am gaining in flesh," he wrote, "and am better than at any time since the summer before the Spanish War; but I'm going to do what the doctor thinks best, as I have already gone to so much trouble and expense." [8]

14. The Crook of the Shepherd

By March, he was off to Georgia, staying at a resort on Jekyll Island with Tom Page and his wife, authors Thomas Bailey Aldrich, Weir Mitchell and J. Pierpont Morgan, among others. Somewhat tongue in cheek, he reported to his mother that "nobody here is in comfortable circumstances — everybody is disgracefully rich." [9] He was not put off by the company he was keeping, however. He was never uncomfortable around affluent people and was just as at home there as he was in a dirt-floored cabin in the Cumberland Mountains. The time away from the constant pressure of trying to meet Scribner's deadlines was quickly working wonders for his health.

In the meantime, the serialization of *The Little Shepherd* in *Scribner's* was drawing favorable comments from all over. Apparently, there was no one who did not like the story of Chad, Jack, Melissa, Major Buford, the Turners and all the others. It was a story that literally captured the imagination of the people who read it, and even today, decades after its first publication, it is the one book of Fox's that is most likely to be found on library shelves and in book stores.

So pleased was Scribner's with the initial success of the story, they had advance copies printed from the magazine plates and bound as a book — once they had received the final installments from John and set them to print — and sent to selected readers. By the first week in September of 1903, *The Little Shepherd of Kingdom Come* hit the book stores all over the country. Scribner's confidence in the selling power of the book was not unfounded. They had ordered an initial printing of 20,000 copies, but quickly changed that to 30,000. It immediately became a best-seller when that phrase was new and little understood. Not only did it sell well, but just possibly became the first novel in the United States to deserve the title of best-seller. By the middle of September, advance orders for the book had already overtaken the initial printing of 30,000 and had gone to 35,000 and had as yet shown no sign of slowing down.

In something of a backward way, he had written the dedication of the book in 1898 when he was pursuing Currie Duke and only beginning to think of writing the novel that eventually became *The Little Shepherd of Kingdom Come*. Now that the novel was finished, he was beginning to have serious doubts about that dedication.

His courtship of Currie had not worked out, and she had married someone else — Wilbur Mathews — but he still had enough real affection for her that he still wanted the dedication of the book to remain as it was written five years before. He was worried, however, and wondered how people who saw his dedication of a novel to a married woman would react. He had been questioned by one or two people who knew him about the dedication, and not being able to make such a decision by himself, he resolved to write Currie and ask her opinion and her permission. In his letters to her, he suggested that it was none of anyone's business excepting three people: Currie, Wilbur and himself, and if Wilbur was as "broadminded" as he thought he was, then there should not be any problem. Currie did not think so either and was flattered with the dedication. The dedication would stay the way he wanted.

He had inscribed the book to Currie Duke, now Mathews, which reveals that he still thought kindly of her, and apparently she of him, although there was certainly nothing improper to be found there. They were friends. They had once been close, perhaps only flirtatious with each other, perhaps more serious than that, but she was married, and now they were friends. "I cannot begin to tell you," she wrote him in October, "how appreciative I am of your great kindness and of my extreme pleasure this delightful book with its inscription; I thank you with all my heart and I feel very proud indeed." She chided him for not visiting them that fall, and signed the letter "your friend, Currie."[10]

By December 1, Scribner's had printed 90,000 copies of *The Little Shepherd*, and had sold 50,000. John was ecstatic. He could finally see "the good times" coming, and this time, he was sure there could be no mistake about it. The old debts he had carried with him since the collapse of the boom in Big Stone Gap had preyed on his mind for years, and he had thought he would never be able to satisfy himself nor his creditors. Now it looked as though things would be straightened out after all. "I tell you," he wrote Charles Scribner in December, "it's pretty cheerful—this seeing daylight ahead. I'd like to start clean with the world on Jan. 1—but I'm afraid that's too much to hope for." [11]

The book was so much different, so much better, than anything Fox had written up to that time that those who read and reviewed it were not a little amazed in his growth as a writer. They always knew he could tell a story better than anyone else when he really wanted to, but this was different. He abandoned much of the sentimentality and saccharin sweetness that had plagued his stories and novels thus far, and brought into this one a more realistic set of characters, with speech and mannerisms to match. He could not abandon all of the sentiment that he had worn out over the years entirely, and it does crop up from time to time in the novel, but not enough to be distasteful, even to modern readers.

It was the best thing he had ever written, and people all over the country read and read again *The Little Shepherd of Kingdom Come*. Children who read the book wrote him letters that he treasured above all others, begging him to write more about Chad and Margaret and Jack. "Please," they wrote, "make it very exciting and don't make the men cry too much," and "We were very sorry Melissa died but she was in Chad's way before." They were excited about it enough to write him and tell him about it, something that one assumes happens rarely today, and they cautioned him not to serialize the sequel. "Don't put it as a serial in a magazine," they wrote, "because it is so tiresome to have to stop in an exciting part." [12] Another wrote that reading the book "makes me wish I had been born in Kentucky instead of New Hampshire." [13]

They could not seem to get enough of the novel and its characters, and even today, those who read it years ago still recall the still, white face of Melissa on her deathbed, and the loyal and trusting dog Jack, sent back home again and again, once from Lexington with a placard around his neck written by a young mountain boy named Chad that read: "I own this dog. His name is Jack. He is on his way to Kingdom Come. Please feed him. Uncle Joel Turner will shoot any man who steels him." [14] And Jack went home, with his own dog's heart broken and bewildered, and fiction or not, one can see him still, waiting outside an old mountain cabin with his muzzle resting on feet that have traveled so far and looking down the dusty road from time to time with old and weary eyes, waiting for a horse and rider coming from the west to take him home.

If the public liked the book, the critics could do no less, and although they occasionally threw in comments that wondered whether the book was indeed "great" literature, they could not deny that it was immensely popular. Even the publication of the book late in the year could not keep it from the best-seller lists of 1903 where it rubbed shoulders with the likes of *Gordon Keith* by John's old friend Thomas Nelson Page and *The Mettle of the Pasture* by another old friend, James Lane Allen. By the next year, both Page and Allen had disappeared from the best-seller lists, but *The Little Shepherd of Kingdom Come* remained, and over the years it still continued to sell a few copies now and again and is one of two books for which John Fox, Jr., is best known.

Even as he was at work on *The Little Shepherd*, Fox did not forget the stories that had made him something of a name in publishing circles, and continued, albeit at a slower pace, to write the mountain dialect stories for which he was so famous. Scribner's wanted a Christmas story for their December 1901 issue, and he sent them "Christmas Night with Satan." The year before, he had sent *The Ladies' Home Journal* a story called "Christmas Eve on Lonesome," and in July of 1902, Scribner's published "The Army of the Callahan," arguably one of his best. They added to those three two others that had not been previously published: "The Pardon of Becky Day" and "A Crisis for the Guard," and issued them in book form in 1904 under the title *Christmas Eve on Lonesome and Other Stories*. The stories weathered well enough that Scribner's brought them out again in 1909, and again in 1911 together with the 1897 *Hell-fer-Sartin and Other Stories*, along with four other stories gleaned from *Bluegrass and Rhododendron*.

With his new notoriety, these other volumes were well accepted, even if much of it was older material in new and improved format, but they could not hold a candle to the popularity of *The Little Shepherd*. It far and away sold more copies than anything he had ever written before, including those volumes of stories published at the same time as *The Little Shepherd*, mainly to ride the coattails of that novel's popularity. For the first time in his life, he found himself free of debt and with enough money to send home to his mother in more substantial amounts than he had sent before. That relief was itself more than he had imagined it would be.

By the first of 1904, and before he left for Japan, he was back in the swing of things on the reading circuits, and found larger and more enthusiastic crowds than before. The money from *The Little Shepherd* continued to roll in, and in his newfound wealth he gave serious thought of traveling to Europe, something he had wanted to do for years. Now he could afford to go, and he would go, but not just yet. "I'm thinking of going to Europe in February, he wrote to his mother the middle of January, with Owen Aldis—a brother-in-law of Mrs. Page. It is a great opportunity. He is an old traveler, speaks languages, is a walking-encyclopedia and he takes a valet who will be at my disposal as well as his." [15]

He was excited about the proposed trip, and after all those years of wanting to be someone who could travel when and where he wanted with little regard for money, it almost made him giddy to think about it. He cautioned his mother not to say anything about the trip since it was still in the preparatory stages, but even so, he felt he had to somehow justify the trip to himself. "I suppose I owe it to myself," he wrote, "to go abroad before I'm so old that I can't absorb things and for the broadening effect it will have on my work." [16]

John would not have long to enjoy his newfound fame, and the trip to Europe would have to wait until later. As in 1898, there were events happening half a world away which had absolutely nothing to do with him personally, but would again sweep him along into a strange land with, (at least to his Anglo-Saxon mind) even stranger peoples than he had witnessed in Cuba.

Well past the middle of *Crittenden*, he had his character, the journalist Grafton, reveal John's own feelings at the time about being a correspondent during wartime, unable to shoot back at those who were shooting at him and vowing, finally, never to find himself in that situation again. "It was his peculiar province to stand up and be shot at without the satisfaction of shooting back.... And it struck him, too, that this was a ghastly business [being a correspondent], and ... unjustifiable, and ... he would never go to another war except as a soldier."[17]

John had written his family back home in the Gap while he was in Cuba, saying the same thing, but by the time 1904 rolled around, troubles in the Far East between Japan and Russia had gotten serious enough to cause a shooting war, and John forgot his earlier vows not to go again. This time, not even Tom Page could dissuade him from going, nor persuade Scribner's from sending him.

CHAPTER 15

Cherry Blossom Correspondents

> We have been met here with a bitter disappointment.
>
> Richard Harding Davis

> Truly the life of the war correspondent is hard in Japan.
>
> John Fox, Jr.

BY THE TIME THE RUSSO-JAPANESE WAR broke out, the era of hyphenated wars was fast drawing to a close. In the future, there would not be enough hyphens to go around, and more concise names would have to be chosen for those future struggles between nations. The causes, however, would still remain abstract no matter what the name of any particular struggle, and this war between Russian and Japan would have as its instigation something as mundane as water that does not freeze in the winter.

The actual fighting had its start some years before when Great Britain invaded China and took over a small village on the coast, and immediately gave it a more colonially-appropriate English name of Port Arthur. In the meantime, Russia, searching for a warm-water port so that it could ship and receive merchandise all year round, looked toward the Black Sea, but saw that it was blocked by Great Britain and other European countries. Not to be discouraged, Russia started building the great Trans-Siberian Railway toward Port Arthur and Korea in the eastern part of Russia towards the Pacific Ocean and the warmer water on that eastern end of their country.

Japan thus began to sit up and take notice. It was and is a small island, especially considering the number of people that it had to support. The nearby

Korean peninsula looked promising for annexation, but not if Russia got there first. Also, everyone thought that China would soon break completely down politically, and anyone with a presence on the Asian continent might have a claim to the spoils. It was worth a chance.

So, in the summer of 1894, Japan sat out to push the Chinese out of Korea, which it did with little trouble, and then, seeing how easy this game of war could be against the right enemy, Japan invaded Manchuria, and in the process took Port Arthur.

Russia was more that somewhat excited by all this as one might expect, since the Japanese occupation of both Korea and Manchuria would make the building of the Trans-Siberian Railway and Russia's goal of a warm-water port quite academic. Russia was not the only one upset by the Japanese successes, and with pressure from Great Britain and others, Japan was forced to relinquish its hold on Manchuria and Port Arthur. The citizens of Port Arthur — those who were left alive, that is — were not sorry to see the Japanese go. Their initial invasion of the city in the winter of 1894 had resulted in one of the great massacres of the nineteenth century.

Later events were like salt in an open wound to the Japanese. After Japan was forced to evacuate Manchuria, Russia promptly leased Port Arthur from the British and extended the Trans-Siberian Railway into that port city. With the Japanese safely out of Manchuria, it was not a big stretch for the Russians to occupy all of Manchuria and then perhaps cast a leering eye toward Korea. There were the usual protests from the Japanese in particular, and not surprisingly the Russians paid no attention. The Boxer Rebellion, which the Japanese always suspected the Russians of starting, only served, reasoned the already suspicious Japanese, as an excuse for the Russians to send more troops for the always popular reason of "keeping the peace." Again, not surprisingly, no one thought to ask the Koreans, the Manchurians, or the Chinese what they thought about all this. Perhaps they knew what the answer would be. John would become troubled by the same thought soon after seeing Manchuria for himself. "I don't wonder," he wrote, "that the Russians are fighting for that land, nor shall I wonder should the Japanese, if they win, try to keep it. But how it should belong to anybody but the Chinaman who has tilled it in peace and with no harm to anybody for thousands of years—I can't for the life of me see." [1]

Eventually, Russian forces crossed the Yalu River and into Korea and began to build fortifications and everything else that goes along with occupying a foreign land. With that, the die was cast, and Japan opened the festivities on February 8, 1904. In a scene somewhat reminiscent of another attack some 40 years later, the Japanese slipped their torpedo boats into Port Arthur and shelled ship and shore before slipping back the way they came.

Not long after the shelling of Port Arthur, the Japanese landed forces in Korea in preparation for an invasion of Manchuria, and the Russo-Japanese

15. Cherry Blossom Correspondents

War was on in earnest. It was about this time that Scribner's thought it a good idea to send their most popular author and sometime war correspondent over there to see what was going on. They asked John to go, and despite his previous vows never to go to any war as a correspondent again, he accepted without much of a second thought. He was feeling good with his most recent literary successes, and a chance to travel to Japan might only come once in a lifetime. He could not refuse. Besides, Richard Harding Davis was going too, and that Davis was taking along Cecil, his wife, bothered John not at all. It would be an adventure.

Most of John's friends did not want him to go, especially Tom Page and his wife Florence. When they heard the news of his going, Florence Page wrote John, "I have a big lump in my throat as I write this, and there is a special prayer for you in my heart night and morning. I would love to run up with Tom to see you and wish you Godspeed, but I don't think I should be happy to see you sail off to the uncertainties of war. You have been our dear faithful friend for many years, dear John," she said, "and I devoutly pray God to keep you safe and bring you back to us free from harm." [2]

John was very much aware of the effect of his going abroad would have on his friends and family, and in the short time he had to decide, it seemed to him that everyone but himself and Scribner's was against it, but there never was any real question in his own mind whether or not he would go. He had written his mother en route to San Francisco that, except for a slight cold, he was feeling better than he had in his life, which, taking into account the number of times he had been sick, probably did little to reassure his mother. "Please don't worry," he wrote, "as I shall see the world, improve in health, make money and not be gone long." [3]

Despite his assurances to his mother, he knew he was going to a strange land where there was sure to be fighting, and he was worldly enough to know that merely being a correspondent for an American magazine was no guarantee that someone from either side might take a potshot at him. He determined, upon the advice of his brother Rector, to make a will before he left.

Knowing John's careless attitude toward almost everything of importance, it is not surprising that the will he left was one of extreme simplicity. It is also not surprising that he never made another, and that after his death that simple document that seemed so plainly written to him would cause trouble for his family. He left everything he had to his mother and sisters, and provided for Horace's newly adopted daughter Mildred of whom, he was very fond. He had it witnessed, named Rector and Horace as administrators, and that was that.

On February 23, he boarded the *China* at San Francisco with a letter of introduction from Theodore Roosevelt and a passport signed by Secretary of State John Hay, and turned his face to the west. At that point, he was excited. In only a matter of a few weeks—days even—he would look back and wonder at that misplaced excitement.

Despite John's assertion to his mother that he had taken a camera to Japan and would send photographs back home, very few photographs of this adventure are to be found. This one, apparently taken on board the S.S. *China*, shows John in a more playful mood than he would be upon leaving Japan in a few months' time (courtesy Duncan Tavern).

The voyage from San Francisco was uneventful. He had been seasick, and the glare from the ocean's surface had affected his eyes which continued to bother him on an irregular basis, but all in all, he felt in tiptop shape by the time he glimpsed Hawaii. On board with him was Richard Harding Davis and his wife Cecil, also headed for Japan, and the time went faster, and the days seemed shorter as these two old friends caught up on gossip and their writing. Neither Davis nor Fox could get over the holiday atmosphere on board the ship. When they had left San Francisco, it had been snowing, but at sea, heading towards the Pacific islands, the weather changed dramatically. "Somehow we cannot take this trip seriously," wrote Davis to his mother. "It is such a holiday trip all though not grim and human like the Boer war. Just quaint and queer. A trip of cherry blossoms and Geisha girls."[4]

Davis had been taken with Fox almost from their first meeting years ago, and their similar situations in Cuba brought them closer together, so that when they both left San Francisco for the Orient to cover this war, they were already close friends, and would remain so ever afterward. Davis would write his mother in July, after many nonadventures in Japan, about his companion and friend. "He is one of the best men when you get to know him," he said, "as we have, of my acquaintance. Out here they always speak of him as the Little Shepherd because he spends his time rescuing drunkards, beach combers and remittance men, and his money too." [5]

When they arrived in Hawaii on board the *China*, John noticed early on the changes that had occurred on that island in the last few years—since 1898—under the United States' "protection," and he was not impressed with what he saw. The island and the people were quickly transforming themselves into a tourist attraction complete with guitar-playing hula-girls who had, he thought, just about as much pure Hawaiian blood in their veins as he had himself. He thought it unseemly and vulgar.

He reserved his biggest shock of culture, at least up to that time, for the obvious mixture of the races on the island that truly offended his southern gentleman nineteenth century sensibilities. "Every possible human mixture of blood I had seen that day, I fancied," he wrote, "but of the morals that caused the mixture I will not speak." [6] It was his first real look at the world outside the confines of nineteenth century America, and he did not approve of what he saw there.

In Honolulu, he took time when they stopped over to refuel to look up a daughter of General Basil Duke, and sister to Currie. She had the unusual name of Thomas Morgan Duke, but everyone called her Tommie, and she was married to Sidney Ballou. The visit was friendly, and they were glad to see someone from home, perhaps more than he was to see them. He did not know it at the time, but almost a year to the day, Tommie would die giving birth to her only child, a daughter. But for now, for a short time, they talked of "The General," of things in America, of Kentucky, and a little of Currie, and when he took his leave of them, they were more than a little homesick and lonesome at his leaving. [7] Eleven days later, they sailed into the harbor at Yokohama, Japan.

With his first exposure to Japan, John forgot, for a moment, his mission, and became the enthralled, goggle-eyed tourist, and he was not the only one. Davis and his wife were equally excited about Japan. Davis had written his mother on their way there that he was "almost hoping the Government won't let us go to the front and that for a week at least Cecil and I can sit in tea houses with our shoes off while the nesans bring us tea and the geishas rub their knees and make bows to us." [8] He would come to regret those lines very much. In little more than a week, his attitude would change, and like most of the other correspondents already in Japan, he would become bewildered at his situation which, he wrote his mother, "continues to remain in such doubt that I cannot tell of it, as it changes hourly.... I may be kept waiting here for weeks and weeks."[9]

On his way to Tokyo and the war he had come to cover, John could not rest nor sleep. Everything was new and caught his eye. Nothing, it seemed, escaped his notice. It was an exotic land full of strange and pleasing sights, sounds and people. Everything seemed so small, even tiny and fragile. He saw "little patches of half-drowned rice bulbs, cottages thatched with rice straw, with green things growing on the roof, and little gardens laid out with an art minute and exquisite, [and] blossoming trees of wild cherry." [10]

Once in Tokyo, he found no more evidence of the war than he had seen when he first set foot in Japan. Davis would write his ever patient mother that "were it not for our own squabbles we would not know not only that the country was at war but not even that war existed *anywhere* in the world. We are here entirely en tourist and it cannot be helped." [11] Except for the bulletin boards all over town that carried news filtered by the Japanese censors, one could convince oneself that there was indeed no war here to be written of, or even seen.

John would see some signs of the war through the noble sacrifices of the Japanese citizenry. "The women," he observed, "let their hair go undressed once a month that they may contribute each month the price of the dressing — five sen. A gentleman discovered that every servant in his household, from butler down, was contributing a certain amount of his wages each month, and in consequence offered to raise wages just the amount each servant was giving away." [12] The servants could not allow him to do so, since to allow that would take the honor away from themselves. All over Tokyo, he saw similar sacrifices, both small and large. He could not help but be impressed by the Japanese devotion and patriotism, especially those mothers who felt so proud when their sons left hearth and home for the far away front. If they were killed, the honor was greater, if anything. "And when he is brought home dead," he learned, "his body is received at the station by his kin with proud faces and no tears. The Roman mother has come back to earth again, and it is the Japanese mother who makes Japan the high priestess of patriotism among the nations of the world." [13]

Richard Harding Davis was among the first of the correspondents to suspect what the Japanese officials — under whose guidance the correspondents found themselves — were up to. The delays that were forced upon them — and kept them from going to the front where the war was being fought — sharply reminded him of his earlier wish that the Japanese would keep them in Tokyo for a little while so they could play the tourist. Soon enough, his temper began to disintegrate. "My temper is vile today," he wrote his mother, "as I cannot enjoy the gentle pleasures of this town any longer and with this long trip to Port Arthur before I can turn towards home. I am as cross as a sick bear." [14]

John, too, was beginning to realize all might not be as it should be, and joined Davis in his concerns. They discovered early that the journalists would not be free to go when and where they wanted — except as tourists — but would be led by, and under the supervision of, the Japanese government. They would be led, it seemed, like children when their hosts were ready for them to go, and be taken to the exact spot from where they were to view the war. "The first column of correspondents," he wrote his family on March 27, "expects to go on April 5. I am slated for the second column but nobody knows when anybody is to go — or what anybody will be allowed to see after he has gone to the front. I've been here 10 days now and I'm already getting pretty tired." [15]

While they were still safely entombed in Tokyo, Jack London, American author and correspondent who was working for William Randolph Hearst,

15. Cherry Blossom Correspondents

This photograph shows Richard Harding Davis and his wife Cecil along with John (right) in Tokyo (courtesy Duncan Tavern).

came through Tokyo on his way home, mad as the proverbial wet hen. He had been in Japan long enough to be disgusted with the whole thing. It was an omen of things to come, though it went by little noticed by the rest of the correspondents. "He is very bitter against the wonderful little people," wrote Davis, "and says he carries away with him only a feeling of irritation. But I told him that probably would soon wear off and he would remember only the pleasant things. I did envy him so, going home after having seen a fight and I not yet started." [16]

Jack London had seen his fight, and had been in one or two himself. London did not like the Japanese, and could not restrain himself from showing it. The Russians, on the other hand, were people more like himself — racially at least — and he felt more compassion toward them. He had gotten into hot discussions with the Japanese who had confiscated his camera, and on another occasion he had knocked down another Japanese for apparently stealing. The details are sparse, but the upshot of it was that London had been detained by the Japanese, and Davis had to intercede with President Roosevelt on London's behalf, and through diplomatic channels London was released and gladly made his way home. Despite Davis's assertion to London that time would heal his bad thoughts of Japan, London did not have that much time left to forget his treatment at the hands of the Japanese. [17]

John and Davis were staying outside of Tokyo now, though not so far as to hear quickly should anything important happen. They were set up at the Fujina Hotel in the mountains of Miyanoshita, and John, at least, was beginning to resign himself to his situation and was taking their circumstances better than was Davis, who could not abide the inactivity, but staying with John he could not stay cantankerous for very long. It was, after all, a beautiful country.

John spent hours with the daughter of his host, loving every minute of it. She spoke English, though slowly and deliberately with an accent that endeared her to him. The days flew by, and before he knew it, he awakened to realize that he had been in Japan for a month. "I'm just as far from the front as ever," he wrote to his mother on April 15, "and it looks as though I'd never see the front. The first batch of correspondents have gone but to judge from the rules that are to govern their actions I think they would as well have stayed in Tokio [sic]." [18]

They visited nearly every shop in Tokyo, buying for themselves and friends and family back home. The prices were so cheap in comparison to the same items back home that they could hardly restrain themselves from buying and spending their money lavishly. In one shop, John bought three suits of silk pajamas, six suits of silk underwear, six silk shirts, six cotton shirts, six crepe shirts, three more silk shirts, and a dozen "turn over" collars, along with a little embroidery.

The next day, at another shop, he bought a tweed suit, and one of white serge. He bought a pair of white pants, another silk suit, and two more silk shirts. For his mother and sisters, he bought in Yokohama six kimonos: three blue, one pink, one gray. He bought one gray silk bed cover, and two silk table cloths, and so on. The prices were too cheap, and it was fun for a while to spend so little and receive so much, but even that soon lost its attraction.

He had been in Japan for a month and had seen just about everything there was to see, excepting what he had originally come to Japan for in the first place. He saw the opening of the legislative assembly, or Diet, had seen and heard the emperor of Japan — that mixture of religion and politics— and had enjoyed the company of both Japanese and American diplomats. He had eaten of the Japanese cuisine which only served to make him miss American cooking all the more. After his bouts of seasickness, he was more than a little wary of "a soup in which floats bits of strange fishes from the vastly deep, unknown green things and an island of yellow custard; to slices of many colored raw fish, tough cocks' combs (real ones) or even to the stewed chicken which at this dinner, at least, had been shorn of everything except bones and tough sinews. The other day I tried it again with no better success." [19]

He had seen geishas and dancing girls and had enjoyed all the nightlife that Tokyo and Yokohama had to offer, in the company of Davis and Cecil. They went to the races and everyone except Davis lost most of their money.

He saw sumo wrestlers and holy shrines and everything else, but eventually the job Scribner's had sent him to Japan to do would have to be done. He had written a couple of articles that did not concern the war very much, and had sent them on, but he could not, in good conscience, continue to enjoy himself at Scribner's expense and have nothing to show for it. He had placed his confidence in the good graces of the Japanese, trusting that all would be well and that he and the others would go to the front sooner or later, and kept believing that everything would eventually work out like he thought it should. He trusted the Japanese, but he was beginning to suspect that it might have been a mistake for him to do so.

So, after the first two dispatches for Scribner's—"The Trail of the Saxon," which really was not a trail at all, and "Hardships of the Campaign," which was not that hard either—he vowed to write no more until he had gotten to the front. He had grown tired of the constant cherry blossoms and polite people. He liked them well enough—admired them really—but the inactivity and the life of a tourist was beginning to drive him insane. After sending that last dispatch to Scribner's he said, "Here goes another, but it shall be my last, and I shall write no more until the needle of my compass points to Manchuria. A month ago the first column got away when the land was lit with the glory of cherry-blossoms. We have been leaving every week since—next week we leave again. One man among us now calls himself a cherry-blossom correspondent. He was lucky to say it first. Clear across the Pacific we can hear the chuckle at home over our plight even from the dear ones who sent us to Japan. If it were not such a tragedy it would be very funny indeed." [20]

Charles Scribner wanted any dispatches he could get from John and Davis, and had received letters describing the difficulties they were encountering in trying to find out the least bit of news from the war. Scribner, however, was less bothered by the whole fiasco than were the correspondents, and wrote to John that he should not worry too much about what he was unable to send back to New York in the way of articles for the magazine. Scribner understood, perhaps more than John and Davis realized, the difficulties they were experiencing with the Japanese, and with a not so subtle jab at the usually unflappable Davis, Scribner suggested that Davis could not expect to know very much about what was going on since he was so far from New York— meaning that the newspaper and magazine writers at home had more information about the war than did those correspondents who were on the scene.

Despite the trouble he and Davis were having in getting to where the war could be seen and reported, John continued to be very much impressed with the Japanese people in general. They had the courtly manner of the Old South. They bowed and were polite, and the most violent thing he had seen thus far was the national sport of sumo wrestling, and even that was rife with polite gestures and ceremony, and if one of the wrestlers were hurt, it was a rare event indeed. "I haven't seen an angry look nor heard an angry tone since I've

John seems to be waiting for something to happen, and enjoying the sights in a rickshaw provided by the hotel in Tokyo (courtesy John Fox, Jr., Museum).

been in Japan," he wrote, but added significantly, "except among the foreigners." [21]

He saw things changing, however, and was reminded somewhat of the things he had seen in Hawaii. There was a vast difference between the new Japan and the old, and he saw that especially in those harbingers of change — the college student. "The most interesting and significant," he observed, "is the Tokio [sic] University student you see [in] the public gardens. He has an intelligent face, looks you straight in the eye, is agile as a panther, and as tall, I believe, as the average college student," and somewhat ominously, at least from a mid-twentieth century perspective, he concluded, "But these students — one can't help wondering what, when they grow up, they will do for Japan and to the rest of the East." [22]

Davis had already written to his mother about John's penchant for getting himself into philanthropic scrapes. They had begun calling him "The Little Shepherd" because of his constant naïveté toward others who needed, or seemed to need, his help. Oftentimes, they did not need his help, and were sometimes puzzled, especially the Japanese, at his insistence that they take it. Children especially unloosed his purse strings.

Occasionally he and the other correspondents in Tokyo would hear of battles being fought, or, more often, they were reduced to getting information from the bulletin boards on the streets, and every week they were told that "next week" they would leave for the front. "I don't believe any correspondent will even hear a bullet or shell in this campaign," John wrote to his worried mother. "We are treated like children, nuisances and possible spies and I'd drop the whole business and go home — if I could, honorably. But I can't and must do the best I can. I still have no more idea when I shall leave for the front," he said somewhat forlornly, "than I had when I got here. I am exasperated and disgusted but there is no help for it." [23]

He had been in Japan for four months now, and some had been there longer, when finally on July 18 they were put aboard the *Empress of China* at Yokohama and were on their way to Manchuria, or so they thought. Nothing could raise their hopes very high after the last four months, but it was, however, with some difficulty that they kept their spirits in check. Two days earlier, John had written a cautiously optimistic letter to his mother saying, "I'm leaving for Port Arthur (I think) ... to-morrow. I don't think we will be allowed to see any fighting except at a distance of 2 or 3 miles." [24] Ten days later, they were not much closer than they were before.

"An explanation has occurred to me," he mused near the end of his third dispatch to Scribner's. "You know the Japanese does nearly everything but his fighting — backward. Of course he reads and writes backward. At the theatre you find the dressing-room in the lobby. Keys turn from left to right, boring-tools and screws, I understand, turn from right to left, and a Japanese carpenter draws his plane toward him instead of pushing it away. Sometimes even the Japanese thinks and talks backward." Following his train of thought, he concluded, "Perhaps then our trouble is that the Japanese tells the truth backward and we can't understand." [25]

That they were at last moving helped his feelings some, but he could not entirely get over the Japanese penchant for mixing up their language with nuances that foreigners like himself did not readily understand. He assumed, as did the others, that if they were told they would be going to the front in two days, then it should not be ten or more days before they even sailed in that direction. He was especially put off by the Japanese use of the word "tadaima," which language dictionaries faithfully defined as "soon."

When asked when they would be allowed to do this or that, or to go here or there, the Japanese would invariably answer, "tadaima." He soon learned that "tadaima" as a word had a lot in common with the Cumberland mountaineers' use of the word "several." "The unwary stranger," he wrote in exasperation, "will be told to-day that it [tadaima] does mean 'soon' and as such in dictionaries he shall find it. But I have tracked 'tadaima' to its lair and dragged it, naked and ashamed, into the white light of truth. And I know 'tadaima' at any time refers only to the season next to come. Early in March, for instance, it means literally—'next summer about two o'clock.'" [26]

The strain was beginning to show on all of them. On the one hand, they had come this far, and to go home in defeat was more than most of them could endure, and so they stayed, hoping that the Japanese would relent, that things would all of a sudden become more "normal," but at the same time, they suspected that the Japanese language held no such word as "normal," or if it did, it would mean something entirely different than westerners could expect.

"In May came the battle of Nansham," John wrote, "and the advance on Port Arthur [and] in June followed Tehlitzu. Both battles any man would have gladly risked his life to see," but they were not allowed to see either. Despite all that had gone before, they were willing to forgive the Japanese and forget all the delays, but only if the Japanese allowed them to see the war firsthand. Now they were aboard *The Empress of China*, and hopes were high, despite their best efforts to hold them in check, and they began to believe that their travails were finally behind them. "Whither we were bound we knew not for sure. But there were three men among us who had been guaranteed, they said, by the word of a Major-General's mouth, that they should see the fall of Port Arthur." [27]

They were moving, and at least, that was something. They stopped overnight Kōbe, a town that was, he said, unremarkable, except that it "might be any town anywhere." [28] Leaving Kōbe, they moved over a sea that seemed to him like molten silver with only small volcanic islands to break up the monotony. They steamed into the Inland Sea, not yet out of sight of land, and into the Shimonoseki Straits that lead to the Sea of Japan and open waters. Past Nagasaki they went, putting in there to await a transport that would, they were assured by their Japanese hosts, take them where they all wanted to go.

When everyone left ship the next morning for the hotel, John stayed behind in the port city, wanting to see all that he could of the surrounding countryside. "I clung to Nagasaki as long as I could," he said, and as a result of his tardiness, he had to make his own way back to the foreign hotel where the rest of the correspondents were making the best of it that they could in that crowded place. Davis had already been there ahead of him and did not like anything about what he saw of the place. Davis could not abide discomfort and would not tolerate it for long, especially when it could be avoided, and so he had gone to another hotel — one more suitable to his tastes — and had left word for John to follow on as soon as he could. "So in a rickety rickshaw I rattled after him through the empty streets," John remembered. "I found [Davis] in a Japanese room as big as the dining-room of an American hotel, covered with eighty mats, full of magic wood-work, and looking out where there were no walls (the walls in a Japanese house are taken out by day) for full fifty feet on mountain and sea and passing transports and sampans."[29]

They kept their expansive accommodations for two nights and a day, but it was expensive by Japanese standards at least. "The gold of the one sunset and the silver of the one dawn were included in the turkey-tracked, serpent-long bill that was unrolled before our wondering eyes," [30] John wrote with just a little sarcasm, but they paid and resigned themselves to the notion that, at least, they had been better housed and accommodated than the other correspondents who had bunked in the foreign hotel in Nagasaki in spite of the prices charged them by their Japanese hosts. John could not help but to be amused by it all. "I cheerfully recommend the method to highway robbers that captain other palaces of extortion," he wrote, "in other parts of the world." [31]

They went back to the harbor at Moji after a minimum, by Japanese standards at least, of waiting around answering useless questions put to them by self-important Japanese officials. They were finally hustled aboard the transport *Heijo Maru* and set out once again, turning north into the Sea of Japan. "We are leaving this base of supplies at 2 o'clock to-day," wrote John just before they left. "Whether we are going to Port Arthur or to General Oku's army inland, we don't know." [32] He might have added that the Japanese were no more helpful than they had been before. They were determined, he said of himself and Davis, to see at least one good fight and then head for home as fast as their feet could carry them. "This has been a wretched, disastrous failure," he complained, "this expedition — and we are both thoroughly disgusted." [33]

Another stopover at the Elliott Islands entailed a delay of three days, watching other transports come and go, and again being told they would be leaving any time now. Time dragged, but soon they were again on their way toward Port Arthur and Manchuria, and as they drew closer, though still miles away, they heard their first "sullen thunder of a big gun." [34] They were still a long way from seeing the battle, but the "thunder" raised their spirits. They put off at Talienwan and straight into the hurly-burly of a city close to the front lines. "We landed among carts," John wrote, "Chinese coolies, Japanese soldiers, Chinese wagons, mules, donkeys, horses, ponies, squealing stallions, ammunition, a medley of human cries. The bustle was terrific. A man must look out for himself in that apparent confusion." [35] This was more like it, and their hopes continued to rise.

Tempers are often shorter in places of noise and confusion, and it was here that the usually unflappable Davis misplaced his. He had loaded a cart with his belongings, only to have a Japanese officer throw them off into the street — camera, clothes, notebooks, everything — in a jumble amid "blistering curses" all around. Reloaded on another cart, they were promptly offloaded again in a repeat performance. It was no use, and Davis never really regained his temper toward the Japanese after this incident, nor regained, what had been at times, hopeful good spirits about the campaign they were sent to cover. Despite everything else, the Japanese had not acted like gentlemen, and that, for Richard Harding Davis, could not and would not be forgiven.

Port Arthur and the war that surrounded it remained elusive, however, and the next day their hopes that had been steadily rising were dashed. They went to bed that night with the sound of the eminent fall of Port Arthur in their ears. "The dream was shattered before we went to sleep," wrote John. "The truth was that we were not to go to Port Arthur at all. Next day we traveled — whither God only knew — with every boom of a big gun at the Russian fortress behind us sounding the knell of a hope in the heart of each and every man." [36]

Davis was angrier than he could ever remember being in his life. Almost at arm's length was the story they had traveled thousands of miles and waited months and weeks to see, and now, through some perverse reasoning of the Japanese, they were not permitted to go any closer. "We have been met here with a bitter disappointment," he wrote his mother on July 31st. Tired of the deception of the Japanese, Davis now remembered London's disgust with the "little people," and felt it like his own. He doubted he would soon forget, nor only remember, the "pleasant" times. "The only mistake I made," he wrote, "was in not going home the first time they deceived us instead of waiting for this and worst of all." [37] This was not the worst, though none of them could know or guess at it then. The worst — the very worst — was yet to come.

On horseback now, which was an adventure in and of itself, they started overland in pursuit of the Japanese general Oku. John had accumulated a horse named Fuji and a Japanese servant named Takeuchi, both of whom shared something of the same characteristics and personality. Both would bear watching, he thought.

Takeuchi's first bill to John for his services thus far overran the correct amount by exactly half. When asked about the discrepancy, the poor man worked for an hour over the figures before finally admitting the figures were wrong. "He had overlooked among other things," wrote John of Japanese mathematics, "one item — the funeral expenses of some relative, which he had charged to me. I made it clear that such an item was hardly legitimate and since then we have had less trouble." [38]

They all felt the disappointment of leaving a perfectly good battle at Port Arthur behind and riding — often on unwilling transportation — off into the wilderness of Manchuria in search of another battle. The roads were as bad as anything he had seen in the Cumberland and Appalachian Mountains back home, and if it were not for the irascibility and downright dangerous personality of the horse he rode, he would have felt more sorry for what the other mounts endured in traveling those roads. "These were the men," he observed, "who thought they were going to Port Arthur and who, with the sound of the big guns at that fortress growing fainter behind them, struck Oku's trail, up through a rolling valley that was bordered by two blue volcanic mountain chains. The sky was cloudless and the sun was hot. The roads were as bad as roads would likely be after 4,000 years of travel and 4,000 years of neglect, but the wonder was that, after the Russian army had tramped them twice and the Japanese army had tramped them once, they were not worse." [39]

The column they were traveling with went on forever, and at any rise of land they saw it stretching before and behind. It contained wagons, carts, men on foot and on horseback, the rattle and clang of metal against metal and even a bicycle weaving itself and its rider through the ruts and mud and dust, sometimes all three at once. That one bicycle and its passenger amidst the clutter and paraphernalia of war made the whole thing seem surreal.

John's horse was giving him trouble, and though he sawed away at the reins, it was only a temporary remedy. The horse seemed to have a mouth insensitive to iron bit, and generally seemed to have only two gaits he was willing to do: full stop and full gallop. "With malediction on tongue," he wrote, "and murder in heart, I sawed his gutta-percha mouth until my fingers were blistered and my very jaws ached, but I could hold him back only a while."[40] Finally, in exasperation, he let him go, holding on for dear life. "Through coil after coil of that war-dragon's length, past the creaking, straining vertebrae, taking a whack with teeth or heels at something now and then and something now and then taking a similar whack at him. The etiquette of the road Fuji either knew not, or cared nothing for." [41]

They stopped over in the first Chinese walled city he had ever seen, rested that evening and night, and left Kinchau early the next morning. He liked what he saw of the Manchurians. They did not seem as posturing as the Japanese — seemed more real, actually — and showed no pretense or affectation. "They seem a good-natured race — these Manchurians," he wrote, "genuine,

submissive, kindly, but genuine and human in contrast, if I must say it, with the Japanese." [42]

Late in the evening of the third day, they arrived in a Chinese village distinguished primarily by its dirtiness and anonymity, and smell. "We left Pa-lien-tan this morning and made thirty-two miles," he remembered. "We took lunch in a stinking Chinese village, and the chicken—well, it was a question which was the more disturbing conjecture—how long it had lived or how long it had been dead." [43]

Fuji remained cantankerous. Mostly, John hung on and was only along for the ride. His horse was a poor pupil and took no instruction from him. "Fuji has not improved," he wrote. "He kicked an Italian on the leg to-day and I've just helped to bandage it. Again to-day I had to let him go. I tried to tire him out by riding him through mud-holes and see-sawing him across deep wagon-ruts. But it was no use. If a horse, bullock, man, woman, child, cat, or dog is visible 500 yards away, Fuji with a squeal makes for it. When the object is overtaken, Fuji pays no attention to it, but looks for something else toward which he can start his squealing way. For brutal, insensate curiosity give me Fuji, or rather give him to anybody but me." [44]

The next day, in company with Davis and the others, he was shocked to see his first strange white face in weeks. The Japanese had captured a Russian soldier and were moving him to the rear, to what purpose or future John could not venture a guess, but he could not be too optimistic for the poor soldier's future. "The thrill was that the man was the first Russian prisoner we had seen—the shock that among those yellow faces was a captive with a skin like ours. I couldn't help feeling pity and shame," he wrote, "pity for him and a shame for myself that I needn't explain. I wondered how I should have felt had I been in his place and suddenly found four white men staring at me. It's no use. Blood is thicker than water—or anything else—in the end." [45]

They began to see more and more signs of the war, but they saw no more captured Russians. In fact, the one prisoner they saw would prove to be the only Russian they saw throughout their entire stay in Manchuria. They were, however, beginning to see sick Japanese soldiers, marching past them, going to the rear, no longer able to fight. [46].

The heat was almost unbearable. Even Fuji felt its effects and was somewhat more tractable, though not much. It was too hot to run, but he thought up new deviltries to plague his master. "I should say," wrote a saddle-sore and over-heated Fox, "that his record in six hours to-day was about this: stumbling with right forefoot—300 times; stumbling with left hind-foot—200 times; neighs—1,000." [47]

Opposite: Fox spent time in Japan and Manchuria covering the Russo-Japanese War. The frustrations that he and the other war correspondents would experience had not yet set in (courtesy John Fox, Jr., Museum).

Finally, they all arrived in Kaiping safe and sound which, considering the hardships of the march, was remarkable. There was poetic justice, however: Fuji had come up lame, but received little sympathy from anyone. John found himself afoot, and was almost grateful for the change from the "hurricane deck" of his oriental transportation. As luck would have it, now that he had been obliged to walk on the way to Kaiping, it began to rain, "and when it rains in Manchuria, it really seems to rain," he said. "I was on foot in a light flannel shirt, and had no coat or poncho. In ten minutes the road had a slippery coating of mud, I was wet to the skin and, as my boots had very low heels, I was slipping right, left, and backward with every step." [48]

On they went, slopping through the muddy roads that had turned to rivers of yellow water and mud, and through endless cornfields so tall they could not see where they were going, nor where they had been. The mud balled up on their feet and made walking miserable, and they needed their full concentration just to steadily place one foot in front of the other and keep from sliding backwards more than they were moving forward. That they were now soaked to the skin did not help matters any. Finally, the city of Kaiping came into sight, and naturally, their destination was on the other side of a waist-deep river, but at this point, they all reasoned they could not get any wetter.

John waded across, and in the process cleaned his boots completely of the sticky Manchurian mud. On the other side, he managed to clamor aboard a mule offered to him by an interpreter. Naturally, once across the river, they were told by the Japanese that they would have to return to the city proper — on the *other* side of the river, of course — and so, back they went, again braving the river. Again on the right side of the river, they sought shelter from what had now become a downpour in a Chinese temple. "I was cold, muddy, hungry, and tired to the bone," John remembered. He, Davis and the others exchanged their sopping wet clothes for dry Chinese togs and sat around writing their dispatches, letters home, and trying to get warm again. There is little doubt of the spectacle they made to any casual observer, but at this point, they did not care.[49]

They next day, with John once again aboard the now un-lame Fuji, they gratefully left Kaiping behind them. Yohatong came and went, as did "dirty, fly-ridden Tashikao," and then Haicheng, recently captured from the Russians. They were now only five miles from the front, but despite the nearness, the war still seemed to be everywhere except where they happened to be themselves. The Japanese, who could give lessons to the Chinese on inscrutability, inexplicably decided the correspondents must wait yet another week.

They were given "sleeve-badges" to wear that would presumably keep any Japanese soldier from mistaking them for Russians, and perhaps keep them from being shot should they wander out beyond where they were cautioned by their Japanese hosts to stay. At this point, getting shot at would at least give them something to write about. "We are to play a week's engagement here in a drama of still life," wrote John disgustedly. "With a sleeve-badge of identification on — the Red Badge of Shame we call it — we can wander more or less freely within the city walls. We can even climb on them and walk around the town — about two miles — but we cannot go outside without a written application from the entire company, and then only under a guard." [50]

It continued to rain, and although they were inside and dry, the constant downpour coupled with their enforced idleness did nothing to raise their spirits. "We still are inside this old Chinese town," wrote Davis near the middle of August. "It has rained for five days, and this one is the first in which we could go abroad. Unless you swim very well it is not safe to cross one of these streets." [51] They had marched north into Manchuria for 11 days, and no battle nor any General Oku in sight. "We are quartered in a walled Chinese town, in a Chinese house and in spite of rain, mud, flies, black fleas and an occasional scorpion, we are quite comfortable and pretty cheerful," wrote John. "You need not be at all alarmed for my safety as the Japanese, I fear, will not allow any of us to get within a mile of rifle-bullet or shell. This is very comforting to mothers," he said ruefully, "but pretty hard on us." [52]

A decisive battle was expected any time at Liao-Yang, no more than 30 miles away, and they were again promised that they would see it. They could not help but go over in their minds the list of previous promises that had been made and broken by the Japanese — battles missed and dispatches unwritten because of it. They were not hopeful, and only halfheartedly wished the Russians would drop a shell on the city where they were, just one at least, so that they would have something legitimate to send back home to their editors.

Piecemeal, the Japanese would offer them some tidbit to keep them from outright rebellion. After nearly two weeks in search of the Japanese general Oku, it was anticlimactic that the Japanese reported the elusive general to be not more than a mile or so from where they had spent the last week, and wonder of wonders, they were — correspondents all — invited to come and meet him. They shined up everything they had that would take on a glow, brushed off their best clothes and in general made themselves as presentable as they could.

Three of the veteran correspondents, including the always dapper Davis, decorated their blouses with ribbons and medals from far away campaigns that they had covered. "I had a volunteer policeman's badge," wrote John, somewhat tongue-in-cheek and poking a little fun at his companions, "that came from the mountains of old Virginia. I was proud of it, and it meant campaigns, too, but I couldn't pull it [out] amidst the glory of those three." [53]

The meeting, solemn though it was, came to naught. They were introduced to Prince Nashimoto and to General Oku, who again promised they would go to the front and see a battle at last. It did not excite them too much. There was too long a history of broken promises behind this most current of promises for them to turn somersaults over one more.

The rain continued off and on — mostly on, which deepened the already considerable mud and intensified the awful smells coming from the town. "We have flies, mosquitoes, night-bugs, that are homelike in species and scorpions that are not," wrote John. "Every man shakes his shoes in the morning for a hiding scorpion." [54] They were quickly losing all patience, and thoughts of going home were practically the only ones they entertained. "We did not come here to sit in temples," wrote a thoroughly disgusted and homesick Davis, "so John and I will leave in a week, battle or no battle. The argument that having waited so long one might as well wait a little longer does not touch us." [55]

Chapter 16

The Backward Trail

I am exasperated and disgusted.

John Fox, Jr.

THEY WERE READY TO GO, and it did not matter very much to them which way they went. A battle would be fine, but they had all but given up on ever seeing anything even remotely resembling one, and by now home seemed more of a pleasant thought than anything that the Japanese were willing to offer them. The Japanese, sensing their disgust and exasperation, thought to mollify their anger by offering them another bone to chew on.

In a day or two, John, Richard Harding Davis and the others were taken by their guides, or guards as it were, out on what John euphemistically described as a "reconnaissance." There they were shown the positions of the Russian enemy ten miles away, and their guides seemed overly fearful that a Russian bullet would somehow find them from that distance. They saw nothing worth reporting and heard even less, and the obviously affected enthusiasm of their Japanese guides was not contagious.

They were lectured by Japanese officers on battles they were not allowed to see, obviously to keep them from becoming too discouraged. The officer drew maps and diagrams for them very seriously, but they meant little or nothing since none of them had any points of reference where this war was concerned. "A certain division, he [the Japanese officer] said of a certain regiment, at a certain time had done a certain thing," wrote John. "It was a perfect lecture except that all the really essential facts were skillfully suppressed." [1]

After this, the weariness began to creep in on all of them, and with it the awful realization that they had been led all over a foreign country by uncaring hosts, and they would very likely never see what they had wasted months trying to find. That search for the figurative "Holy Grail" would, they thought, prove as fruitless as the search for the real thing would have been. "We are getting mighty tired now," John wrote, "[and] several of us concluded up at the monastery to-day that we would go home pretty soon unless there was a change." [2] The realization that they were thousands of miles from home and that it would take weeks if not months for them to get back there, even if they left right away, did nothing to lift their spirits.

In the end, they were taken out of their minimum-security prison of Haicheng and were on their way, or so they supposed, to see a battle at last. They *did* witness a battle of sorts, but they were so far away they could not tell one opposing side from the other. "Two hours we marched," John reported, "climbed then a little hill ... crawled over the top to where the battle was raging — some ten miles away." [3] It did not take long for the correspondents, many of whom had been in the Spanish-American War only a few years before and were used to having bullets buzz and whistle around their ears, to get bored. Some, to the horror of their Japanese guides, lay down and took a nap, while others simply turned their backs to the "raging battle" and read newspapers that were too old to provide much except casual entertainment.

Their guides protested that the correspondents had complained for weeks that they had not been allowed to see a battle, and now that the Japanese had relented and shown them one, they did not have the courtesy to watch. One reporter excused himself, saying he did not realize that a battle was going on. "The only thing about that battle of which you were certain," wrote Davis, "was that it was a perfectly safe battle to watch. It was the first one I ever witnessed that did not require you to calmly smoke a pipe in order to conceal the fact that you were scared." [4]

Davis, never at a loss for words when he was angry, summed it all up by saying that "in order to see this battle we had traveled half around the world, had then waited four wasted months at Tokio [sic], then had taken a sea voyage of ten days, then for twelve days had ridden through mud and dust in pursuit of the army, then for twelve more days, while battles raged ten miles away, had been kept prisoners in a compound where five out of the eighteen correspondents were sick with dysentery or fever, and finally as a reward we were released from captivity and taken to see smoke rings eight miles away!" [5]

To make up for their seeing a nonbattle, the Japanese said they would show them the real thing the next morning, but they would have to get up very early to get there before things started to happen. In the blackest dark before dawn, they were roused by their Japanese guides, and sleepy-eyed, they stumbled out of the compound, checking their pockets for notebooks and cameras, tobacco and pencils, and anything else they might need to survive the Manchurian night and the battle they were sure to see. "Nobody said a word," wrote John, "and the silence and mystery of the march was oppressive as we waded streams and ploughed through mud between walls of dripping corn ... a more stealing, mysterious, conspirator-like expedition I have never known." [6]

Davis, however, was not optimistic. "Either that we might not miss one minute of it, or that we should be too sleepy to see anything of it, we started in black darkness, at three o'clock in the morning, the hour, as we are told, when one's vitality is at its lowest, and one which should be reserved for the exclusive use of burglars and robbers of hen roosts." The confusion was such that only 18 men and an equal number of horses and mules in darkness can make. "Concerning that hour," continued Davis, "I learned this, that whatever its effects may be upon human beings, it finds a horse at his most strenuous moment. At that hour by the light of three paper lanterns we tried to saddle eighteen horses, donkeys, and ponies, and the sole object of each was to kick the light out of the lantern nearest him. We finally rode off through a darkness that was lightened only by a gray, dripping fog, and in a silence broken only by the patter of rain upon the corn that towered high above our heads and for many miles hemmed us in." [7]

If anything, it was worse, when daylight came, than it had been the day before. This time, they did not even see any smoke — nothing. The Russians, they were informed, had slipped away during the night and were in full flight. At any rate, there was no battle. All night long, they had waited in the rain, sleet and cold wind for this. Their spirits, already low as they thought they could possibly be, sank even lower. Eventually, and soon, they would hit rock bottom, and no amount of promises from the Japanese nor anyone else would make them rise again.

They followed their guides thinking they might catch up with the retreating Russians, and at least see something, but not one Russian soldier did they see, only signs of where they had been. Except for the mysterious caution of their Japanese "protectors" and the smell of horses and men that lingered long after they were gone, there was still little sign of warfare in this muddy, fly-infested countryside. Disgusted, and with no end of delays and broken promises in sight, they determined to send General Oku their own promise. "We had a serious consultation that night," said John, and one can almost see them all gathered around a smoking fire, like conspirators plotting an assassination. "The artists couldn't very well draw what they couldn't see," John wrote. "Some of us, not being military experts, and therefore dependent on mental pictures and incident for material, were equally helpless. Thus far the spoils of war had been battle-fields, empty trenches, a few wounded Japanese soldiers, and one Russian prisoner in a red shirt. So, hearing that General Oku feared for our safety, we sent him a round-robin relieving him of any responsibility on our account, and praying that we should be allowed to go closer to the fighting, or our occupation would be gone." [8]

It did no good. General Oku sent them word that the Russians were apparently in flight, and that they probably would be for some time, and that the likelihood of a battle any time soon was very remote. Even if there was a battle soon, they were told, the correspondents would not be allowed within four miles of the firing. This, according to John, was the final straw. They had finally had all they could endure. They had stood more than anyone had the right to expect of them. Now, "on a bright sunny morning," no closer to the war than they had been since first setting foot on Manchurian soil, "Richard Harding Davis, Melton Prior, [9] the wild Irishman, [10] and I sat alone in the last dirty compound, with the opening guns of Liao-Yang booming in the distance." [11]

They would be told that the Russians had abandoned Liao-Yang without a fight and that the Japanese had entered the city already without firing a shot. There would be no battle, they were assured, for the city of Liao-Yang. It would be to their everlasting regret that, this one final time, they believed the Japanese were telling them the truth.

They were left alone in their decision to turn back. They were uncertain if they were making the right choice, but in their misery they felt that they had no other choice to make. The few Japanese that had constituted their guards and guides had disappeared along with a few of the more hopeful correspondents. John sold his horse and wondered at the nerve of the man who bought him, "for the price, though small, was big for Fuji. I pulled the vicious stallion's wayward forelock with malicious affection several times, and watched" as his new owner rode away on a horse that had proved to be "a more dangerous fate than any danger that war could hang over him." [12]

They went back over the same route they had struggled along for days and weeks, and it had an anticlimactic quality about it. They felt like school boys who were skipping school and cutting classes, and resisted, as much as they could, the urge to look over their shoulders. It was as if a huge and deafening noise had passed through the land, and the silence that took its place was, in and of itself, deafening. "Our first halt was at Hai-Cheng," remembered Davis, "in the same compound in which for many days with the others we had been imprisoned. But our halt was a brief one. We found the compound glaring in the sun, empty, silent, filled only with memories of the men who, with their laughter, their stories, and their songs had made it live." [13] They kicked around the deserted village, finding and sorting through the trash, but found nothing of real interest.

They hurried away from the spot, as if escaping ghosts they could not see nor hear. They got lost and wandered through fields of corn so tall that they lost much of their sense of direction, and the perverse Manchurian mud was just as contrary as before, pulling at their boots and slowing them down no matter what direction they traveled.

They met Japanese troops, who were not so enthusiastic about their going to the front. They heard rumors of a foreign hotel in New-Chwang, and in their uncomfortable condition grasped at the straw, and John rode ahead on a borrowed horse to reconnoiter the situation.

The hotel was there; the rumor was true. Davis and the others were supposedly right behind him, and John made arrangements of food, rooms, beer and even champagne and servants to stay up and await the arrival of Davis and the rest. "Then I went to bed," he wrote. "About two o'clock there was a pounding on my door, and a little Japanese officer with a two-handed sword some five feet long came in and arrested me as a Russian spy." [14] If the Japanese officer had not been so serious and his sword so long, John admitted that he would have hung out his window and laughed himself breathless. "But I had ridden into that town on the biggest white horse I ever saw," he wrote, "and I looked like an English field-marshall without his blouse. I had gone to the Japanese headquarters. I had registered my name and the names of my three friends on the hotel-book. I had filled out the blank that is usual for the passing stranger in time of war. I had added information that was not asked for on that blank. I had engaged four rooms, had ordered dinner for four people, and had things to eat and things to drink awaiting for the other three whenever they should come. I had my war-pass in my pocket, which I displayed, and yet this Japanese officer, the second in command at Newchwang and a graduate of Yale, as I learned afterward, woke me up at two o'clock in the morning, and in excellent English put me under arrest as a Russian spy." [15]

Davis had not shown up until midnight, and finding John already in bed asleep did not waken him, and so, through all his troubles with the Yale officer, John was unaware that Davis and the others had already arrived. The next day, the lateness of his companions whom he thought to have been right behind him was explained. Almost as soon as John had left them, Melton Prior's cart had overturned and the mules had lain down in the mud, contrary creatures that they were, and refused to get back up. Finally, they were forced to abandon the stubborn animals and load Prior's belongings on another cart. By that time, it was dark, and they still had three hours of traveling through the mud and dark before reaching New-Chwang.

George Lynch had decided to go on to the hotel alone, and promptly got lost. That, and Melton Prior's decision to wait until morning before going any farther, was really what got John arrested in the first place. Lynch stumbled around in the darkness outside the city until he fell in with a litter of Chinese pigs, who were no more pleased to be awakened by Lynch than John would be shortly by the Yale man posing as a Japanese officer. The pigs were not arrested as spies, but Lynch was. "This had one advantage," wrote Davis, "as he [Lynch] now was able to find New-Chwang, to which place he was marched, closely guarded, arriving there at half-past two in the morning." [16]

Lynch swore to his captors that if they would only take him to the hotel where his friends were now staying — he had no idea that Prior was not there, and that Davis had slipped in unnoticed — they would vouch for his credentials as correspondent. What they found, in this comedy of errors, was a sleepy-eyed, indignant John Fox, whom they had promptly arrested as a spy as well.

It was unfortunate for them that all their papers, even including Davis's, were left with Prior some miles to their rear. Of course, Prior showed up the next morning with papers aplenty for all, and Fox and Lynch were freed, but not before John had given the Yale officer a good piece of his mind, which must have been quite colorful, since John declined to repeat what he said in print.

They were mad to get out of Manchuria and start for home, and were stymied nearly every way they turned, trying to go through regular Japanese channels. Finally, taking matters into their own hands, they gathered up everything they had and boarded a British steamer to Shanghai, hoping to catch passage there to Japan and then home. John was still mad over being arrested, and from the guard-rail of the British ship he "hurled through the captain's brass speaking-trumpet our farewells to the Japanese." [17] Those curses must have been something to see and hear — really good ones *can* be seen — for the British officers begged him to desist with his farewells until they were under the protection of a nearby British man-of-war.

"To avoid floating mines," he wrote, "we anchored that night outside the bar, but next morning we struck the wide, free, blue seas, with an English captain, whose tales made Gulliver's Travels sound like the story of a Summer in a Garden. Without flies, fleas, mosquitoes, or scorpions, we slept when and where we pleased and as long as we pleased. Once more we wore the white man's clothes and ate his food and drank his drink, and were happy." [18]

Three days later, they arrived safe and sound in Shanghai having learned two days previously in Chefoo that a big battle was now going on at Liao-Yang, the very spot the Japanese had told them they had already taken without firing a shot, and that there was not anything there that would interest the correspondents at all. John wrote that upon hearing the news, and the realization that they had all missed this one last opportunity to see a battle, "a melancholy of which no man spoke set in strong with all of us." [19]

No one spoke; they did not have to. All knew what they had lost by leaving when they did, but who among them could have known? They had again been tricked by a lie and by believing that lie, they had missed what turned out to be one of the greatest battles, in sheer numbers at least, of the twentieth century. Two hundred thousand men on each side, Russian and Japanese, fought for ten days, and they had missed it. Davis was physically ill, and distraught almost to the point of tears of frustration. "So, our half-year of time and money, of dreary waiting, of daily humiliations at the hands of officers with minds diseased by suspicion, all of which would have been made up to us by the sight of this one great spectacle, was to the end absolutely lost to us. Perhaps," Davis wrote sadly, "we made a mistake in judgment." [20]

But it was not entirely their fault, and they all thought, including Davis, that if faced with the same circumstances again, they would not have done much differently. "Our misfortune lay in the fact that our experience with other armies had led us to believe that officers and gentlemen speak the truth," wrote Davis, "that men with titles of nobility, and with the higher titles of General and Major-General, do not lie. In that we were mistaken." [21] From Shanghai, they sailed into Nagasaki and spent the next three days in crowded train cars on the way to Yokohama.

Finally, in Tokyo, insult was added to injury when the Japanese asked them if they would change their minds about leaving, and would they "consider going back to Port Arthur?" [22] They refused, but the Japanese, according to John, were insistent, and asked them to please think it over. They played along, told the Japanese they had changed their minds, and informed them that they would indeed go to Port Arthur. Of course, the Japanese then said they could not go. They were not surprised.

"All my life," John wrote near the end of *Following the Sun Flag*, a collection of those dispatches he sent to Scribner's from Japan and Manchuria, "Japan had been one of the two countries on earth I most wanted to see. No more enthusiastic pro-Japanese ever put foot on the shore of that little island than I was when I swung into Yokohama Harbor nearly seven months before. I had lost much — [in those seven months] — but I was carrying away in heart and mind the nameless charm of the land and of the people."[23]

He had liked the people and the country well enough, but of the experience of war correspondent, he was not so enthusiastic. "After seven months," he wrote, "my spoils of war were post-mortem battle-fields, wounded convalescents in hospitals, deserted trenches, a few graves, and one Russian prisoner in a red shirt." [24] As soon as he returned to Big Stone Gap, he retrieved his good humor, and could even smile and joke a little about his non-adventure, but while it was happening to himself, Davis, Jack London and the others, it had been no laughing matter.

By the first of October, they were all back in the United States, and never did a homeland look so good to two weary travelers as it did to Davis and Fox. Davis went on to Philadelphia, to his mother's and father's home, where the latter was seriously ill and would soon die. Fox stopped over in New York and decided to remain there for some days, visiting friends and taking care of business that had accumulated in his absence of seven months. "I thought you would prefer that I saw the people up here," he wrote his anxious mother near the end of October, "and came home to stay rather than that I should go home only for a few days and then have to come to New York to sign the Scribner contracts—since you know how much longer I am apt to stay up here when I do leave home." [25]

John got back to the United States before the balance of his articles on the Russo-Japanese War were published by Scribner's. "Trail of the Saxon" had appeared in *Scribner's* magazine in June of 1904, and "Hardships of the Campaign" had appeared in July, but he was long back in America when "Making for Manchuria" was published in December. "On the War Dragon's Trail" appeared in January of 1905, "White Slaves of Haicheng" in February, and

Fox made friends everywhere he went, and language barriers did not seem to get in the way. After returning home, he continued to receive letters from those men and women he had met "following the sun flag" (courtesy John Fox, Jr., Museum).

lastly, "Backward Trail of the Saxon" in March. "Lingering in Tokio" was written to round out the collection, and the seven articles were gathered together by Scribner's, and published in book form in April 1905. It was given the title of *Following the Sun Flag* with a subtitle of *A Vain Pursuit Through Manchuria*, but it was not, as he thought, a "vain pursuit," but one that generated some of the finest writing he ever did. He dedicated the book, somewhat cryptically, to "The Men of Many Wars," and congratulated any of them who might have been more successful than he thought he had been.

Following the Sun Flag was some of the best writing he had ever done, even if it was not a novel nor a collection of the mountain tales he loved to write. There is no way it could stand against *The Little Shepherd of Kingdom Come*, which continued to have phenomenal sales, nor the soon to be written *Trail of the Lonesome Pine*, but in its own genre it is one of the best he ever wrote and is vastly superior to the Spanish-American War articles he had written for *Harper's* only a few years before.

His writing had matured fantastically, and coupled with the publication and, at least at that time, the unheard of sales of *The Little Shepherd of Kingdom Come*, it was a completely different writer that had emerged into the twentieth century.

CHAPTER 17

A Noble Profession

I know that all of you will love her.

John Fox, Jr.

THE COUNTRY THAT JOHN RETURNED TO that late fall of 1904 did not seem to him to be the same one he had left only a few months before, and he was reminded of the way he had felt upon his return home after the Spanish-American War. In the Gap and other small towns across America, life went on at a sedate pace, little changed over the years. A tennis court for the young and eligible had been built in the side yard of the house in the Gap, and for a short, sweet time, the troubles of the world seemed far away. The summers were hot and the winters long and cold in Big Stone Gap; old people, and sometimes the young, got sick and died, and there were marriages and births to replace those that were lost. To John, now a writer of unquestioned promise and importance, the town seemed too quiet for his now more cosmopolitan tastes, and even seemed more dull than usual, and when the town was held up against the attractions of New York or Chicago, or even Louisville and Lexington, Kentucky, there could be no comparison.

None of the rest of the family was overly fond of the Gap now that all hope of a fortune to be made was gone, and in the summer of 1904, while John was still out of the country, there was an abortive attempt to find a suitable home for them all, back in Bourbon County, Kentucky, near their old home at Stony Point or perhaps nearby Paris. James and John Sr. went to look things over and walked around a couple of prospects, but John's mother refused to go, and eventually the whole scheme died for lack of interest. Now, for better or worse, like it or not, the Gap was their home for the rest of their lives.

In the years following the publication of *The Little Shepherd of Kingdom Come*, John's money problems were alleviated by the huge sales of the book. By extension, the money problems of his family were also eased somewhat.

His newfound wealth and his unaffected generosity placed him in the tacit position of family leadership, usurping, to some degree, that position from his father and older brother James. All his money problems for the present and future did not vanish entirely, for John liked spending money too much, when he had any. He was spending more and more time in New York, supposedly to be closer to his publisher. As a result, his money, substantial though it was, flowed freely from his purse. It could not last forever.

He was generous to a fault with his family, and although his letters to them would become more and more infrequent as the years went by when he was away from home, nearly all of them enclosed checks for his mother to share with the others. "Dear Mother," he would almost invariably write, "I enclose a cheque for $200," or $100, or $150, sometimes as much as $300, but always with no strings attached. "Use it for yourself and the girls," he would suggest, or, "Let May [Horace's wife] have a certain amount," and that was all. If their money ran short between letters, he would always tell them to get it from the bank, and he would take care of it the next time he came home. [1]

For nearly a year after returning from Japan and Manchuria, he did very little writing beyond the additions he made to the articles for *Following the Sun Flag*. He managed to "misplace" a Christmas story he had written for Scribner's and had to rewrite the whole thing. While he was in Manchuria, Scribner's had thought to capitalize on his current popularity as a writer by publishing a collection of stories entitled *Christmas Eve on Lonesome and Other Stories*, and it sold fairly well, considering that three of the five sketches were not new to followers of Fox's writing and had been published previously.

Then, in a long delayed burst of writing activity, he wrote, in one week, a novelette of just under one hundred

Fox in middle age, a successful writer who had at last found his profession (courtesy Duncan Tavern).

pages, and Scribner's published *A Knight of the Cumberland* in three installments, beginning in September 1906. Immediately after the final installment in November, *A Knight of the Cumberland* was brought out in book form in October. Three years later, Scribner's would combine the novelette with the previously published, and not much longer, book, *The Kentuckians*, and bring that combination out in 1909.

Throughout the rest of 1906 and 1907, he alternated between short visits to the Gap and longer, more extended stays in New York, Maine and Washington, D.C. He stayed with any number of friends, especially the Davises and the Pages, enjoying the nightlife of the cities and his own notoriety as a bestselling author. His success may not have gone to his head, but who could blame him if it had? In New York, there were plenty of parties, balls, galas and the like, and plenty of drinking and smoking which did nothing to improve his health. When he drank, he did not write very much, nor very well. He received a double hit from his New York visits: He drank and smoked too much, and the distractions were too many.

In October, he wrote Tom Page from Big Stone Gap that he was "still hammering away" on his new novel: *The Trail of the Lonesome Pine*, adding one chapter after another. "I never was in such good health," he wrote. "As usual, I don't drink here. I smoke very little and I walk these mountains in a heavy sweater two hours every afternoon. And," he added cheerfully, "I get up before 9 A.M.!" [2]

Despite his claim that he "never was in such good health," these interludes of relatively good health were short-lived. Since his return from Manchuria late in 1904, he had been plagued by everything from bad teeth to poor eyesight, and many of his old complaints, perhaps intensified by his campaigns in Cuba and the Orient, returned in force to attack him. He tried every cure he could think of, and nothing seemed to work past occasional and impermanent relief. He dosed himself periodically with hot water to clear his throat and sinuses. Various home remedies failed him also, but so did more professional cures that doctors prescribed for him. A stay or two in northern hospitals only offered a temporary respite. From now to the end of his life, he would seldom be a well man, despite his occasional assertions to his friends and family.

Mineral springs were all the rage at this time and people from all over the country flocked to those sites to "take the waters" with often dubious and uncertain results. It was inevitable that he should try them himself. But he was bored just sitting around drinking water. The entertainment value in those places was nonexistent, and being surrounded by sick people who were more than willing to talk of their complaints for hours on end was not something that such resorts wanted to advertise. "This continues to be," he wrote from such visit at French Lick Springs Hotel in Indiana to his father on Valentine's Day, 1907, "about the loneliest place I ever saw and up to date, I've spoken but to one person in the hotel." [3]

John Jr. at home with one of the family's many dogs. All were especially fond of their dogs, even to the point of having framed photographs of their favorites in their home, and a section of the backyard set aside for those dogs that had died (courtesy John Fox, Jr., Museum).

He thought, however, that he was getting better and "getting in great shape and," and said, "I'm doing a great deal of work on 'The Trail of the Lonesome Pine.'" [4] He had the idea for a novel about a mountain girl and an outsider from the Bluegrass for some time, and even while he was working on *The Little Shepherd*, the plot of this new novel continued to take shape in his mind. It was a theme and an idea that he had used before, perhaps most notably in *A Mountain Europa*, but this time he intended it to be bigger, better, more colorful and more realistic than its literary forefather.

One story suggests the new novel he was working on actually had its beginnings at the old Imperial Hotel in Tokyo. "It became the custom with us," wrote one correspondent from the Russo-Japanese War, "to gather in one of the little rooms off the lobby of the old Imperial Hotel to drink our after-dinner coffee and liqueurs, smoke a cigar or two—and talk. One night Fox told us a story. It was a sweet, beautiful story, told only as that gentle Southerner could tell a story. It was about a girl named June and interwoven in it was something about a lone pine tree." The next morning, or so the anonymous author says, Richard Harding Davis, who apparently was intrigued by the story himself, told John that he should "make it into a book."

"'Maybe I will one of these days, but I haven't time now', Fox replied.

"'Which means that you will never do it,' was Davis's rejoinder."

Just to show Davis that he was wrong, or so the story goes, John dictated to the author of the newspaper article for three hours, finishing the first two chapters, which the author of the tale claims John never changed when he actually finished the book, some four or five years later. Even Davis was impressed. [5]

By the end of July 1907, John was confident enough that he would soon be near the end of the story that he wrote his mother from Loon Lake in New York that he was in good health, for the time being, and he said, "I hope to get the book so near its final shape that I shall be able to call on Scribner's for more money. Just pull along," he told her, "and charge things until the middle of August and then we will all be all right again." [6]

The money from *The Little Shepherd* was not yet gone, and royalties continued to come in from what continued to be a very good seller for Scribner's, but John was spending that money like water. One estimate of his earnings from that book says that he made $100,000 from serialization and various editions of the book that was published by Scribner's and other secondary publishers, including overseas publishing houses, but a great deal of that money was already gone. There had been old debts to be settled, both for himself and the debts of his family which he took as his own. In three years after the initial publication of *The Little Shepherd*, he found himself having to count his nickels and dimes again. The crook of the shepherd proved no match for the city and John's free-spending ways.

The money he unresistantly sent home continued to drain his already dwindling finances, and during these and the years that followed he found himself writing for the money, which anyone knows is a poor instigator for an author. He was, to be sure, spending money on himself, but it appears that his real and perceived responsibilities to those family members still in the Gap was a constant concern for him. He was placed in the uncomfortable position of continually asking — begging really — Scribner's for advances upon royalties for work not even finished, and sometimes not even started. Naturally, when the work was done, those advances were deducted from any future royalties, and even though he did not actually lose any money by taking those advances, the method he employed getting them did not seem quite honorable to him, but he could not bring himself to do otherwise, and his publisher did not complain.

In the fall of 1907, another distraction came into his life, and this one more serious than money. In New York, he was invited to attend a dinner at Delmonico's to honor an opera star from Vienna, Austria. It was here that he met, for the first time, Fritzi Scheff. He had heard of her — who had not heard of the girl who had turned New York's opera lovers on their collective ears? Scheff was at the peak of her career; she had made her American debut in *Fidelio*

John Jr. strikes a senatorial pose in the backyard at Big Stone Gap, Virginia. While writing *The Trail of the Lonesome Pine,* he often took leisurely breaks in his bath robe (courtesy John Fox, Jr., Museum).

at the Metropolitan Opera House in December of 1900, was enjoying a magnificent success in *Mademoiselle Modiste* in New York, and would soon be taking that operetta on the road, where crowds of hundreds, if not thousands, would be brought to their feet by her beauty and talent. He had seen her photograph, but nothing had prepared him for seeing her in person. He arrived late, as usual, and was instantly taken with her. Even by today's standards, Fritzi Scheff was a beautiful woman, and had charm, wit, talent, personality and, John thought, the most captivating foreign accent he had ever heard. She was also married, though separated at the time from her husband.

John arrived in sharp contrast to the others already there. He had dressed not in formal attire but a simple suit. She was instantly drawn to him, and pleasantly surprised at his manners and the way his self-confidence shone through. He was handsome enough, she thought, though not in a startling, pretentious sort of way. His profile, she would later say, reminded her of one of those American Indians that Fritzi's mother would later be so afraid would scalp her in the "wilds" of Virginia, but it was his personality that carried him through to her and impressed her the most that night.

Fritzi Scheff was taking the New York stage by storm at about the time this photograph was taken. At a time when many photographs left much to be desired of their subjects, Fritzi was a beautiful woman by anyone's standards. It is no wonder that John fell head over heels in love with her (Dover Publications and James Camner).

After their initial meeting, it was some time before their paths crossed again, but at that second meeting, at a party given by a mutual friend, both were practically oblivious to the other guests and spent most of the evening in each other's company. John showed off his ability at the piano while Fritzi sang "Kiss Me Again." [7] It may not have been love at first sight, but neither could deny the attraction that they had for one another.

John had done much over the years to avoid any kind of permanent attachments, but this time he would not be able to avoid this entanglement, and he obviously did not want to very much. By the spring of 1908, they had become an "item" for casual gossip among the New York crowd who knew them both very well.

By the time May rolled around, Fritzi's marriage, already technically over when she met John at Delmonico's, now ended in divorce from Baron Fritz von Bardeleben, an officer in the German army. Her divorce would cause new problems, real and imagined, for the couple, especially John, whose overly sensitive feelings about divorces in general and his desire to protect Fritzi from all gossip in particular would drive a temporary wedge between himself and his friends.

Now, at the age of 44 — Fritzi was 17 years younger — he found himself openly pursuing a divorced woman, and although divorce was becoming less and less rare at the turn of the century, it was by no means an accepted circumstance in all communities and in all households. It troubled him and initially had an effect upon where he thought she and he would be welcomed, though it appears that he was overreacting quite a bit, and if it bothered Fritzi she certainly did not let it show.

It was his best and truest of friends, the Pages, that he avoided most during the first months of their courtship. The Pages came from, and were steeped in, a conservative Southern culture that preferred to ignore the realities of divorce and had gone so far as not to invite divorcees into their home. John was certainly aware of their feelings and likely thought it to be better not to place his dearest friends in an uncomfortable and awkward position. It is very likely, given their true affections for him, that the Pages would not have sacrificed a treasured friendship for a principle that was fast becoming antiquated.

Also, the Pages were themselves not immune from "progress" in marital affairs. Minna Field, Tom Page's stepdaughter, had eloped with Preston Gibson near the end of January 1900 and married, before leaving for the Gibsons' family home in Kentucky. The marriage, however, did not last, and the Gibsons were soon divorced. By 1907, Minna had remarried to an English gentleman which was, by the way, something of a fad among young American ladies at the time. Like many families before and since, the Pages were forced to reevaluate their family's personal standards in light of new and sometimes unwelcome developments.

John and Fritzi would, no doubt, have been more than welcome at Oakwood, the Pages' home in Virginia, or at York Harbor, or anywhere else the Pages might be, but gentleman that John was, he could not allow himself, for several months, to intrude, not because of his feelings, but out of concern for theirs.

The Pages were understandably a little hurt by his absence from them, but as always they knew him better than anyone else, knew his reasons, and forgave him all. They had written him in April that he need not worry about Fritzi's former marriage and to hurry to them as fast as he could, but he was too embarrassed by his lengthy and unexplained absence from them to go, and did not answer. The summer came and went, and another letter from Tom went unanswered also, until November, when John could stand the alienation no longer.

He had carried both letters with him, he said, when he finally replied to Tom Page, "in the hope that I could get the courage to answer both." There had been plenty of talk, gossip and innuendo about their highly visible courtship and Fritzi's divorce, so that even the usually unflappable Fritzi was beginning to show some surprise at the intensity of the gossip, and she was predictably hurt by it all. "I was almost heartbroken over the tears," wrote John, "the humiliation and the distress caused an innocent woman at the time, and you can imagine how my engagement to her retrospectively made my pain the more keen." [8]

After their second meeting in New York, where he played the piano and she sang, and he playfully told her he usually did not care much for foreign women, she had begun running into him nearly everywhere she went. At an engagement in Louisville, she had only just arrived when she and he "accidentally" met. He would later confide to her that their meeting that day was no accident at all, and that he knew she was coming to Louisville; he had arrived earlier, and had walked around the Louisville streets, killing time, until their "chance" encounter.

She liked him, too, and their relationship matured throughout the coming summer months while he alienated himself from the Pages and his other friends. By August 1908, they were engaged. No one but his family and close friends, like the Davises and, three months later, the Pages, were aware of their secret, but such things have a way of getting out. Before long, talk of secret engagements gave way to talk of public marriage.

By the time their courtship was beginning to take a serious turn, he had finished *The Trail of the Lonesome Pine*, and Scribner's had begun its serialization in the January 1908 issue of their magazine. Not surprisingly, when it came out in book form later that year, it was dedicated to "F.S.," which fooled absolutely no one. It would prove to be his last great work. The promise of *The Little Shepherd* was fulfilled with this latest effort, but as quickly as his literary star rose, it would descend almost as rapidly. There would be other stories and other books, and occasionally the old spark could be seen, but never again would the quality be there, nor the intensity of purpose that had been there in his two previous, best-selling novels.

The Trail of the Lonesome Pine was peopled with characters from real life, as was the best of his writings before 1908. Not that they were all living, breathing people on whom his characters had been based — though many were — but they were believable enough most of the time, and to the readers of *Lonesome Pine* they became real, and made the book a success.

The people and places, and the time of *Lonesome Pine*, provide perhaps a clearer picture of what life was like for the Fox family after their move to the Gap in 1890. John had written about their misadventures before — several times, in fact — but it is in the fictionalization of those people and times that seems most real. The story of Jack Hale and June Tolliver, the Gap and a huge towering pine tree, that god-like, was given power of protection over the star-crossed lovers, captured the hearts and imagination of readers, and quickly made the story a favorite and again placed the name of John Fox, Jr., on the best-seller lists.

Much of the novel's unromantic content had been previously published in *Bluegrass and Rhododendron* (1901). "The Red Fox of the Mountains," "The Hanging of Talton Hall," and "Civilizing the Cumberland" were all used, sometimes transcribed directly into the storyline.

Strangely enough, as popular as the human characters of the novel were, it was the third principal character of the book that garnered as much attention as Jack Hale and June Tolliver and the rest. Almost immediately readers began searching for "the" lonesome pine that figured prominently in the story. [9]

Fox denied most of the public's supposition about whether or not this character or that was based on real or imagined people, but many were so obviously drawn from actual persons that that denial did little good. He denied the existence of the Lonesome Pine, though in a few years he would come to believe that there might actually be such a tree, but he never saw it himself, either before or after he wrote the novel. But almost immediately after publication of the book, and for years afterward, newspaper articles, complete with "actual" photographs of the tree, pretended to "solve" the mystery.

The Trail of the Lonesome Pine proved wildly popular, sold thousands of copies, and placed John firmly back on the best-seller lists. Now, he thought he had everything: a beautiful woman he loved almost to distraction; money enough to enjoy the noble profession he always wanted, and now seemed to have; and a future as an author that seemed as bright or brighter than he ever imagined that it would. And for a time, short though it was, he was happy.

CHAPTER 18

An Imperfect Union

> John has not kissed me for three days.
> Fritzi Scheff Fox

PERHAPS, IN THE END, HE WANTED HER more than he should have. Most who have written or said anything of John Fox's marriage to Fritzi Scheff, either at the time it was going on or in the years after it was over, usually preface their comments with "it was doomed from the start." Few marriages can be so neatly categorized, and neither can this one.

He should have known better; both of them should. He had written, in his stories and novels, time and time again of two people from totally different worlds who loved each other without compromise, and of the problems they encountered because of their differences in background. But it is only in fiction that love becomes enough. In reality, it takes a great deal more than love alone to keep a marriage afloat.

In the meantime, however, the doubts that either of them might have entertained were nowhere to be seen in their relationship, and neither John nor Fritzi suspected any future other than a happy one. In the meantime, however, now that he was seriously involved with Fritzi, he found one more reason to keep himself away from home and was, as usual, apologetic to his mother and father about his absences. "But you will understand," he wrote to his mother in September 1908, "that it has been on account of Fritzi that I have stayed [away]." [1]

When he was out of Fritzi's sight, which wasn't often, he thought of her, constantly. When he was by her side, he could not bring himself to leave her, even for short periods of time. It is a wonder, knowing his sometimes erratic writing habits, that he was able to write, revise and finish *The Trail of the Lonesome Pine* at all, when he was at the same time courting her, but he was able to do so.

John and Fritzi soon after their marriage, and, for a time, smiling and happy (courtesy John Fox, Jr., Museum).

In November of 1908, one month after *Lonesome Pine* came out in book form, he brought Fritzi to the Gap for the first time, and it was thought that they might marry soon after their arrival. "I think," wrote John Sr. to James, "[that] Johnnie and Fritzi contemplate paying us a visit on or about Nov. 23rd, and possibly marry while here on the 25th but this is not settled yet." [2] Nothing would have pleased the old man and John Jr.'s mother more than to have their son, who they were seeing less and less of now, marry where they could attend, but it was not to be.

The visit to the Gap was a short but successful one. They welcomed her with open arms, and by all accounts, they were as charmed by her as she was by them. The family was amazed at the number of trunks and baggage she brought with her. That amount of luggage seemed more appropriate for an extended stay than a visit of a few days. William Cable Moore, who had only recently been married to Elizabeth, John's youngest sister, remembered that "on her first visit to the Gap ... shortly before she and John were married ... she brought her companion, her personal maid, eight trunks and a number of bags, including one important bag or case in which she carried her many beautiful jewels." Usually, it was even worse. "When on tour," continued Moore, "she usually traveled in a private car and took with her in addition to her companion and her maid, her agent and occasionally a guest or two. Her auto, chauffeur and footman she usually sent overland to meet her at the next stop." [3] The Gap had never seen anything like it.

Even John's father was fascinated, and completely won over. Apparently, there was still a little romantic fire left in the old schoolmaster. "They came on the late train from Louisville," he wrote to James, who had not been at the Gap during all the excitement, "[at] 9 o'clock, last Sunday night a week ago— Johnnie, Dick [Richard Fox], Fritzi, her traveling companion and a maid. We had dinner for them about 10 o'clock that night. Fritzi was very friendly and agreeable all the time, and insisted that we call her just 'Fritzi,' and that she call me 'Father' and Mother 'Mother' all which we did." [4]

They all slept late the next day, having stayed up nearly all night. Guests and friends flocked in to see a New York opera star close up, including Elizabeth and Cabell Moore and Horace Fox and his wife, May. Fritzi took it all in stride, and even gave an impromptu recital for the guests. "I think they were the merriest company I ever saw," said John Sr., flushed and pleased all at the same time, "and enjoyed themselves as much as if they had been at the Waldorf." [5]

John had been concerned and not a little nervous about whether or not everything would come off all right, and he hovered around Fritzi throughout their stay, and hardly left her side for a minute. He loved her so much himself that he could hardly imagine that anyone would not feel the same way about her, but he needn't have been worried. Fritzi captured them from the first moment they saw her on the train platform. A few days later, on Tuesday, in a flurry of activity, the entire party left for Bristol, Tennessee, on their way back to New York. Dick dropped off in Cincinnati, and the rest went on alone.

It would be interesting to know what John's mother thought of Fritzi after that visit in November, but whatever his mother thought, John's father was very pleased with his prospective daughter-in-law, and John Jr.'s mother was probably satisfied as well, at least as much as most mothers are about losing their sons to another woman. "We found Fritzi very sociable and agreeable," wrote the senior Fox in his diary for November 23, 1908, "not spoiled by flattery, and seemingly very natural and childlike and seemed pleased with us all." [6] It was one of the few times that he failed to mention the weather in his diary.

The marriage would not happen in the Gap, and the whole family was disappointed, but Fritzi's engagements could not let her remain in one place very long, and wherever she went, John went with her. Soon after they left the Gap, they settled on a date for their marriage, but even then it had to be a tentative one. John wrote to the Tom Pages on December 11 that everything was in place. "I am going to be married on Sunday morning at my brother Rector's house at Mt. Kisco," he wrote. "If possible, we shall keep the wedding from the newspapers for a week. I have got tired of letting this consideration and that delay it, and so Fritzi and I have decided on Sunday." [7]

The marriage took place as he had told the Pages it would, at the home of his brother Rector Fox at Mount Kisco, New York, on December 13, 1908. They had corralled a local clergyman, and with John C. Hunt, a friend of Rector's, as best man, the deed was done almost too quickly. It seemed to John that anything as serious as a marriage should take longer, but nervous as he was, he was glad to have it over with. [8]

After the wedding, to which no guests were invited because of the sickness of Rector's children, Fritzi returned to the Hotel Gotham in New York where she had been staying while appearing in *The Prima Donna*, but she returned alone. Thinking, perhaps, to keep the marriage quiet, John returned later and took a room in an apartment nearby the hotel. That night, Fritzi appeared on stage and sang as usual, as if nothing had happened.

It was too good to last, however, and immediately questions about marriage followed both of them, but especially Fritzi. She would say nothing, but finally her personal representative, Charles Dillingham, announced, when asked, that the wedding had indeed taken place, and perhaps foretelling of future troubles for the bride and groom said, "I do not know where Mr. and Mrs. Fox will spend their honeymoon or what their future arrangements are. All I know is the Mme. Scheff will go on singing and carrying out her part of her contract." And somewhat prophetically, he said, "The honeymoon will have to come later." [9]

John had never been married before, and had never really come close to the altar. He had, however, been romantically linked with several young ladies over the years, most notably Currie Duke. None of those dalliances resulted in anything more than the extremely platonic courting of the day. Most of John's friends and family thought that he had missed the matrimonial boat, so to speak, when he let Currie Duke get away from him, thinking he would have been happiest with her. They were a couple with much in common, but for one reason or another the match did not take, and John ventured off to Cuba and the Spanish-American War just when their romance seemed to be taking a more serious turn. Currie married someone else, and John met the star of the New York stage, Fritzi Scheff, while Currie entered into what has been characterized as not the happiest of marriages. By 1908, when John and Fritzi married, Currie would have only two more years of her marriage left.

Fritzi, on the other hand, had already been married to Baron Fritz von Bardeleben, a lieutenant in the German army, and had borne and lost her only child. They had divorced early in 1908, but had apparently been separated for some time. John was blind to all these circumstances, and there is no reason why he shouldn't have been. It appears that his "friends" were more concerned with his choices than he was.

His family, however, was supportive. John's new brother-in-law, William Cable Moore, was impressed with Fritzi from the start. "She was ... well educated," he said some years after the marriage was dissolved, "spoke several languages besides German and English, was a fine needle woman and a good cook, and she had a neat, trim figure and knew how to wear her clothes. She was considered, I believe, one of the best dressed women in America in her day. Her wardrobe, it was said, consisted of a hundred pairs of shoes and slippers, a hundred hats and dresses and suits and furs." [10]

Critics of the marriage were unrelenting, and often seemed joyous at any real or perceived crack in the marriage. Ballard Thurston, the geologist-photographer, among other professions, and supposedly the basis for the character Jack Hale in *The Trail of the Lonesome Pine*, wrote that "John and Fritzi stopped in Louisville on their bridal trip and were the guests of Mr. and Mrs. J.D. Stewart. Mrs. Stewart was my oldest niece so I saw something of them. We went out to the golf links for a round. Before starting, Fritzi said with a petulant tone, 'John has not kissed me for three days.' I was not surprised when I heard of the separation." [11]

As the years went by, there was no letup in those who psychoanalyzed the marriage and generally laid the blame for the failure at Fritzi's doorstep. They also blamed her for John's lack of real work during and after his marriage to her. John Townsend, who gave radio addresses in the 1930s on Kentucky authors, thought that Fox's ill-fated marriage to Fritzi did much to destroy his later efforts at writing, conveniently forgetting that one of his best works, *The Trail of the Lonesome Pine*, was completed while he was courting her. Townsend also thought the marriage was "another melancholy tale of too much artistic temperament, both sea-sick with fame, with incompatibility again serving as the chief reason of their early separation." But he was more concerned with the loss of John's writing that the marriage supposedly caused. "Five important years in Fox's career," he complained, "were almost lost through his marriage; and when he did come back to the writing wars, his old punch showed a very decided charley-horse. He limped and squeaked at every joint." [12]

Others would see their happiness and be happy for them, and would realize that there can seldom be one evil genus that brings a marriage down. John's writing habits before he married Fritzi were uncertain at times, and could be erratic at best, and while his distractions with a new wife can be readily understood, that alone cannot be blamed for any lack of writing on his part.

Madison Cawein, the old poet friend of John's, wrote in February 1910 of having seen Fox and Fritzi in Louisville. "John Fox, Jr., and his wife," he wrote to R.E. Lee Gibson, "are here also, and we had a little affair at our home that stirred up the natives. Fritzi Scheff, as you know, is Fox's wife; she is a charmer as well as a singer. Last night we had a great time at the theatre, and after that an elaborate dinner at the Rathskeller of the Seelbach Hotel where at much champagne was drunken. I like these little diversions; they make one forget one's troubles." [13]

At any rate, what exactly they saw in each other is a little hard to say, but there was definitely an attraction. As it turned out, the initial attraction could not sustain the years that followed, but it is also unlikely that the marriage was "doomed from the start." They came from backgrounds so dissimilar that even the most imaginative scriptwriters would seldom have matched such characters, believing, and rightly so, that the public would not believe any such scenario. Fritzi was so unlike her new husband that even his friends commented, certainly not to his face, about the incompatibility of the match. Socially, she was more like Fox's friend Richard Harding Davis, who was beginning to have marital troubles of his own, and both Fritzi and Richard Davis ran in similar circles in New York. She was what she was, and it wasn't her fault, and she offered no apologies. She could not change, and neither could John, and the marriage was bound to suffer for it.

The original house in the Gap was too small to accommodate the new couple and had to be enlarged. Indoor plumbing was added as well, and the attic finished and turned into a bedroom for them with a large porch off the bedroom facing the north and the mountains around the Gap. They spent their summers there, but mostly they were gone off together, following up Fritzi's singing contracts. On the rare occasions when John did not go with her, his sister Minnie went along as a sort of traveling companion and would continue to do so as long as John and Fritzi remained married. Apparently, Fritzi and Minnie were the closest and traveled together a great deal, but mostly it was John who was her traveling companion. They were in Texas, Cincinnati, Louisville, and all cities in between. They went to California and New York City, and as they traveled John became more and more enthralled with his wife. "She is like a child about Xmas," he wrote to his mother from Los Angeles near the end of their first year of marriage, "and it's fearful to think of the lonely Xmases she has had. She is devoted to Minnie & to all of you and she has no greater pleasure than in doing things for other people.... Monday Fritzi starts in to work again. I shall try to work, too." [14]

Fritzi loved the Gap and at times even seemed to love it more than John, who more and more wanted to spend his time in New York. Big Stone Gap was too small for him now, and he continued to look for more excuses to get away, even more than he had in the past. In New York, or on the road with Fritzi, or even in the Gap, his writing came slowly and more difficult for him. He had years ago taken up the game of golf, and like with many who venture into the game it became, at times, all consuming. Oftentimes, Fritzi would practically have to force him to come inside the house in the Gap and write, when he would rather be knocking golf balls around the back yard of the house in the Gap.

Rumors of John's drinking and jealousy started to surface not too long after they were married. Fritzi was a beautiful and charming woman and naturally attracted men to her no matter where she was, married or not. There was nothing serious about any of it, and not the least hint of anything improper, and most times John was not too upset by the attentions his wife

John and Fritzi's marriage in 1908 was probably a mistake for both of them. Their eventual divorce in 1913 did not surprise very many of their friends, but there were happier times, as this photograph shows (courtesy Southwest Virginia Museum).

was receiving. It was when he drank that the worst came out, and his jealousy got the better of his good judgment.

Fritzi related years after the fact that John had once threatened to shoot a man if the man so much as looked at her again. Of course he had been drinking, and she, embarrassed almost to death, had run upstairs to their room. When John didn't show up, she went back down and discovered him with the threatened man, drinking and talking as if nothing had happened. There were not too many such incidences of that sort, however, and none that seriously jeopardized their marriage, so far as is known, but she didn't know what to make of his jealousy and mercurial temperament. At those times, she was, of course, frightened at what he might do and the predicaments he might get himself into, but she wisely thought it best not to try too hard to change him too much, too suddenly.

She made more money than he did, and that too was a source of irritation to him. Being the Southern gentleman he was, he could not bring himself to take her money, even though she saw nothing wrong in a married couple sharing their incomes, and thought it a little strange that he would be so stubborn about it. He was doing fairly well as a writer, and making more money than he had before, but his output was sporadic at best and the checks did not come in regularly, and even if they had his earnings could in no way match the two or three thousand dollars a week that Fritzi was bringing in.

In their travels, he insisted upon paying everything himself, which taxed his finances more than he could afford, and he found himself uncomfortably strapped for cash from time to time. Fritzi, on the other hand, was extravagant in her buying sprees and saw no reason why she should not be. She bought more hats, dresses, shoes, gowns and jewelry, and John couldn't help but be somewhat embarrassed by it all, especially when he knew he, her husband, could not afford to buy for her the things she wanted. It was not a good omen for their marriage, but he still adored her, and she loved him.

She tried, however, perhaps harder than anyone had any right to expect her to. The cultural shock of the Gap as opposed to the lights and excitement of the big cities must have been quite an obstacle for her to overcome, but in a short time she did overcome those obstacles, even to the point of bringing her mother to the Gap for a visit soon after she and John were married. Accounts say that her mother was scared nearly witless at the thought she might be scalped in the wilds of Virginia. "Her mother was quite a character," said William Cabell Moore, "knew no English, was a bit timid, thought America and the Gap a wild, rough country, and apparently expected to be scalped any moment." She got used to it nearly as quickly as her daughter, however, and to the delight of the townspeople they both joined together on the porch of the house on Shawnee Avenue and sang their hearts out. "She had a beautiful contralto voice," said Moore of Fritzi's mother, "and she sang beautifully. She and Fritzi often sang German opera songs together, sometimes on an upstairs porch off Fritzi's rooms, much to the surprise and delight of the neighbors. Sometimes they sang German folk songs and sometimes John played for them and sang too." [15]

Typhoid, that old enemy of frontier towns, reared its ugly head once again in the Gap, just after John and Fritzi began to make their home there, and everyone who was not too ill or too afraid to help plunged in with a vengeance. Fritzi did her part when no one expected her to. At other times, she donated Christmas trees and furnished a trophy cup for the popular baseball teams in the Gap, and did just about everything she could think of to make herself fit into what had to be a very foreign environment for her. She was successful, and the townspeople took to her without reservation, but even that was not enough.

After the first year or so of marriage, John gradually lapsed into his old bachelor habits, much to the disappointment of Fritzi. He would disappear for days at a time, never telling her where he was going, nor when he would return. His habit of avoiding any decisions played heavily upon her nerves. He would excite her with plans for a vacation or trip, set a date for them to leave, and then promptly forget all about it, until perhaps two weeks or more later, and then suddenly remember the plans and announce to a startled Fritzi that they were leaving immediately.

He became less and less interested in writing during these years, though his output of material was respectable enough. He was working in the shadow of *The Little Shepherd of Kingdom Come* and *The Trail of the Lonesome Pine*, and both were hard acts for him to follow. He was beginning to think that his best work was behind him, and that nagging doubt served to make him morose and moody. He had changed. When they had first met, Fritzi was charmed by the Southern manners of this writer from Kentucky and Virginia and that was what she fell in love with. But there were two sides to John Fox, Jr., and there had been two opposing sides for years, long before he met Fritzi in New York City. If he wasn't on the move — and it often seemed that his destinations didn't matter in the least so long as he was going somewhere — he wasn't happy. Unfortunately, the change in him was just another in a long line of problems for the marriage, which was on shaky and unsteady ground almost from the beginning. They didn't fight and they didn't argue any more than any married couple, but the little slings and arrows of their difficulties lay like a wound that is bound too tightly and festering underneath, and by the time 1912 rolled around it seemed it would be just a matter of time before the final break would occur.

John's output of writing for the years that he was married to Fritzi are admittedly small, but for the time involved, it would be difficult to make a case that he should have written more. He was just coming off the publication of two best-sellers, and perhaps more is expected of best-selling authors than should be by readers and publishers alike. In the four years he and Fritzi were married, he managed to write four magazine articles for *Scribner's* and one for *Collier's*, and finish one novel, which really is not that bad a production for an author. But many wanted more, or thought he should have done better. When it didn't happen, Fritzi became an easy target for what his friends and critics perceived as the reason for his not writing more and better than he did.

In December 1909, *Collier's* published "Christmas Tree on Pigeon," and the next December, Scribner's published "Christmas for Big Ame" in their 1910 Christmas edition. Previous to this, Scribner's had published, starting in August 1910, "On Horseback to Kingdom Come," "On the Road to Hell-fer-Sartin," and in October, "On the Trail of the Lonesome Pine." Through it all, he worked on a new novel, and although it didn't come easy, he kept grinding away at it one page, one chapter at a time, until it was finished and published in book form, the year after he and Fritzi separated.

Over his writing career, John had written of exotic places and people, and he had visited and knew firsthand most of the places and people that had appeared in his stories and books. But in the mountains, the country was a lot different in 1910 than it was only a few years before when he had begun writing. Nearly thirty years had gone by since he had seen his first mountain in eastern Kentucky, and the mountains that had barred civilization and "furriners" from entering those isolated valleys were being torn down and circumvented, and no longer held back anyone with the least bit of determination to cross them. Naturally, any exposure from the outside could not but help change the inside, and now, he wanted to see those changes, if any, "with the eyes of [his] body instead of [his] mind," before the alteration of his beloved mountains became unrecognizable. [16]

And so, one day in June, perhaps one of those times when Fritzi didn't know where he was, he started out in search of Kingdom Come, in nearby Kentucky, searching for the old stamping grounds of the fictional Chad Buford and Jack and the Turners, Melissa and the locales of his first best-seller, *The Little Shepherd of Kingdom Come*. He had hardly gotten started on horseback before the changes were apparent. In Letcher County, Kentucky, where mere trails had once snaked their way through the mountains, now roads ran — wagon wide — through the hills. Where once roads ran, now railroads supporting huge smoking steam locomotives chugged their way from mountain town to Bluegrass city, and beyond, and back again. Even the hills and forests were no longer the same. "The hills showed now stripped of trees, fire-scalded, and covered with dead logs, and not a seedling visible to take their places: it looked not only like criminal waste but a disregard of the unborn that was little short of fiendish." [17]

He was surprised that no one asked him his business. Thirty years ago, he would not have gotten anywhere without questions being asked that ranged from where he was going to his place of birth to when he was coming back, and nearly anything else a lonely mountain family could think of to ask. Now, it seemed that strangers were common enough so that no one cared very much what they were doing or where they were going. There were other, more dramatic changes. "The log-cabin was no more," he noticed. "The houses were tidy, weather-boarded, and painted; men were ploughing industriously in the fields; I passed children in the road, no longer in tatters, and with schoolbooks in their hands, and not a soul asked who I was and what I was doing over there. Evidently the stranger was no longer a rare bird along that creek; curiosity was slack and suspicion was gone." [18]

Finally, stopping over for the night in Whitesburg, he was recognized only after identifying himself, and was pleased that his host for the night had read his books, and thought he was "all right." "This was cheering news," he said, "for it had not been always like this in the olden days. I recalled having no little trouble over my first book about the mountaineers, of just escaping a 'rough house' at the hands of some students of a mountain college, and of being often charged by educated mountaineers that I had not done them justice, and by "furriners" of having given the mountaineers credit for more than was their due." [19] It was this host in Whitesburg who related to John that when the "Guard" was watching over Talton Hall back in 1892, and fearing that a gang of his friends from Kentucky would try to break him out of jail and save him from the hangman's noose, a stray cat had run wildly through the jail causing enough commotion to startle the guards into thinking the Kentuckians had arrived. John naturally denied remembering anything like that happening. He did learn, however, that there actually was a halfhearted attempt at rescue, but the Kentucky boys, upon hearing about the strong guard around Hall, decided the prisoner wasn't worth saving as much as they had originally thought and had turned back toward Kentucky.

Now, it seemed, even the feuding had died out, and now killings were isolated ones and usually had nothing to do with family squabbles. It was apparent to him that law and order — that which the Guard had brought to the Gap so long ago — had taken root in neighboring counties and states as well. The next morning, "fortified with accurate directions," he wrote, "I turned the little sorrel and, everybody from curb-stone, doorway, and porch regarding me curiously and silently, I rode down the street and through a flock of geese to the river." [20]

Finally, he gained Kingdom Come Creek, and found that the stream served as part of the road, and forced him to ride distances, sloshing through the shallow waters. "There were quicksands in that river-bed and rocks to make the sinner stumble," [21] he noticed, but still he wandered on until he found what he was looking for. He found logs, cut just like the Turners of *The Little Shepherd of Kingdom Come* had cut long ago, waiting for higher water — perhaps all winter — until next spring when the real flooding would begin, and in his mind's eye, he was able to see the Turners and Chad tying those logs together and Jack leaping from one to the other in anticipation of a trip to

Frankfort that, for him, would have to wait until much later. He could visualize Melissa and the Old Woman and the cabin and the rose-bush where Melissa had preserved Chad's last footprint before he left her with a little board placed over it. "Indeed, so real was it all that, when I started on, I found myself with a heartache as keen as though I had come back to a place where I had once lived, and from which had gone forever real people whom I had loved there many years ago." [22] There was a lump in his throat and a sadness in his heart, and he realized that he had grown too close to those characters he had created in his mind, and was a little troubled that they, even if imaginary, affected him so. He knew he had been away from the mountains too long.

Everywhere he looked, he found images that reminded him of those places and characters in *The Little Shepherd*. They were not the same, of course, but they were familiar enough at times to give him a start every now and then. He found what served as the old Turner homestead, with a clapboarded house now, and more clean than any cabin had ever been or could be in the old days, and wonder of wonders, he found a young lady that would have served as a model for the long-dead Melissa who was friendly and beautiful but not overly impressed with who he was. "Her throat was a proud column of alabaster," he wrote, "and her eyes were frank and straight-forward, big, and violet blue. Melissa I had never imagined lovelier than this mountain girl standing before me at the head of Kingdom Come, and Melissa would have been lovelier had I seen this girl first." [23]

A young boy who could have passed for Chad, he thought, played the banjo after their noon meal at the "Turner" home. The banjo was a new one, John observed, but the songs and tunes were the same ones that Chad had beaten out on his rough homemade instrument years before. He had found what he was looking for, though not without effort and some imagination, and as he turned his horse back toward Pine Mountain and the Gap, he was satisfied in his own mind that the places that Chad, the Turners, and all the rest from *The Little Shepherd* had lived, loved and died, still existed and those characters he had created out of his own mind still lived, though changed, in Kingdom Come.

As he rode back toward the Gap, he couldn't get the changes that had come to the mountains in the short span of a few years out of his mind. Some of those changes were good, but others did not seem as worthwhile, and his mind was torn between a feeling of success over his trip to find Kingdom Come and an uncomfortable longing for what he did not find there anymore. "Down I dropped then by a trail that was steep, winding, and obstructed by stones, roots, and fallen trees; down at last to the head of a little creek, down through a sylvan dell that was heavy with wood smells and riotous with laurel and rhododendron — down until below me, coke ovens [of the Gap] flamed like a pit of hell." [24]

When he got back to the Gap that hot day in June, he was already planning another mountain trip and a search for what might or might not be there,

All his life, John attracted two kinds of people: children and beautiful women. Here he is shown on the steps of the house at Big Stone Gap with an unidentified companion (courtesy John Fox, Jr., Museum).

but before he would go he sat down and with two fingers pounded out "On Horseback to Kingdom Come" on a No. 10 Remington typewriter. It is reminiscent of his *Sun Flag* dispatches from Manchuria, with clear pictures of a time and people replaced, and not always unwillingly, by the progress that comes to all frontier regions sooner or later.

The first story that really gained him any sort of national recognition had been "On Hell-fer-Sartain Creek," and so, traveling by train from the Gap to Jackson, Kentucky, in Breathitt County, he began his search for that place, and he found things there less changed than on Kingdom Come. The old feuds had only recently died down, but it was apparent to him that the embers of that deadly fire still smoldered and could be fanned into a flame with little effort. There was law, he observed, in Jackson, but it was of a tenuous sort, though accepted gratefully by most of the citizens there. But those who wielded the law, he saw, still were wont to use that law to their own ends, which was not much different than it had been in years past.

Directions to Hell-fer-Sartain Creek were plentiful, if conflicting. "Already I had learned that I could get a horse — a 'good horse'— and many people told me as many ways of getting to Hell-fer-Sartain, which ranged according to opinion from forty to sixty miles away." [25] The horse they loaned him looked as though it couldn't make four miles, much less 40, and certainly 60 would be the death of them both, but one bystander suggested that the old mare was "willing," and that he had ridden her 25 years before, and she was all right then. Of course, the mare had the use of both eyes that time, and hadn't been blind in at least one, but she'd be just fine, he was assured. John was not convinced, but lacking any other offers, mounted and headed toward Leslie County, Kentucky, which now seemed farther away than anyone had estimated.

The old mare proved to be everything her mountaineer advertisers said she was, and more. "Most of the time the road was the creek bed," John remembered, just like the one into Kingdom Come, "sometimes rocky, and sometimes of solid gray slate into which the wheels of heavily-laden wagons had worn ruts but a trifle wider than their rims, sometimes eighteen inches deep. The old mare was an expert mountaineer. She had but one gait on level ground — a swinging pace. Uphill she would go at a quick walk, and down hill stiff-legged, letting the force of gravity do her work, and making the avoidance of vertebral dislocation my work." [26] John was reminded of Fuji, his old war-horse from the Russo-Japanese War, and he grew quickly fond of the old mare, dubbing her "Old Faithful."

He saw many of the same changes here that he had seen on his previous trip to Kingdom Come that same summer, but these changes here were of a rougher sort, he thought, and the people he met seemed more suspicious of his motives. The first night he stayed with an ex-sheriff of Jackson, Kentucky, Ed Callahan, who had recently been wounded from ambush, the latest victim in an ongoing feud in Breathitt County, and he was, John saw, practically a prisoner in his own home, being afraid to show himself either before open window or door, and having a covered passage between his home and the store he ran for a livelihood. Callahan would seldom leave his home, John wrote, "appearing outside only when he must, and never knowing when a bullet would sing at him from the bushes; never standing in a doorway or before a window with a light behind him, clothed in black to give no aim to a rifle, and when he had to leave home, slipping off at night by some unfrequented way, and always — no matter where he was — in momentary danger of death." John felt his skin crawl a little while he sat at the table eating, unable to see through the darkness outside, wondering who or what might be looking in at either of them over the sights of a rifle outside in the darkness. [27]

On past Buckhorn, toward Hazard, John and "Old Faithful" went the next morning, after leaving the doomed ex-sheriff, starting well before daylight. He was mistaken for a preacher along the way and finally encountered Devil's Jump Branch. Following that for an hour over some of the roughest terrain that was ever created, he eventually stepped into the creek that bore the name of Hell-fer-Sartin. "As I rode down," he said, "I was politely told the name of the creek by a man and by a woman, each without a smile and each correcting my pronunciation to Hell-fer-*Certain*—for the present generation of mountaineers is losing its dialect fast." [28]

He heard them before he saw them. He could hear preaching that only preachers in the mountains can deliver, or care to. What they lacked in originality of message, they more than made up in length and volume. The preaching was going on in a nearby school house, which often happens since schools are always closed on Sundays, and the building stands waiting and wanting on those days. "The bellowing tones of the preacher issued from the little frame school-house," he remembered, "and the windows of it were suddenly filled with curious faces regarding me." [29] The people, he noticed, wandered in and out of the services at will, and little or no notice was taken, and having heard sermons of hell-fire and damnation before — almost exactly the same — he too left after a few minutes.

Finding the real church house on up the creek, he found that someone, obviously seeing a necessity to change the name of a church that bore such a name as "Hell-fer-Certain Church," had changed the name to something less remarkable. He noted that the church, under its new name, was for some reason closed that day, and he turned back toward Jackson.

Back he went, and nearing Jackson, nightfall overtook him and "Old Faithful." He began looking for the ford of the river, which was rising and getting deeper than when he had crossed it only a day or two before. He could hear its roar as a constant companion, and in the darkness that roar of swift water seemed more terrifying than it should have been to him. He was alone, and although his direction back to Jackson was sure in his mind, he also knew he would have to cross that river sooner or later. In black darkness, he felt his way along, not too troubled about running into anything. "Old Faithful's" gait forbade anything even remotely approaching a reckless speed. "At several houses I shouted inquiries, and at each house I got the cheerful news that at each I was the same distance away from the ford, so that Old Faithful and I were at least holding our own." [30]

Suddenly, it occurred to him that perhaps his horse was more knowledgeable than he was about the road they were traveling and the ford they were looking for. "It occurred to me," he said, "that during the twenty-seven blushing summers of her young life she had been over that road before and that she had not perhaps forgotten how to swim. So I let the reins loose on her neck and while I saw, as it seemed, many fords, I let her alone, and we ambled on through the dark for two hours and, it seemed, many more. And I did well to trust her, for without warning she suddenly turned down a steep bank where I could see nothing and boldly entered the river. The swirling yellow struck the middle of my saddle skirts but I had the faith of ages in that old mare now and she was soon climbing the other bank precisely at the mouth of the creek." [31]

At the front of the hotel in Jackson, he left his trusty steed with a great deal of "regret and affection," and tenderhearted as he always was toward animals and the less fortunate, he entertained for a short time an idea of buying the old mare and putting her out to graze for the rest of her life. But he could not take her home with him on the train, and realized that as soon as he was out of sight, any money he used to buy her retirement would be pocketed by her owner, and "Old Faithful" would be hired out again and again.

He had been to Hell-fer-Sartain, and as he was leaving on the train, bound for the Gap and home, he knew he would never willingly go back. He had been glad of the trip to Kingdom Come, and had been glad of this trip, but he had not been as impressed with Hell-fer-Sartain as he had been with Kingdom Come. The places and the people seemed different, and he left with regret for what he had expected to see, and had not. "On the way to the train," he remembered, "I saw Old Faithful plodding along the road with a stranger on her back, starting out for another toilsome trip. I wish," he wrote, "I had bought freedom for Old Faithful." [32]

He went back to the Gap, worn out from the trip. His trip from Jackson to Hell-fer-Sartain had taken him 46 hours to travel there and back, and all but 19 had been spent in the saddle. He was tired, but as usual, such outings and exercise did more to better his health than anything he was able to do all his life. After a bath and food and rest, he felt rejuvenated, and again, sitting down at his desk in the back of the house, scooting up closer to the typewriter, the story of "On the Road to Hell-fer-Sartin" unfolded.

Two years before, he had written *The Trail of the Lonesome Pine*, and almost immediately the Gap was visited by tourists looking for the "real" June Tolliver, the "real" Jack Hale, the "real" this or that of the novel, but most of all they looked for the "real" Lonesome Pine. Denials by himself to the contrary did little to change their minds. The tree had to be there. The characters in the novel had been too real, too alive, including the lonely old pine tree, for it all not to have been real. As in the directions he had gotten in Jackson on his way to Hell-fer-Sartain, the location of the pine was many and varied. One would avow that it was here, another there, and another would swear to have seen it somewhere else. Train car conductors were known to point out a certain pine to the passengers, swearing that this was the "real" one for sure. In time, John began to think there might be something to their romantic searches, though the characterization of that Lonesome Pine originally came straight and only from his writer's imagination.

He started in March, when the first signs of spring were coming to the Gap, and the weather, though changeable, had moderated enough to allow some degree of comfort. Down Shawnee Avenue he rode, away from the house, away from the center of town. Up the hill on the other end he went, past the last houses of the town, and on toward Wise, Virginia, where the "Red Fox" and Talton Hall had been hanged, so long ago, it seemed. There, he turned into Pound Gap, where Doc Taylor, alias "The Red Fox," had ambushed and murdered the Mullins family with the help of Calvin and Henan Fleming— the crime for which Taylor was hanged and Calvin killed in a shootout with the law in West Virginia, and Henan received a new hat.

Part of the way, the road was as good as one could find in that part of the mountains, but eventually gave way to the more primitive trail that had not changed in decades. Stepping from wide road to primitive trail was like going back in time. The trees and rocks were the same, and the winding trail through them was the same as it had been when Talt Hall had fled over it, escaping, he thought from his murder of Enos Hylton, and finding only "twittering birds" to foretell his future. He rode past the "killing rock" where the ambush of the Mullins family had taken place. The bones of the horses killed nearly 25 years before were still to be seen scattered along the side of the road in the weeds.

As soon as he and his traveling companion and guide got close to the gap between Kentucky and Virginia and rode down the other side of the mountain toward the town of Jenkins, they began to see drunk men sitting on the side of the road, and others propped up against anything available while the moonshine did its work. "Now I had supposed that moonshining in the mountains was on the wane," he said, "and so it is generally: but in that region where local option had cut out competition and the opening of mines had made a flush market the trade seemed to flourish and the trail I followed was slippery with the stuff." They even tried a taste of the wicked brew themselves, but, John admitted, "A wet tongue was enough for me." [33] Later they would find other homemade liquid fire that was more pleasing to their taste, or perhaps they quickly got used to the rough liquor.

They were caught in a fierce storm, with wind and rain whipping their faces, and John was reminded of his trip back from Hell-fer-Sartain onboard "Old Faithful"—he couldn't see a thing in the darkness, and the noise of the storm precluded his hearing very much. "Literally for an hour we traveled by the flashes of lightning," he said, "going as far as we could see by one flash, stopping to wait for another one, and then going ahead again. Meanwhile as a trivial detail the rain was pouring and the wind swished it this way and that as though cackling witches were slinging water at us with wet brooms." [34] Soon they were both soaked to the skin, their outer clothes weighing heavily on their shoulders, and their inner clothes wrinkling and creeping uncomfortably against their clammy skin.

Finally, a light twinkling through wildly waving tree branches showed them the way to a cabin where they found shelter and food that consisted of the usual rough—but good to a hungry man—fare. John was beginning to acquire a taste for the moonshine they had picked up on their way. "After supper," he remembered, "I took another dram of the moonshine—to help dry out my clothes—and the agent [his traveling companion and guide] was right; nobody could tell just what the stuff would do to him. Ten minutes later I stepped out of the house with the agent and reeled into his arms, deathly sick, and I went at once to bed." [35] The next morning, he observed, he was all right again.

His hosts were more talkative than ever after their meager breakfast, and the subject of entertainment came up. There wasn't much around the mountains to do for fun, they told him. The boys generally got drunk, and the girls got married as soon as they could, some as young as 13, and some as old as 17, but they were lighthearted about it.

Later in the morning, having taken their leave of their hosts of the night, they found their first rumored site of the Lonesome Pine. "There was a lonesome cove down there and a lonesome cabin," John wrote, "but there was no towering pine anywhere, nor had the occupant of that cabin, a typical black-haired mountaineer, never heard of it, but he had known the Red Fox well." [36] A leisurely talk with the old mountaineer revealed another source for the ever evasive pine. "Ef thar's anybody who knows whar it is, hit's Uncle Hosey Bowlin," [37] and off they went again, in pursuit of the holy grail.

Uncle Hosea looked awfully old, John thought, but his white beard may have made him look older than he actually was. He still had his wits about him, however, and they even got sharper after a sip or two from their deadly bottle of moonshine that John and his companion had thoughtfully brought along with them. It also made Uncle Hosey's blue eyes sparkle a little. No, he had never heard of a pine tree called the Lonesome Pine, but he did remember long ago that hunters knew of one that they used to call the Lone Pine. "That was close enough for me," wrote John. "I had run that pine to earth at last. It was on top of the mountain, Uncle Hosea said, and stood at the foot of some great cliffs right on the State line." [38] But, said the old man, they would not be able to find it by themselves—he would have to guide them (and their bottle of moonshine) there, and since he wasn't feeling too pert, they would have to wait until he was feeling better.

Two months later, John made the journey again, this time in the company of an artist who was to illustrate those three sketches he had written for Scribner's. The dogwoods were in full bloom at that time of year, and it was almost as if the mountainsides were covered with a late snowfall. Curiously, they hardly saw anyone on their journey, and then found that the revenue agents had been active in John's absence and had arrested moonshiners and destroyed moonshine stills left and right. It was a dry trip they made that May, but a sober one.

Uncle Hosea was still living when they saw him again that spring, but "still ailing and genuinely distressed that he was yet unable to take us where we wanted to go. Once more, then, I had to turn back on the trail of the Lonesome Pine," he wrote. "I am still waiting for Uncle Hosea to improve, and if he ever gets strong enough, I shall strike that trail again. It may be, however, that the pine which lives only in the memory of that one old man will, as far as any man can know, die with him. But though no mortal eye may see and know it again, may it still stand, 'catching the last light of sunset'." [39]

Those 1910 articles for *Scribner's* were some of the best writing he did after the publication of *Lonesome Pine*. In them, he returned, literally, to the roots of his greatest successes, and, always a keen observer of those around him, he wrote convincingly of the moonshiners that he saw and did a little business with himself. They still seem to be running off a "dram" or two around Black Mountain. The boy whose banjo playing reminded him of Chad's still seems to pick a tune recognized from long ago, even if it hasn't been heard before, and "Old Faithful" still plods along the road outside of Jackson, carrying strangers to the oddest places and bringing them back safely again in the darkest of nights, and Old Uncle Hosea still sits under his apple tree, trying to get over his "ailing" long enough to guide a curious stranger on a perilous trip to find a lone pine tree, and, perhaps finally, be satisfied.

These years were happy ones for him. He and Fritzi were having their differences from time to time, and money continued to be in uncertain supply, but he was getting accustomed to that, if not used to it. But none of their problems seemed overly serious as far as marital problems go. He looked around himself and could not help but think that now he had it all: a beautiful wife of whom he can readily be excused for being jealous, a home, and a family, though neither had children of their own. He had written two novels in the last ten years that had defined the term best-seller in America, and he had settled in — notwithstanding frequent trips north — to the respected position of resident author and local star of the Gap. But such happiness, it seems, must always come with a price, and the years to come would not always be easy or peaceful ones for him. There was terrible sadness on the horizon, and before the next few years were finished he would begin to think, with some justification, that his happiest days were behind him.

The year 1912 would prove to be the darkest year of his life.

CHAPTER 19

The Eyes of a Father

> Our Golden Anniversary is next year ... if we live to see it.
>
> John Fox, Sr.

It has been said from time to time that the dying often have premonitions of their own deaths. It is possible, however, that too much credence is put into such things. When John Sr. wrote to his son James on the last day of January 1911, he sounded a little more dejected than usual, but soon he was in a more cheerful mood and forgot his wondering thoughts of anniversaries and of the races that are run between time and life.

Throughout the rest of 1910 and all of 1911, things continued much as they had before for John Jr. and Fritzi. His and Fritzi's trips to the Gap and the remodeled house were mostly confined to the summers, with their winters being spent in New York City, though they switched back and forth constantly, and it would be difficult to say that they stayed more in one place than they did in the other. In the summer, the Gap was filled with distractions.

Baseball, along with golf, was one of John's favorite pastimes, and although he would be the first to admit that he was not much of a hitter, his enthusiasm for the sport never wavered. He loved baseball, but it was golf that was his affliction. In Lexington, Bristol, Knoxville, and even in the backyard at home, any bright sunny day would find him and Fritzi, and friends from near and far, playing somewhere. A golf course in Lexington named one of its particularly difficult greens "Hell-fer-Sartain" in honor of the story and its author. Like many of the pastimes and interests that seized him throughout his life, the game of golf became his passion, almost to the exclusion of any other interest, especially in the summers. Fritzi noticed that it even interfered with his writing.

Fourth of July celebrations were always special in the Gap, and in those times of unabashed patriotism the celebrations of that holiday took on epic proportions. Entire towns turned out and closed up for the festivities. There were baseball games and jousting tournaments, and even airplane demonstrations, to the wonder of all who saw these strange and wonderful contraptions. But mostly, there were the mountains and peace and quiet that attracted John back to the Gap, and leisure time to do whatever he wanted, or do nothing at all. He wrote best while in his study at the Gap, usually with one or more of the many dogs they kept, sleeping nearby. The summer droned outside in the backyard just behind him, while his two-fingered style of typing could be heard all through the house. It was a sleepy time in a sleepy little town, and mostly he wanted to be outside, but he plugged away, working on a new novel that he wondered, in those times of endless summers, if he would ever finish.

On January 28, 1912, John Sr. and his wife Minerva celebrated their 50 years together. It did not seem so long ago to him that he had brought her to his home that winter over muddy roads to meet his three motherless sons. He could still remember how well she had taken the hardships of that trip, and remembered how impressed he had been with her that day. That feeling had never waned over the five decades they had been married and the seven other children they had between them. Nearly all their children had been on hand for the celebration. Everett and all his family had been there; John was there without Fritzi who was singing somewhere and could not come; Horace and his wife May; Minnie, still living at home and unmarried; and Elizabeth, in from Washington, D.C., with her husband Cabell Moore. Rector and Richard had come with their families all the way from New York, and it was especially good to see them and their children, whom they did not see as often as they would have liked. James had not been able to come, and neither had Oliver, and that had disappointed them only a little. Sidney's widow, Polly, who had remained close to the family all those years since

This photograph of John Fox, Jr., was used as the frontispiece for the 1911 edition of *Crittenden* and shows the stiff formality that many of his later photographs reveal (Charles Scribner's Sons).

Sidney's death, did not make it either. But all the noise and excitement in the now crowded house, made their absence, if not unnoticed, at least easier for them to overlook.

On Monday, however, most of them left except Everett and his family, and the next day the last of their guests and children had gone away again, excepting Minnie and John Jr., who stayed until the first of February, "and now we are feeling lonesome," John Sr. wrote in his diary. The house seemed so big and empty that Monday and the days that followed. The children and grandchildren had come in like a whirlwind, and all of a sudden lights had burned well into the night. There had been laughing and talking and remembering and stories to tell, and it had been so much like he wanted it to be every day, so that when they left, as he knew they must, tears gathered in the corners of his eyes.

By the middle of February, John Fox, Sr., complained of not feeling very well, but marked it down to the cold and nasty weather they were having. On the 18th, he wrote in his diary that "I have had an unpleasant day, a sharp muscular pain in my left side, and constipation of the bowels all of which is very distressing." [1] His aches and pains did not keep him from his regular rounds about town and the house, and he carried on his business much as usual, if at a slower pace, and staying indoors only when the weather was at its worst.

John Jr. came home again, but only stayed a few days, and was off again. By the end of February, John Sr. was still not feeling well, and wished the winter would soon be over. There was nothing like an early spring to raise the spirits of those afflicted by aching bones and muscles. By March, he was better, though weak, and a week or so later, the weather had moderated enough so that he was able to plant early crops in his garden, prune apple trees, and gather and pile rocks out of the fields.

John began spending a little more time at home now, leaving only on those times he needed to meet Fritzi in Cincinnati or Kentucky or go to Chicago to see the opening of "The Trail of the Lonesome Pine" stage play. James came by, as did Everett, and as always they were glad to see them. Late snowstorms made it seem as if spring would never come, and the land around them lay bare and brown when the snow was gone. "The grass is very backward," John Sr. noted in his diary. [2]

March went, and April came, and he still felt unwell and it seemed to trouble him more than before. To his family he was silent about it, and said very little, but those that saw him every day, especially Minerva, noticed he was not well. By the end of April, he made a short notation in his diary. "A partly cloudy day," he wrote for April 28. "Not feeling very strong. I have been closely around home all day," [3] and by the first week in May, "I am dreadfully weak and can scarcely hold up my head." [4] Two days later, he was able to leave the house for a short walk up Popular Hill, but his legs trembled beneath him, and he was glad to get back home and sit down, breathless and weak.

Through the month of May, the entries in his diary followed a similar pattern, only progressively worse. On May 4 he wrote, "Am feeling much more weak to day," and on the 5th, he did not leave the house all day. On the 8th, he complained that "my condition does not improve and I am dreadfully weak," and on the 9th, he admitted again that he had to "stay very close to the house." [5]

Elizabeth came to see him on Sunday, the 12th, and he could tell she was worried about him, and he tried to reassure her that he would soon shrug off whatever was ailing him, and be good as new again. When she left, she did not seem convinced, and neither was he.

On Sunday, the 19th of May, he wrote, "I feel a little stronger to day," but the next day his entry was short and terse: "The same as yesterday," he wrote. [6] Nearly a week later, on the 26th of May, he wrote simply, "Sick," and for the next five days, the entry remained the same, and then the pages went blank and unwritten. He had kept his diary faithfully for 60 years, every day, in his fine script, detailing the weather and short descriptions of those who had visited and those who had not. He had written of crops and grass seed gathering in Georgetown and Paris and Lexington, from where he had written his wife Minerva that he missed her very much and would rent a horse to ride back to Stony Point to see her, but the cost was too dear.

He had written of the birth of all his sons and daughters. He had written of the death of Kitty, which broke his heart, and the marriage to a new wife two years, which that made his heart strong again. He had written of Sidney's death in New York, of friends and neighbors whose troubles bothered him also. In those pages he had revealed his gladness when spring came and delight in how much his garden had grown that particular season, and how terrible the roads were in the winter, and how it was often the worst weather he had ever seen. Through all those slim volumes, his handwriting never faltered in telling of his life in the coming and going of the seasons of the earth.

On June 28, 1912, at 7:25 in the evening, John Fox, Sr., died in the house he and his sons had built, near the end of Shawnee Avenue, in Big Stone Gap, Virginia.

They had been called to his bedside, and all but Everett, Oliver, Rector and Richard made it in time. John and the others were there, and when the old man finally slipped away, John Jr. felt a terrible cold and clamminess run through his body, and he thought his own heart would cease.

By the time they were ready to take their father back to Paris, all the rest of the family had gotten home. They took him home to Paris and buried him in the family plot near the middle of Paris Cemetery that Sunday, and stood around with friends and family, some of whom still lived nearby, not wanting to leave or go back to that empty house in the Gap. Finally, someone made to leave, and the rest, seeming almost grateful for the guidance, followed down the hill and away from the death and loneliness of that place.

John Fox, Sr., in the backyard at Big Stone Gap with friends. This photograph was likely taken shortly before his death in 1912 (courtesy University of Kentucky).

Now, John was more alone than he had ever been in his life, and the sad truth was that no one could think of anything to make him feel otherwise. They knew, through their own grief, that there was nothing they could say or do. John had depended upon his family for everything, especially the unquestioning and unwavering support they had always given him, but his father was the one he had been closest to, the one he wanted to be most like, the one whose character and personality he most admired, and duplicated when he could. He saw in his own face the lines that belonged to his father, and when he looked into a mirror, into his own eyes, he saw those same eyes he had trusted as a child. He felt inside himself the beliefs and emotions that he and his father had both shared with each other. So much of them both were the same, and when the old man died, John felt the reality of his own age. He was surprised at the suddenness of his father's death, and he was never the same afterwards. Elizabeth would remember that their father's death affected John seemingly more than the others. "His grief," she would say, "was pitiful to see." [7]

He could not bear to stay in the house, nor even in the Gap. Everything reminded him of his loss; everything he touched that had belonged to his father seemed more precious to him now than it would have been otherwise. By the middle of July he and Fritzi were gone away, and so was his mother. Over the next few months, their mother and Minnie traveled to New York to stay a while with her sons Rector and Richard, and down to Washington, D.C., to visit Elizabeth and Cabell. Anywhere they went was better than staying in the Gap, and the rambling old house stood empty except for Horace and May who came by from time to time to check on things and see that the dogs were all right.

Nothing was the same around the old house on Shawnee Avenue after the death of his father. The house was too quiet, but also too noisy with memories of the man who had been their strength and a steadying influence for all of them for so long. A few years before, they had entertained thoughts of moving back to Kentucky, and had even made a trip or two to investigate homes there. He had wanted a big house so all of them could be together, and a few acres, he said, to keep him busy and perhaps make a little money for them. But the plan had fallen through. His father had never really felt entirely comfortable in the Gap, and had never really felt at home since leaving Stony Point so long ago and moving into Paris and opening his school there. The move to the Gap in 1890 only worsened things for him. John Sr. had often felt abandoned with John, and James and Oliver off somewhere doing whatever it was that they were doing. Horace was busy with his own concerns, and Richard and Rector were hundreds of miles away. His family had splintered and scattered, and as he grew older he had missed them more and longed for those old times at Stony Point when they had all been together.

With his father's death, John grew more and more protective of his mother, and the more concerned he became with his mother's well-being the less patient he became with Fritzi, and she with him. Mostly, he worried whether his mother and Minnie had enough money or not. "Please let me know, mother," he wrote her at Rector's in Mt. Kisco, where she was visiting, "when you need any money, or get it from Rector and ask him to draw on me for it." [8]

He always felt responsible for both his mother and father, especially in the last decade, and now that his father was gone all that attention fell on his mother, and to a lesser extent, though not by much, upon his sister Minnie, who had never married and never would.

By the end of July, he had returned to the Gap, leaving Fritzi in New York where her engagements were keeping her as busy as ever. His mother was still with Rector, and he encouraged her to stay there as long as she wished. Staying away from the Gap would be good for her, he was confident, and for the time being she did as he wished. "I hope you are well and cheerful," he wrote from the Gap where he was now staying in the house alone, "and having a good time and I want you to stay just as long as you please, go to Washington and Chicago and come home only when you get good and ready. Please let me know when you need any money. And don't worry the least bit about us. We miss you and will be mightily glad to see you but we are happy that you are with Rector and Hilda and your grandchildren. Furthermore," he continued, "you have given up your whole life to others and now we want you to consider only yourself, for God knows you have earned that right." [9]

In Washington, D.C., now, with Minnie and Elizabeth and Cabell with her, his mother was beginning to be lonesome for her home once again. She had never been one to travel very much, and had enjoyed, up to a point, seeing Richard and Rector and the grandchildren, but now she felt it was time to go back. Her husband was gone, lain far away in Paris, and naturally she would miss him and look for him puttering around the yard and getting underfoot in the house. She could smile a little now without crying when she thought of his habits that sometimes annoyed her. He did not have a great deal of humor about him, had always remained a little like the stern school teacher he had been all his life, but underneath she knew he had been a good man, and she had loved him very much, and so she went back home and began to plan the rest of her life without the only companion she had ever known.

John Jr.'s domestic affairs continued to deteriorate. There is some suggestion that the climax of his marital problems came when his father lay dead in the house in Big Stone Gap, and may have initiated the final break over some real or perceived slight or backhanded insult. Whatever brought the whole thing to a head, by the time late summer and fall of 1912 rolled around, he and Fritzi were separated and both apparently wishing for a speedy divorce, and one without complications, if possible.

New York, where the divorce papers were eventually filed, would tolerate no other grounds for the dissolution of a marriage but that of infidelity, and John allowed Fritzi to file for divorce on those grounds, though it is almost a certainty that it was untrue, and only an accommodation to Fritzi. In time, the divorce would be granted, but in the meantime, no matter how anxious both of them were, they would have to wait until that slow process worked itself out. [10]

His divorce still unresolved in 1913, he went everywhere he could think of that he would be welcome, everywhere but home to the Gap. There he could not go. The memories of his father, though somewhat lessened by time, kept him away as did the unhappiness of his failed marriage. The rambling house had been enlarged and remodeled for Fritzi and himself, and everything done, every enlargement could not help but remind him of that. He never liked reminders of things unpleasant, and usually avoided them as much as he could. Years before, he had written of another man, stricken with love and in danger of losing it all. "I wonder if you can know what it is to have somebody such a part of your life," he had Crittenden say in the novel of the same name, "that you never hear a noble strain of music, never read a noble line of poetry, never catch a high mood from nature, nor from your own best thoughts — that you do not imagine her by your side to share your pleasure in it all." [11] The sad truth was that he missed her, missed her more than he thought he would now that she was gone, but not enough to go looking for her and try to get her back.

His latest novel was finished, finally. He had worked on it for the entire period of his marriage, and only now was it finished, and it, like his marriage, promised much, or so he thought, but in delivery turned out to be much less than he desired.

In *The Heart of the Hills*, he returned to what he knew and did best, with the exception of his nonfiction reporting. He returned to the mountains and mountain people. In his writing, they had always been faithful to him. The few times he had deviated from that theme, he had not been so successful. Now, even that would not work for him. He was able to carry on the local color stories long after others had given them up. Until the time he would pen his last word, he would always find a market — a willing, albeit a smaller and smaller market — for his mountain stories. That market might not always be enthusiastic nor as large as before, but it would always be willing.

The mountains, however, had changed by 1913, and the people in those mountains had changed as well. His trips to Hell-fer-Sartain and Kingdom Come, and his search for the one true Lonesome Pine, had revealed to him just how vast and irreparable those changes had been, and it had also been a reminder to him that he and his brothers were in no small part responsible for some of those changes. It was a part of his life that he was beginning to see more clearly, and to regret. He had seen what the unrestricted buying and selling of land and coal and the blind harvesting of both had done to the people and the mountains, and he had never admitted to himself that what he and James and the others were doing was a bad thing. John salved his conscience by constantly reminding himself that if they did not do it, someone else would. But now, seeing the evidence of that destruction all around him, he could see that it was something that he was uncomfortable at having been a part of. He had thus determined to write something of the way he felt in the pages of this new book, to try and explain what had happened to his mountains, and why. He thought, this one last time, he would be able to write about his beloved mountains, and people would understand why he loved so much those mountains and those people who lived in them.

All in all, reviews of the book were decidedly lukewarm at best, and most suggested, either directly or indirectly, that Fox needed to move on to other topics, if he could. He had mined his "richest veins" of material about the mountains and mountain people. He had written about almost nothing else, and the sameness of the stories was beginning to tell upon his success as a writer, and also his popularity. [12]

He would make one final effort in 1917 to sift the "tailings" from the mountain story before moving on to something radically different. Despite the New York *Times*' assertion that he had no talent for the historical novel, he would try to change the course of his writing and pen a masterpiece that would, he hoped, bring him back to the success and fame of *The Little Shepherd* and *Lonesome Pine*.

In the years immediately following the publication of *The Heart of the Hills*, he saw no peace of mind. His successes in writing popular novels had created problems he had never imagined, and much of his time after 1913 was taken up in fighting various infringements upon his books, especially *The Trail of the Lonesome Pine*. Almost immediately, the book was made into a stage play and a successful one at that. The legal maneuvering and royalty payment negotiations took up too much of his time, and his dealings with producers, directors and their lawyers did nothing calm his nerves. He was often angry and disgusted, but he found himself stuck and could not let go.

His last two best-selling novels were so popular it was almost inevitable that someone would want to make them into stage plays, and the growing popularity of motion pictures opened up new and intriguing possibilities. He had worked on ideas for plays ever since his New York days, and had started one or two, but had never finished anything.

A song was written by Ballard MacDonald and Harry Carroll that took the name of "The Trail of the Lonesome Pine," and he had to bring suit against the publishers for use of the title without his permission, in an attempt to gain royalties. *The Little Shepherd of Kingdom Come* was to be made into a stage play, and he signed a contract that fall of 1913, but the play was not produced until three years later because of mismanagement and everything else that goes with producing plays and motion pictures. *The Trail of the Lonesome Pine* was almost immediately produced as a successful play, and film versions of both books, and the resulting negotiations, were a drain on his temper and patience, and the agents and lawyers he hired to help him sort through the mess were of little help. [13]

After their divorce early in 1913, Fritzi's career began a downhill slide that was probably as inevitable as their marriage. She would, in time, gain additional success and fame in vaudeville acts, but in-between times, money became very tight for her. With his worries about royalties, and the lack of real success with *The Heart of the Hills*, John's monetary income once again left him counting his nickels and dimes. He had to caution his mother, to whom he continued to send money, to move more carefully with her finances, and, if necessary, let some of her bills go unpaid for a time, until he could get himself in "better shape."

By the middle of 1913, John was 50 years old, and often tired. The things that interested him in his youth no longer concerned him as much as they once had. Middle age was upon him, and had it not been for what he knew middle age foretold, he would not have minded the release that particular age gave him. By that time, he could not help realize that his future was in his past, and he wondered what he had and had not accomplished, but his thoughts seemed to dwell longest on what might have been.

With the wrapping up of *The Heart of the Hills* and the settling of his New York business, he hoped that he had worried the last he would ever have to worry about Fritzi and his domestic affairs. He had no new ideas about what to write, and suddenly found himself completely free of any responsibility as far as deadlines and lawsuits were concerned. Taking Horace Greely's advice somewhat belatedly, he determined to go hunting out west that fall of 1913, to get as completely and as far away from the Gap and New York as he could possibly get, and still remain in this country. Perhaps, he thought, when he returned, things would be all right again.

CHAPTER 20

West and East — North and South

I am ... anxious to get home.

John Fox, Jr.

JOHN'S FIRST TRIP ACROSS THE MISSISSIPPI had been in the 1880s when he had gone with the family whose children he was tutoring. He had been fascinated then by the open spaces and the absolute vastness of that part of America. In 1913 he made his way west again, looking for adventure and, perhaps, peace of mind. The West was just as he remembered it from his times with Fritzi, criss-crossing the country, and from his train excursion so many years before with the Kemp family. "Even on the way," he would write of this last trip west, "you are prepared for the vast reaches of earth and sky in the west but you can never know the spirit of the land until you ride between them — alone." [1] He and a few close friends had been planning their trip for weeks, and the train ride across the rivers and plains only served to heighten their excitement. It was cold when they arrived at Jackson's Hole, Wyoming, but their guide did not seem to mind, going around in clothing best suited, or so John thought, for Fourth of July celebrations back home in the Gap. On horseback, they entered Jackson's Hole just as darkness was falling.

The next day, John, the complete dude, went shopping for his western outfit. "I have never seen anything like myself outside a western drama on a moving picture screen," he remembered after trying on his purchases. "I had an eight millimetre Mannlicher, a sombrero, high boots and midnight chaps that would have made a bear-dog howl with joy. Several inhabitants grinned delightedly while Dismal Dick helped encase me. These chaps were stiff as boards and I walked perforce as wide as any cowboy on the range. Other inhabitants including several girls were hanging with expectant grins over the fence around the corral which enclosed our horses and the pack train." [2] He thought, however, that after his experiences aboard Fuji in Manchuria, anything the West had to offer in the way of cantankerous horses he could handle quite well.

He need not have worried. His mount proved to be the gentlest of the bunch, which was a good thing, since he supposed that, once on the ground in his new stiff clothes, he would be like the proverbial turtle on its back, unable to get up by himself. That night, they slept under the western stars for the first time, and like everyone who sees those stars for the first time, all other thoughts are lost in the sheer number and wonder of stars that seem to be hanging just beyond reach.

They had been steadily climbing the day before that first camp, and continued climbing the next day, though it was not very perceptible to John, the rise of the country being so gradual, but the next morning the higher altitude was noticeable and troubling to him. "Next morning I was light-headed and a cigarette made me dizzy. It was the altitude of course and I wondered how long the discomfort would last. Bearface told me it had taken him two years to get over it [and] I began to wonder how I was going to climb hills when I couldn't walk comfortably on level ground and how with such palpitation I was going to shoot straight." [3]

The second morning in camp dawned bitterly cold, with two feet of snow on the ground, but they did not wait for it to get warmer. Off they went, and soon found "bunches" of elk grazing and dozing on the hillsides. They began moving toward the elk, some distance away, mostly downhill, and then uphill. "We slipped, slid, crunched and banged down a few feet at a time," he remembered, "clutching now and then a friendly little tree and by the time I was peering through a little opening in the trees down at that resting elk, my legs were quivering like quaking aspens, I was out of breath and my heart was beating a lively tattoo against my ribs ... and I raised and fired. To my utter amazement that elk rose and before I could fire again disappeared." [5] He had missed and could not understand how it could happen, since the animal had been so close. They had killed an elk the day before with little or no trouble at all, which had made such things seem easier than they actually proved to be. His guide said nothing; he had seen it all before, but John was, as he said, "mortified" at his marksmanship, or rather the lack of it.

The next day was almost a repeat of the previous day's debacle. He shot two or three times at running elk and missed them. "If a Zeppelin air ship had come out I should have missed it," he said of his shooting that day. On horseback they galloped over terrain that looked dangerous to walk on, but at breakneck speed, they chased the game across mountain, hill, valley and anything else that got in their way. Finally, they caught up with the herd, "and when we reached the valley there was that magnificent bull two hundred yards in the rear alone — his mane shaggy and black as midnight his noble head uplifted —

sole guardian of his thousand wives. I fell off my horse and fired," and missed again. [5] His next shot was better, and the elk went down. "When I walked up to him in the red snow," remembered John, "I could take little pride in his downfall." [6]

The next day, he killed another, and was done. His license had allowed him two elk, and he had been successful (the first had technically been killed by his guide). The hunt ended fine. He was tired, dirty, worn out and needed a shave worse than at any other time he could remember, but he was happy. "On Saturday morning," he wrote, "we started back to Jackson [Jackson's Hole].... For ten days we had had neither bath nor shave and when we reached Jackson late the next afternoon we got a shave," [7] but for some reason, if they had baths, it was not mentioned.

From Wyoming, they turned south into Colorado. They crossed over Teton Pass, boarded a train, and rode in style, relative style anyway, to Glenwood Springs, Colorado, on the trail of bear and mountain lion.

Out of Glenwood Springs on horseback once again, and alternately followed and led by a pack of hounds, he found his "moving picture chaps" that had seemed so comical back in Wyoming were becoming more and more useful. The terrain had changed from open vistas to more brush and thickly growing aspens through which they were obliged to follow the dogs. Finally, in desperation, he and the others let go of their reins and used both hands to keep the low-hanging limbs from scratching their faces, tearing off their hats, and possibly dismounting them in the most inopportune of places. He was beginning to wish he had "moving picture chaps" from head to toe. He likened the experience to "swimming" through those trees.

The hounds quickly struck a trail, and it was uncertain whether it was the trail of a bear or mountain lion, but since they were after either, they followed the best they could. The chase reminded him a little of his earlier escapades hunting fox and rabbit from horseback in the Bluegrass, but then there had been only fences and low rock walls to contend with. Here, travel was more difficult, and it seemed that the dogs took whatever trail would be most difficult to their followers. "Then up we went — through a grove of big aspens whose trunks were scarred by the claws of young bears and up the same steepness, same brush — twisting, stumbling, crashing — to a grassy knoll. But the music of the hounds was growing fainter — they were going over the ridge and into the next valley beyond." [8]

The dogs treed neither bear nor mountain lion, but a bobcat which they soon shot and skinned, and moved on. They were in some of the best hunting country in the world, and some of the most beautiful. Deer ran through the middle of their camp, and it was debatable who was more surprised: the deer or the hunters. At any rate, the deer escaped unscathed. "I had hunted him for four days in Wyoming," John wrote wonderingly, "and in Colorado—where it was against the law to shoot him," [9] he ran right up, and practically tried to shake hands. John was learning a great deal about the vagaries of and unpredictable nature of hunting.

They struck a bear's trail and then began a madcap rush on horseback to overtake the hounds and the bear. A branch tore John's glasses completely off his face, throwing them into a snowbank; he retrieved them, and hurrying to catch up with the others, put them in his pocket for safety. The bear "treed." "There was a small limb just under his shoulder and I had to graze that limb above or below," he wrote. "I shot above it and the bear climbed two or three feet up the trunk and stopped. Good Heavens, could I have missed that huge black mass, or struck the limb!" [10]

He had not missed, though he took another shot, and the bear fell like a wet sock. Apparently, he was a better shot without his glasses, or at least, no worse. It would prove to be the last shots of the hunting trip. That night, a heavy snow covered any tracks and trails that they might have followed, and they spent all day in camp, eating, drinking coffee and talking over old hunts in great detail and planning future ones.

"That day, too, was the last of the hunt," he wrote a little regretfully, "and next morning we started back to the railroad through the magic beauty of the snow-covered woods and hills. We made Glenwood Springs before dark and straightway under cover of that dark we were swimming in our altogethers in a big pool of natural hot sulphur water — the first bath since we left New York." [11]

When he left Colorado and Wyoming and the West behind him, he looked out on those hills and mountains from his train car window with not a little regret. He had seen the West before, but always it had been from a place similar to where he now sat. It was a big country, these United States, he thought, and nothing could prepare anyone for the sheer size except actually seeing it first hand. The far reaches of the western plains and mountains was something that no one could imagine from books and magazines; even photographs did not do the land justice. It had to be seen to be believed. "You can never know," he would later write of his trip, "the spirit of the land until you ride between" the earth and the sky of the West. [12] He could not but help think that he was leaving it too soon, though he treasured the short time he had spent there, and it was a sincere regret to him as the land passed behind him and out of sight. He would never see it again.

He spent Thanksgiving that year in Washington, D.C., with his sister Elizabeth and Cabell Moore, and his mother and Minnie had come up from the Gap to join them. He left his mother and sister there, and headed back to the Gap, and they, in turn, went north to spend some time with Rector and Richard in New York.

Back home in the Gap, he found things not much changed from before he had gone West. He had begun to see less and less of his old friends, especially those who did not live in or around the Gap where he was spending more and more time now. He had not seen Richard Harding Davis in a while now, and the Pages did not see him as often as they would have liked. His circle of visits and travels had narrowed considerably, but what trips he did take proved to be longer excursions and to more strange and exotic places.

He had always wanted to see England and Europe and had been cheated out of a trip there just before the Russo-Japanese War in 1904. Now, it appeared that he was to have a second chance, and this time he was determined to go. He was just over 50 years old that year of 1913, and he thought if he was going he had better go soon. It was a pretty good bet, he thought, that he would not have another 50 years left to do the things he wanted to do. His divorce was still painful to him, and although his trip west helped, he still felt a little ill at ease around his friends who knew both him and Fritzi and of their marital problems.

John could not bring himself to write anymore. There were tentative efforts here and there, but the interest was gone. He began staying home at the Gap more and more, and perhaps enjoyed living there after his divorce more than ever, but at the same time he found that he was restless. It was as if, all at once, he wanted to go everywhere and do everything. All through 1913 and the first part of 1914 found him hunting and fishing and playing golf when the weather permitted. He had gone to Wyoming and Colorado, and in February 1914 he started on his long-delayed trip to England.

February was a good time to leave for just about anywhere, and they made the crossing of the Atlantic on a German liner in less than ten days. He had some apprehension before he left that his old problem of seasickness would return, but it did not. He had been sick going to Cuba and had turned every color of the rainbow on his way to Manchuria during the Russo-Japanese War. But this time, he had been pleasantly surprised. "I haven't been sea-sick at all," he wrote home to his mother on February 27, "my cold is almost gone (it has been the worst I ever had) and I am in fine health and spirits." [13]

He was in good spirits, and despite the death of his father in 1912, which had devastated him, and his too recent divorce from Fritzi, and he found himself living the life he had always enjoyed living. There were, on board the liner, several members of that great extended family of British royalty, and John soon found himself joining them in pleasant and casual conversations as they made their way across the Atlantic toward England. Naturally they liked him — everyone did — and by the time they docked at Southampton, his pockets were literally full of invitations.

Thomas Nelson Page's stepdaughter Minna was living in London, having remarried after her divorce from Preston Gibson, and John stayed in her home for a night or two and was introduced to the English sport of fox hunting. He had seen, participated in, and written about hunting from horseback before, but this, he thought, was the real thing, and nothing like using mules to ride after rabbits over rough Bluegrass farms. Here there were real foxes, and the outfits the riders wore were immaculate; everything was picture-perfect, and he was, for a while, his old self again. "You should see me," he wrote home to his mother. "I wear a top-hat (padded) a frock-coat (bell-shaped at the hips) pink-topped riding boots, a stock and carry a riding-crop with whip attached. Wallah!" [14]

There was something to do every day and most of the night, and he admitted to exhaustion at the end of the day, and sleep came nearly as quick as his head touched the pillow. The next morning, it started all over again. But it was all too soon over. By the middle of March he was on his way back home on the *Olympic*, striking New York on the 25th. The trip, from start to finish, had been a short one for him, and he had memories galore to bring back with him, plus a few souvenirs for his mother and the others. But like a dash of cold water, his troubles were waiting for him when he stepped back onto American soil in New York.

Negotiations for the movie productions of "Little Shepherd" and "Lonesome Pine" were hitting snag after snag, and in his darkest moments, he almost wished he had never penned the first line of either. He wanted to go home, but could not. He was beginning to think that the arguments and contentions over something that seemed so simple as film rights and royalty payments would never end. By September 1915, he had spent almost the entire year in negotiations and really could not see that he was any closer to a settlement. "I am sick to death of all this delay and bickering and anxious to get home," he wrote his mother again from New York, "but there's no use leaving and having to come back the next day." [15] Finally, he thought things were going well enough in New York that he could leave and let the lawyers and producers get along well enough without him. With his usual stopover in Washington, D.C., to see his sister Elizabeth and her husband, he finally turned his eyes toward the Gap and home.

From 1913 until 1916 he alternated between New York and Virginia, concerned primarily with movie rights and copyright infringements. He wrote little or nothing during those times, not even of his England trip. He no longer had anyone around him that cared to encourage him to write. Fritzi was gone, and her influence upon his writing had always been questionable at best, but now, even what little influence there might have been was no longer there. His father was gone, and the others — Tom Page, Richard Harding Davis, and even his oldest mentor, James Lane Allen — had their own problems and worries, and John's own writing languished.

He had fallen into that ready trap of staying busy with this or that and fooling himself with an illusion that to stay busy meant he was working. It was a dangerous illusion, for as far as his writing was concerned, he was doing very little or nothing.

He continued to be restless, and he did not want to hang around the Gap that winter of 1915, sitting inside, watching the weather turn everything outside varying shades of gray and black and brown. His body ached from the cold weather more and more as he grew older, and he was finding that he had less and less patience with any weather that made him uncomfortable. He was not doing anything anyway, he reasoned, and a trip south, possibly into Florida, possibly to do a little fishing in the warm waters there, might be just what he needed.

He loved Florida at first sight. The ocean was warm and sparkling in the sunshine that was such a rarity this time of year back home. The water was clear at the beach but quickly became opaque and then deeper green farther out as far as his eye could see. In the distance, along the horizon, he could faintly see the curve of the earth. The cool of the early mornings quickly vanished to a soul-warming heat from the sun in a clear sky, and he missed Big Stone Gap not at all, and was not even too sympathetic to the cold dreariness the folks back home were enduring.

He was surprised and fascinated by the size and number of fish in these waters. Back home, the streams of the Cumberlands were losing their fish to the progress that was being built along their banks, but even before progress those streams had held nothing like these Florida behemoths in number or in size.

The first day out, they caught one huge—at least by Kentucky and Virginia standards—tarpon, and the rest promptly disappeared, and they caught no more that day. The next day, under a glaring sun—hotter than the day before—John got his first strike. "Just then," he wrote euphemistically , "I caught my hook in the State of Florida. It was very firm at first and immovable—the State of Florida. Then an earthquake started, and Florida seemed to have started suddenly for the Gulf of Mexico. My line whizzed. Before my amazed eyes a column of polished silver was catapulted from the water—straight and solid it seemed. Then with a kind of grunt it came to a life of writhing terrific energy, and with a sidelong, shaking leap fell back whence it came." [16]

He was not prepared for the sheer strength of the fish, and all of a sudden his heavy rod, reel and even the boat he was on seemed not to be up to the job of capturing an animal of such size and determination. "It was like trying to stop a motor by putting your foot on a tire," he remembered, "like stopping a hurricane with a handful of feathers." [17] The tarpon, however, was having its own problems that were about to get worse. This way and that, it streaked under the water and out of sight, chased by something they could not see. The line began to move frantically toward the shore, no longer trying to escape the hook and line, and just under the surface there was a "frightful, convulsive struggle. The tarpon disappeared and at the spot the water turned red with blood—a circle of blood six feet in diameter—and my line went seaward steadily, irresistibly, and snapped. With and within that shark," he wrote regretfully, "my tarpon was gone." [18]

After the shark attack, he could not help but wonder — looking into the deep water that now suddenly seemed both dangerous as well as beautiful — what could have happened had he and the tarpon somehow changed places. He was not sure he cared for such thoughts so far from the shore that now seemed farther away still, and was grateful for the safety of the houseboat that all the fishermen in their party gathered on in the evenings. It was still a boat on the water where more sharks might be at that moment swimming underneath them, however, and somehow, after that day, no boat would feel safe enough. But there were distractions to keep his mind from what he believed to be realistic fears of the deep waters. "Life on that house-boat was mighty hard," he wrote tongue in cheek. "There were ice and mineral waters and things to go with them aplenty. There were lounging-chairs on deck, and a card-table. There was a cabin as big as an ordinary drawing-room, and in it there were a piano, a graphophone [sic], and books and magazines. Below were quite commodious staterooms, a dining-room, and a bathroom. And, incidentally, there was a most excellent cook." [19]

Again, the next day, he hooked "the State of Florida," and after what could only be described as a battle, he landed the huge, gasping fish. He was no less breathless himself. His arms trembled and his hands were numb from the exertion. "The tarpon was six feet long, said the guide, and would weigh one hundred and fifty pounds," he wrote. "I neither weighed nor measured him.... It had taken me one hour and fifteen minutes to land him, and it was a question during the last quarter of an hour which one of us would die first, and, as a matter of fact, during that time I didn't much care." [20]

In the next few days, he caught more, but none as large, and too soon he turned back north where winter still raged. He returned to the Gap, tanned and healthy, but the radical change in climate was still a shock, so much so that he could not stay. John made trips to Charleston, South Carolina, to hunt ducks, and in January 1916 he returned to Thomasville, Georgia, where he observed, "It is as warm as May at home." [21] There was a cold wind coming, however, and it was one that would chill his very heart and make him colder than any winter the Gap had produced since 1912 after his father died.

He had not seen his old friend and Frankfort roommate Robert Burns Wilson for years and really had not heard from him on anything approaching a regular basis for some time now. They had roomed together in Frankfort, in a grand old home at the corner of Ann and Clinton streets, just down from the old state capitol, and both were the life of any party they attended. It was a lively affair, indeed, when both attended at the same time, which they often did.

As friends — even good, best friends — are wont to do, they had moved in opposite directions from one another, and in doing so they had lost the close contact they had once enjoyed. Wilson had been nearly 13 years older than John, but the age difference had not mattered, and through all their Frankfort days they seemed, even to others, of the same age and more like brothers than good friends.

Wilson, like John, had married late in life and had moved to New York to make his living as an artist and writing the occasional poem, but he found New York a great deal more critical of his work than Frankfort had been. For the next 15 years after moving there, it had been a financial battle for the simplest things a family must have to survive. His painting could not provide everything they needed, but he was too stubborn to give up the thing that to him made his very existence worthwhile.

Finally, they said, Burns Wilson was reduced to painting postal cards for Wanamaker's Department Store for tourists, and the day he took that job, out of desperation, his poet's heart must have surely died within him. After a lengthy illness, he had wound up in the charity ward of a hospital in Brooklyn, but someone, remembering him from his salad days in Frankfort, and hearing of his plight, had sent money, but it was too late. The money arrived in time to move him to more comfortable quarters, but on the last day of March 1916, Robert Burns Wilson died a long way from the hills and streams of Kentucky. He was brought back to Frankfort, as he wished, and the same day that he arrived there he was buried overlooking the Kentucky River that flows past the capitol.

When John received the news, he was stunned. All the old memories of Frankfort in those younger, more innocent days came flooding back, and he could see in his mind's eye the face of his poet friend, laughing and spouting poetry, and sparking with the girls as they went arm-in-arm to a seemingly endless procession of parties and balls. He hated the feeling that came with the news of a friend's death. It was a restless feeling, and one that he was at a loss to know how to deal with. With each death of someone close, it was as if his past was being cut away from him, and at the same time, he could not help but feel his future getting shorter. But there was worse to come.

Less than two weeks after Wilson died in New York, John's very best friend, Richard Harding Davis, died at his home in Mt. Kisco, New York, hard by the home of John's brother Rector. Nothing since the death of his father four years before had hurt him as badly as the news that Richard was dead, and coming upon the heels of Robert Burns Wilson's death, John felt the coldness of a winter's wind blow through his body, chilling his heart and soul.

He and Davis had kept in touch, from time to time, but their paths had also diverged, through no conscious effort by either. Davis had remarried after his divorce from Cecil, and he and Bessie McCoy, his new wife, had produced a daughter, Hope, to whom John would dedicate *In Happy Valley* the next year. [22]

Davis had aged over the last few years, and photographs of him a few years before his death show a man whose youth had finally left him. Those famous good looks that made him as well-known as his writing can still be seen, but there is a resignation about his face, a tiredness, and perhaps a knowledge that even he could not live forever. He had written dozens of books, sometimes more than one a year ever since almost anyone could remember. He had covered war after war and had done some of the best reporting ever to grace the pages of any newspaper anywhere. One can still feel his excitement when he exclaimed, happily when he was just starting his journalistic career and getting paid by the word, about getting paid 50 cents just for writing "for instance."

It was Davis who coined the oft-quoted phrase on one of his many ventures covering wars all over the world, "The Marines have landed. The situation is well in hand," but few remember the source. Now, he had just returned from yet another adventure in Europe, where he saw the beginnings of World War I already going full tilt toward American involvement, and was trying to think of an idea to write another story that would bring in desperately needed money, but could think of nothing to write about. His health had started to break, and this time he knew it was not a passing affliction. "I am tired all over," he had written in his diary on April the 8th, "and have had a sort of a warning that I am getting on." [23]

Since returning from Europe, where he was almost shot as a spy, he had been in and out of bed, but his feeling of tiredness would not leave him for long, and the New York winter did little to raise his spirits.

On April 11, 1916, the house at Mt. Kisco was quiet, and both Bessie, Richard's wife, and Hope were in another part of the house. She had not heard Richard for some time, but knowing he did not like to be disturbed, and knowing how desperately he was trying to find and hold an idea that would turn into a novel, she left him alone.

Sometime around midnight, she slipped through the house, looking for him, and found him slumped in the hallway, the telephone in his hand. Richard Harding Davis was dead, just nine days shy of his 52nd birthday.

John received the news of Richard's death and was affected so intensely that he at first refused to believe it. Hard on the heels of that first disbelief was the realization that it must be so. No one would be so cruel as to say that Richard Harding Davis was dead when it was not true.

He walked that day and paced that night, and the whole house could count his grief for his friends in the number of footsteps that echoed throughout the house in Big Stone Gap. For days afterward they saw him, but he spoke little, no more than he had too, and spent much of his time staring off into space, into Cuba, into New York, into a cemetery in Philadelphia, and into another in Frankfort, Kentucky, where his friends were.

Soon after "Tarpon Fishing at Boca Grande" was published in *Scribner's* in February 1916, he began to write again, and in the process he found that he was again spending more and more of his time in Big Stone Gap. He returned to the life of the Gap almost gratefully. The hectic hustle and bustle of the last few years could not be found in the little southwestern Virginia town. About the most exciting thing that happened — with the exception of the Fourth of July celebrations — was someone's cow wandering off and getting lost, or a new shipment of fancy toilet articles arriving down at Shelton's Drug Store.

The peace and quiet that he found there was therapeutic. He fished the mountain streams, took his mother for rides in an automobile he had acquired, adopted every stray dog that came his way, and played golf with his old friend Bascom Slemp almost every day. Any reporter, or anyone else for that matter, looking for him would more than likely find him knocking around on the links at nearby Bristol or Knoxville, completely oblivious to deadlines and responsibilities. But he was writing again, and within the next year, 1917, he would publish ten stories in *Scribner's*, not counting his Florida fishing story.

Before 1916 would gratefully leave, he would receive the news that James Whitcomb Riley, with whom he had shared the stage and a suit of clothes, had died in July. Some said that Riley drank too much, but usually people who did not drink said that about anyone who did regardless of the amount involved. John had even had people say the same thing about him, and he would rather that they minded their own business, but that did not seem to have very much effect upon them and their gossip.

From January through October 1917, *Scribner's* published those ten stories — one each month — that he had begun writing the year before. He could not bring himself, however, to abandon the mountain source of his inspiration, despite the lackluster success of *The Heart of the Hills*.

In January, *Scribner's* published "Courtship of Allaphair" and followed with "The Compact of Christopher" in which St. Hilda, of *The Heart of the Hills*, makes an encore appearance. "The Angel from Viper" came in March, followed by "The Lord's Own Level," "The Marquise of Queensberry," "The Pope of the Big Sandy," "The Goddess of Happy Valley," "The Battle Prayer of Parson Small," and finally, in October, "The Christmas Tree on Pigeon." The stories were immediately gathered into one volume by Scribner's, and thinking of his old friend, he dedicated the collection of stories to "Hope: Little Daughter of Richard Harding Davis." *In Happy Valley* joined other volumes on the shelves of bookstores, and there it languished.

Critics again drove home the point that such stories as these were out of date, and belonged to another era — another century — and not in a century that was already experiencing the realism of World War I. Reviews of the book were mostly favorable, however, if not ecstatic. The stories were not bad at all; actually they were quite good, and had they been published 20 years before they would have been received with a great deal more excitement than they were in 1917.

Early in that year of 1917, it was apparent to just about everyone that America would soon be fighting in Europe, and John could not do any less than volunteer. He was turned down naturally, and it was not any great surprise to him that he was. He was too old, they said, and that, though they seemed quite cavalier about it, hurt him more than the not going, but he hid it well. He knew he was too old to fight a young man's war — a lot of people were — but the patriotic fever was upon the land, and he spoke on behalf of the government and the war effort, urging people to buy war bonds, and on behalf of the American Red Cross, urging donations of all kinds. He could not go to war to fight, that was clear, but, again going against his earlier vows not to go to war as anything less than a soldier, he volunteered to go abroad and read his stories to the troops. American's involvement in the war was short-lived, however, and he never had the chance to go.

There is some question as to when he actually began writing his last novel *Erskine Dale: Pioneer*. His sister, Elizabeth, wrote years later, that "the story had been written six years before [1913] and was about to be destroyed when he was urged to send it to the publishers." [24] This is almost certainly wrong, since he was still receiving answers to his letters concerning research on the book after his death. It was not unlike him, however, to begin taking notes, assembling research, and perhaps writing a page or two long before the actual writing commenced. If he did in this instance, those notes are as yet carefully hidden.

Erskine Dale would be his best work, he thought, and might even be a complete turnaround for him. His literary stock had slipped dangerously in the last ten years or so, and this change of pace just might, he hoped, put his name back where it once was. The writing did not come easy for him, and during that fall and winter of 1918 it barely moved at all.

Earlier that year, in January, his half-brother James died, and for a while, John made no pretense of writing. There had been no one, with the exception of his father, to whom he was closer. It had been James who had given him money to go to the old Kentucky University and sent him money while he was at Harvard to buy a suit so that he could go to graduation and not be ashamed of his clothes. It was James who brought him to the mountains for the first time and told him of a girl he had seen riding on an ox with a bag of meal in front of her. It was James who trusted him in business, who chided him for his faults of forgetfulness and lack of ambition, who never gave up giving John advice, even when it was not wanted, and who loved him more than any brother had the right to expect.

He remembered the times he had been hateful and short with his brother, and the times he had told James to mind his own business, but those times — those semi-arguments — did not last very long. James had been the one safe rock of his existence. He loved the others, but James was the one he could go to for help and not be turned down, although there might be a lecture on the value of a dollar before the help was dispensed. He could smile at those thoughts now, even while his heart broke.

By the middle of that summer, his writing pace had quickened, and through the fall and winter of 1918 he continued to place the words on paper that would be the last novel he would write. The plot was there; he knew where the story was going; but still, the words came painfully slow. Still, harkening back to Tom Page's advice of another time and another novel, of simply writing one page at a time, he kept at it, and by the first of the year, 1919, he thought he just might finish it after all.

CHAPTER 21

Last Dance

It is sad to see the change but there
is no life without death, and spring
returns.

John Fox, Sr.

I wonder how we will live without him.

Minnie Fox

 HIS MIND HAD BEEN ON KENTUCKY for some time that spring of 1919, and on his solitary walks around his home in Big Stone Gap, Virginia, he could just see the tops of those mountains he loved so well. He had often heard others say that they thought that spring would come early this year or that year, but for him, it could never come early enough. He thought he would be satisfied if the seasons were all combined into one and that it would always be spring.

He was working on a new novel, and he thought it was good. It would be different from anything he had ever written, and he was excited about its prospects. He was not sure about the title but that could be changed into something better, or maybe just left alone. There would be plenty of time for that.

In another time he would have asked his father or his older brother James what they thought of the new book, but not today — not even tomorrow or next week or ever. Both were gone. Even his old friend Davis was gone.

It had been three years this spring since he had received the news of Davis's death, and he could hardly bring himself to admit his loss even now. To be sure, they had not seen each other too much in the months and years before his death, but that did not keep him from missing him now. It was a long way from Big Stone Gap, Virginia, to New York, but he thought of Richard and those good times that they had together so many years before in a swirl of memories that were so good this time of year. He had, in later months after receiving the news of his friend's passing, expect to hear from him about a new adventure or new book he had written. It did not seem possible that great heart, that great infectious personality, could be gone for always. But it was true, and he knew it to be so; even while he was denying it, subconsciously he knew it to be so. And he thought of his friend often.

He had seen 56 springs come and go since that long ago December in 1862, and each one had seemed to hold a promise of better times and fresh starts that were seldom entirely forthcoming or fulfilled. He remembered the times he had written to his mother and father in his youth that prosperity was just around the corner. The mining ventures in Jellico and Big Stone Gap would make them rich forever they thought, but it was not to be. Vast riches never came, and the times he could honestly say he did not have to worry too much about money or how much he spent were few. Now, the end was close enough to make him uncomfortable, though he would almost certainly have years and years left. Perhaps he might reinvent himself with this new novel and regain something of the status of a writer that he had ten years ago.

As he walked along, he unconsciously named the wild plants he saw and invisible creatures that he heard scurrying here and there, just as he used to do when he took these very walks with his father, and a pain cleaved his very heart at the memory. Now, the habit born of those precious walks with his father had ingrained itself so that he caught himself doing it when he walked alone. And he remembered that Jack Hale had done the same with a young and illiterate June Tolliver almost in the very spot where he now stood, and that too seemed like ages ago.

He turned and started back toward the house he had once shared with Fritzi, who in her own way was his June Tolliver, and he remembered that she used to come out of the house and force him to come inside and write when he would rather be knocking a golf ball around in the back yard. But she was not here to do that anymore. And so, heavy with thought of the past and the book that was almost done, he turned and headed back. And as he walked back to his home out of the early spring chill, he noticed that the daffodils had begun to bloom.

Things had settled into a routine in his later years. He could not seem to write the way he used to. He had always, after the first few stories, had difficulty in writing, or rather, making himself take the time to do so. When he could, he did not want to and felt guilty about it. Now that he wanted to he found it almost impossible. It was hard for him to get his mind in order. He was doing what he swore he would never do: settle peacefully into middle age. He was 56 years old now and remembered telling James about how old their father seemed when he was that age, and feeling sorry for him. Now he was the same age as his father was then. Was he? Were they ever so young as to think they would never be so old?

Through the rest of that winter of 1918-1919, he worked on *Erskine Dale*, the new novel, to the exclusion of just about everything else in his life. He lived now in the house at Big Stone Gap, and thoughts of New York and of going there seldom crossed his mind. His mother and sister, Minnie, were with him, and his brother Horace and his wife May, who lived in their own house just across the backyard, came by often enough, but it was still a lonesome and restless time for him.

His sister Minnie, who became his protector before his death, and more so afterwards, remembered in those last weeks and months that he had lost all interest in the things he used to do for pleasure. Even golf had little attraction for him. When she would catch him walking away, or sitting with the light just so, she was sometimes startled at the resemblance he had to their father, and she, no less hurt by her father's death than the rest, found that at such times there were tears in her eyes for her — their — loss.

He was restless, and during the days he walked, almost pacing, thinking, not speaking. In the night Minnie could hear the floor in the old house creaking and squeaking from his walking, until she dropped off to sleep, trying to visualize where he was and what room he had walked out of or into. Occasionally, after the weather had turned warm, she would hear the outside door open and close as he went out, restless, walking.

No, he would tell them, he was not sick, even though he had been terribly sick in March with influenza, and yes, he felt fine, and no, don't worry at all about me. But they did, and watched him closely, and the little things that they ordinarily would not have noticed became larger, more important than perhaps they should have under other circumstances.[1]

And so, it was a relief to them all when summer came, and he seemed to regain his old bounce and vitality. The book was nearing an end, and that had always lifted his spirits, and with the coming of warmer weather he thought a short fishing expedition would be in order, and, he thought, he knew just the right place.

The city of Norton, Virginia, lies just north of Big Stone Gap, a little more than ten miles away. That summer of 1919, it was not a large city, and would never become a large metropolis, and the citizens pretty much liked it that way. Prosperity came to Norton as it had to the Gap and other towns nearby, but over the years it had been a process that had its ups and downs. John enjoyed Norton and had gone through the city on his way to Kentucky and New York on more than one occasion, and although the city shared many traits with its southern sister, it was just exotic enough, just far enough away, even in 1919, to make a trip there interesting.

Erskine Dale was, for all practical purposes, finished. It only remained for him to tie up a few loose ends and bring the story to a close. Certainly, he thought, that amounted to no more than a chapter or two, a few pages at most, but that could wait a day or two. It was too nice outside; the weather had warmed, and summer was undeniably here.

21. Last Dance

His sickness in March had caused him to lose weight and strength when and where he could ill afford to lose any at all, and he was still occasionally a little groggy on his feet. But the sunshine felt good, and as he basked in it he could almost imagine that he was gaining strength directly from it.

And so, to Norton he determined to go. There would be a carnival going on — it was the 4th of July, after all — and, no doubt, a mentally and physically wasted man could find his diversions there. The waters of the mountain streams had warmed somewhat, and after an evening or two of lights, noise and sawdust, a lucky fisherman might hook into a fish large enough to be mistaken for "the state of Florida," or, in this case, the state of Virginia.

It was just such a cure as the doctor might have ordered, and leaving the last page he had written of his new novel on top of the growing stack of other pages on his desk, he gathered up his fishing tackle, hurriedly threw a few clothes together, and headed north for a well-deserved break from his labors.

On July 5, the family at home in the Gap received word that he was too sick to go fishing. He had gone the day before to see a Negro baseball game, and had gone to the carnival afterwards, both of which would wind up a week of 4th of July celebrations that were once so popular. The music was loud and the dancing as boisterous as only mountain dancing can be, and amidst such gaiety, he could not contain himself, and saw no reason why he should. He was among friends, and the need to relax and enjoy himself was too strong to resist.

The dancing tent and the music naturally drew him, and there he danced himself into near exhaustion. The girls who worked the dancing tents were not always treated too well by the rough mountaineers and the men from town were not always on their best behavior either, but, they said, John Fox knew how to treat a lady and was always the perfect gentleman, even with dancing girls.

He danced with an energy that had grown uncommon in him in the last few years, and his turns on the dancing stage were monopolized by the so-called queen of the carnival, but he saw no difference in their social station and treated his partner with the same deference he would have extended to any lady. "They talked as gentleman and lady," wrote Bruce Crawford for the Dearborn *Independent*. "Others leered, and yodeled inane songs, but Fox was courteous, gentlemanly." [2]

His demeanor had an effect upon his dance partner who under other circumstances could be as brash and forward and as coarse as her usual clientele normally were. "And surprising to all was the girl's quiet recognition of the gentleman in her partner. His gallantry had brought to the surface unsuspected refinement in her. Again and again she returned to him after dancing with others." [3] Until late in the night they danced, rested and danced again, and finally, John and the rest stumbled off to bed, happy but tired.

The next day they had planned to go fishing, but he awoke not feeling well, and actually feeling quite sick. He ached and was feverish. A doctor was called, and, at first, it was thought to be nothing very serious—a summer cold, perhaps—or possibly a leftover vestige of his more serious sickness of the past March. With startling quickness he worsened, and the doctor discovered an alarming amount of congestion in his lungs, and diagnosed pneumonia.

In between the first and second visits of the doctor the family had been called, and his worried mother sent a car to bring him home. When they came for him, they were taken aback at how sick he really was, and concern etched their faces, and they tried not to reveal to him how worried they were.

For the second time in his life, he had to be carried inside the rambling old house. They put him to bed and raised his head and shoulders with pillows to help him breath more freely. That night, the sound of his labored breathing could be heard nearly everywhere in the house. A gloom settled in, as it had seven years before when his father lay in the same house, slowly but inexorably sinking away.

The next day, he was no better, and throughout that day he grew steadily worse. One lung had filled with fluid, and the other was rapidly following suit. The day lasted forever. He was in and out of consciousness, delirious when conscious, fitfully moving his head back and forth when not. He did not know how sick he really was.

They carried him outside on the screened-in porch, thinking the warm July air would help him breathe, and for a short time they thought it might, but as the evening came there was only a steady decline in his condition, and all their hopes faded. At about 5:30, in the shade of the trees that had been allowed to grow around the old house, and in sight of the mountains he loved so well, John closed his eyes one last time.

At first they thought he was just sleeping again, and then the realization that he was not hit them all like a physical blow. One minute he had been alive, and the next minute he was gone. They were not prepared for it, and the surprise and shock of his going left them numb and wondering what they would do without him.

On Friday morning, the Louisville & Nashville northbound train number 34 pulled into Paris along the same route John had traveled so many times before. He was taken to a private home on Mt. Airy Avenue and lay there the rest of that day while friends, family and curious filed past. At six o'clock that afternoon, the casket was taken the short distance to the western edge of Paris and carried up the low hill to where his father and brother James lay already. They sang those old songs that are seldom heard anymore—"Cast the Burden on the Lord" and "Christian's Good Night"—as the casket was lowered into its own grave by the side of his father and brother. [4]

They milled around for a while, not wanting to leave. It was not the first time they had climbed that hill to leave one of their own, and it would not be the last. Finally, the boys took their mother in hand and walked away from that place. "It was John's own day," Elizabeth would write some years later, "a day of blue sky and heavenly breeze." [5] They had sung one last song before they left, leaving him, she said, "in the bluegrass country he loved so well." [6]

The day that he died, his friend Tom Page and his wife Florence had just returned from Italy where Tom had been serving as U.S. ambassador. They were unaware that anything had been wrong, and when they got the news of John's death, they, like nearly everyone who knew him, could not believe. In August, Minnie wrote them the details of his death and their shock and disbelief that was with them also. "I cannot realize that this terrible sorrow is true," she wrote. "I feel that John is away on a visit, and I cannot let myself think that this time he is not coming back to us." [7]

She wanted to send them some little remembrance of his, but, Minnie said, "there is so little. No one ever had so much given him, and no one ever lost more than dear John.... I wonder how we will live without him." [8]

At the end of his hunting trip to the West in 1913, he had written of elk and mountains and just the pure pleasure of being away from those people and those things that seemed constantly to pull at him. "And the two memories that abide with the deepest thrill within me," he had written of that trip to the West, "are of an eagle riding a gale with motionless wings." [9] Perhaps, in the end, trying to gasp one last full breath from the screened-in porch of his home in Big Stone Gap, Virginia, he saw that last eagle of the Wyoming mountains one last time and went with it, far away.

Chapter 22

Surviving

> That sure was a long time ago, but in some ways does not seem so long.
>
> Oliver Fox

THE BOOK THAT HE HAD WORKED ON SO HARD did not prove to be the success he had hoped it would be. *Erskine Dale: Pioneer* has a hurried flavor to it and in places seems little more than expansive notes arranged in chronological order. But even so, it is not entirely a bad book.

Sometime early in July of 1919, John was near the end of the manuscript. He had his main character for whom the book is named find his dying Indian stepfather. The old man, dying, speaks to his adopted son one last time: "'My son', said he, 'I knew your voice. I said I should not die until I had seen you again. It is well … it is well', he repeated, and wearily his eyes closed. And thus Erskine knew it would be." [1]

These would be the last words that John wrote, and it would be left to his sister, Elizabeth, to finish the book. Scribner's, who had been John's publisher for years, thought the book would need a couple of more chapters to bring it to a tidy end, but Elizabeth was able to do it in only one chapter of less than three pages. The book itself was so close to being finished that another hour or so would have done it, but likely, John relished the anticipation of finishing the novel and wanted to put off those final words for a day or so, knowing that the writing of the novel — the major part of it — was well behind him.

The book did not sell well, but, perhaps, it was not entirely the fault of style or passing fads in literature. In the end, John's inspiration had gone, and as hard as he had looked, he could find it nowhere.

He had known that inspiration was a relative thing anyway, and although it did seem to leave him during and after his marriage to Fritzi, he never said so himself and never attempted to blame anyone but himself for his writing, or the lack of it. A few years before, he had told Alice Rohe, a reporter for The Denver *Times*, "I confess I am not fond of working," he said, "and I am afraid this belief in inspiration is often an excuse for writers putting off their work. Everyone, I believe, understands the dread of beginning anything, and before I commence my writings I call up every possible thing I can do as an excuse to postpone writing. When I do get started," he confessed, "I work like mad." [2]

He would not force himself to work. When he had a good idea, and felt like it, he would write, but otherwise no force on earth could make him write when he did not feel like it. "The few times I have forced myself to work when the mood was not on me I have not only worked in vain but I have ruined an idea forever — an idea that might have been developed into something good had I waited until I was, as some people say, 'inspired' ... I write when I feel like it," he said, "which...is not often." [3]

Long before that summer of 1919, John saw the change that had come to the mountains from which his "inspiration" had come, and on which he had built his reputation as a writer. He was saddened and a little resigned at what had happened. "In the mountains," he wrote in *The Heart of the Hills*, "the people had awakened from a sleep of a hundred years. Lawlessness was on the decrease, the feud was disappearing, railroads were coming in, the hills were beginning to give up the wealth of their timber, iron, and coal. County schools were increasing, and the pathetic eagerness of mountain children to learn and the pathetic hardships they endured to get to school and to stay there made" his heart ache. [4] The times and the people he had based his best stories on were no more to be found, and the inspiration that he had gained in knowing of those people and those times had gone with them.

Near the end of his life, he spoke of himself and his writing. "'I don't like that word artist', he exclaimed. 'And why should the artist be any more than the businessman or lawyer or doctor or financier sit up and bore people with recounting his puny successes? Why should I be sitting here talking to you about myself? I haven't done anything worth talking about and why should anybody be interested any more in what I have done than in what any one else has done?'" [5]

In the weeks, months and years that came after John's death, things in the Gap settled back into a routine of days, but it was an uneasy routine, and seemed out of balance somehow. Those who should by natural laws have died first still lived, and those who should have lived the longest, by the same contrary law, left much too early.

Minnie never married, and after her brother's death and that of her mother a few years later became the tacit head of the family, and until her own death in 1962 was a well-known and beloved citizen in Big Stone Gap, and reminder of a more genteel time.

The letters stopped coming and going from the Gap with John's death.

None of them, with the exception of his father, ever wrote any more letters than John and his father did, and now that both were gone the writing came to an abrupt halt. There was just not that much left to write about anymore, and those few letters that came from Oliver, Horace, Elizabeth, Rector and Richard have a sad longing about them that is impossible to ignore.

"Do you remember," wrote Oliver to his mother from Harlan, Kentucky, "the little trundle bed that Elizabeth and I used to sleep in, that by day was pushed under your bed in your room in the old house that was burned down at Stony Point? That sure was a long time ago, but in some ways does not seem so long. I am getting along all right and working hard. I was out in the mountains every day for a week," he wrote, "and at noon [I] would build up a fire and toast our lunch ... which made me feel like old times." [6]

Afterword

IN THE SUMMER OF 1856, John Fox, Sr., along with his wife Catherine and their two small boys, James and Sidney, moved to Sharpsburg, Kentucky, where he would teach school for a short time, and where Everett would be born. When "Kitty" died in 1860, he took her back to Bath County, just north of where they had lived that summer, and buried her there just north of the little town of Bethel. The cemetery there has suffered over the years. Stones that once marked graves are scattered and piled on top of each other, some broken, some gone entirely. But a few yards from the highway, a small obelisk and an ancient wrought iron fence watch over Catherine Fox's grave. It has weathered well, and in a small circle, carved just above her name and midway up on the obelisk, John Sr. had them also carve, simply, "My Wife," and he never forgot her.

South and east of Harlan where Oliver Fox was still trying to make money in the coal fields that first December after John Jr.'s death lie the lands around Jellico, and the mountains around the little town have not changed too much over the years. Coal mining still goes on, but there is not the fervent activity that there once was. "The new," as the mountain people might say, "has worn off the coal business."

The old Proctor Coal Company where James and John and Horace worked is hard to find now. Not many people can direct the curious within a mile of its original location. In later years, the lands around Proctor were stripped and reshaped until it is unlikely that John himself would have recognized the country where he first saw the mountains of Kentucky. When one does find the spot, there's not much to see. Bits of coal — good, black coal — litter the ground, and where the mine openings once were are deep looking pools of stagnant water where in the still hot summer frogs and snakes like to make their living. Bits of broken glass lie here and there, along with empty cans of various sorts.

The old tram road where John and Horace worked clad only in their trousers, and their hired help worked in a lot less, is grown up in trees and brush now, and there is little sign of the labor those two brothers did that hot summer. There is still coal being mined around Jellico, but it is mined in more efficient, less romantic ways now, and there is much room for pick and shovel and mule-drawn carts.

Miles away to the east, Big Stone Gap, Virginia, has settled into a comfortable middle age. Coal trucks rumble through the town from time to time, hauling the coal that the Fox brothers knew to be there, and knew they would make their fortune in, but that fortune never materialized. There is no boom going on in the Gap today, but the people are nice. The road that John Jr. and his father loved to walk and think on is paved now, and runs within a stone's throw of the old house over on Shawnee Avenue. Tourists come by every now and then, mostly older folks who remember reading John's books when they were younger and perhaps a little more sentimental.

Over on Clinton Avenue, close by the South Fork of Powell River, is the house where Jack Hale brought a young June Tolliver when he first brought her to the "settlements" for an education and nearly lost her in the process. Every year, the locals and the tourists get to see June and Jack and all the rest in an outdoor drama that, thankfully, bears at least some resemblance to the original story. There is action and drama enough for everyone, and they flock to see if Jack Hale and June Tolliver will really find happiness. They do.

There is no volunteer police force in the Gap these days, and it's not needed, and there has not been a hanging in or around the town for decades.

The old Kentucky University in Lexington where John spent his first days at a "real" school has finally settled on Transylvania University as its name, and has expanded somewhat since the 1880s. The faculty still manages to put a student to sleep every now and then, but not so much so as you would notice. They still teach, and occasionally the students learn.

Lexington has grown up around the old college, and John would not know it now. The spot where volunteers camped and trained in expectation of going south and whipping the Spanish in Cuba is probably one of the prettiest places in town now, and John would be gratified at that. Henry Clay's home is still there and is still being visited by tall mountaineers, but now they bring their wives and children to see the home of the Great Compromiser and mostly leave their guns at home. Most never know that just out from the main house preparations for the Spanish-American War were handled by green militia and volunteers so long ago, and that their great-grandfathers met there, and a few never got back home from Cuba. And that a fledgling correspondent for a magazine most have never heard of was there to see it all and write about it.

Back down south toward Jellico, the road turns left through London toward Hazard, that town that John Fox, Jr., rechristened as Hazlan when Rome Stetson and Martha Lewallen of *A Cumberland Vendetta* were falling in love, and Old Gabe's mill was still grinding corn for his neighbors. If, before reaching Hazard, one might turn from the parkway that breezes by Manchester, Hyden and Hazard, through a maze of twisting and turning and narrow roads, the traveler, if he cared to look, would find that water still flows out of Hell-fer-Sartin Creek, and the mountains it flows from are still as steep and rugged and unsuited to foot travel as they were when John and "Old Faithful" made their way there in 1910. Few who live on that creek know that a famous book once took the name of their little creek for its title. Most libraries no longer have it in their stacks, but the creek is still there, and in places it is not much different than it ever was.

In Frankfort, the house at the corner of Ann and Clinton streets where John roomed with the poet and artist Robert Burns Wilson is still standing, and just behind the house, the old capitol building where Goebel was shot on a cold day in January 1900 is also still standing, and for years served as the state's museum. The state house where the fatal shot was fired is still there as well, and a plaque marks the spot where Goebel fell. A larger-than-life statue of the martyred governor was placed in front of the new state capitol building that was built just after Goebel was killed. In 1962, state workers moved the statue early one morning in December to the grounds in front of the old capitol building where Goebel was shot. The statue stands with its back to the old capitol, and near the bottom, Goebel entreaties his friends to be loyal and kind to the common people. There is no mention of oysters, good or bad. When the statue was moved that early morning in 1962, no one really noticed.

A similar statue was placed over Goebel's grave in Frankfort's cemetery, and sometimes, when it rains, a visitor can almost see the funeral procession coming up the hill on a cold day in February. Just behind Goebel, and a little farther up the hill, John's old friend and roommate, Robert Burns Wilson was buried in 1916. He who wrote of streams and meadows and grass and sunshine lies in a neglected spot. None of the things he wrote about grace his gravesite, and very little of the sunshine makes it through the trees. It is a poor place for a poet.

Within sight of Wilson's grave, and also that of Governor Goebel, the river that brought Chad, Caleb Hazel and the Turners of *the Little Shepherd of Kingdom Come* to Frankfort that long-ago spring still flows by and through the town. In the spring, "tides" still sometimes come, and it is something to see the fierce brown water carrying trees and debris downstream in a swirling haste. However, no one rafts logs on the Kentucky anymore, and the number of those who remember those huge rafts of logs grows less and less as the years pass.

The Old Tampa Bay Hotel, where the "rocking chair period" of the Spanish-American War kept both correspondents and generals from doing what they had come south to do, is still standing, having survived hurricanes and the twentieth century penchant for tearing down the old for no real good reason, except age. The outside still shows some sign of looking as bizarre as it did when John first saw it that summer of 1898. It is the "home" of the University of Tampa now, and it is rare that any journalist, or general for that matter, ever stops by for a visit.

Richard Harding Davis, whom no one could think of as mortal, not even himself, died in his home of a heart attack, still writing, still making plans, still trying to live the life he wanted to live, dying with a telephone in his hand in 1916. That same year, Jack London, who had tried to slap some sense into a Japanese out of frustration during the Russo-Japanese War, and was saved by the intervention of Davis, died at his home in California, was cremated, and his ashes buried under a huge boulder.

Tom and Florence Page returned from Italy in 1919, arriving in New York on the very day John, their best and dearest friend, was gasping his last breath in Big Stone Gap, Virginia. They too found that much had changed over the last few years, and many of their friends were gone. Florence died on June 6, 1921, on the anniversary of her wedding to Tom. A year later, Tom Page fell dead of a heart attack in his garden at "Oakland," in Hanover County, near Richmond, Virginia. Both were buried in Washington, D.C.

Currie Duke Matthews lived as a widow for the rest of her life. Arthritis crippled her hands soon after her husband Wilbur died in 1910, and she could no longer play the violin, but she remained a lady, though times were sometimes hard. In the mid–1930s, she moved to Virginia to live with her married daughter and son-in-law, and lived happily there until 1946. In the spring of that year, being 79 years old, Currie became sick with influenza. Over the next five days she got no better, and as so often happens with people who have seen their best years, her illness quickly sapped her strength and what little health she had remaining. On March 2, late in the night, Currie Duke died, ironically of the same sickness that had weakened John so much that spring of 1919, nearly 30 years before. She was cremated and brought back to be buried near her daughter and son-in-law's home in Ware Church Cemetery, near Mathews, Virginia. Around her grave are flowers that bloom every spring, the result of bulbs planted by a loving grandson.

In 1954, Fritzi Scheff died alone in her apartment in New York City of what the medical examiner described as "old age and natural causes." In the years after her divorce from John Fox, Jr., she had remarried and divorced again. During the Great Depression, she lost everything and was reduced to working as a hostess in Waterbury, Connecticut. She made a modest revival of her career in the 1950s. When she died, she left an estate valued at $476. She was buried in the Actor's Guild Cemetery, a part of Kenisco Cemetery at Valhalla, New York, and no one goes looking for her grave anymore. The marker is of the simplest sort, bearing only her name and the dates 1879 and

After John's death in July 1919, several family photographs were made, including this one. Pictured from left to right, front row, are: Hilda Seccomb Fox (Rector's wife), Horace Fox, Mildred Fox (Horace and May's adopted daughter), May Worth Fox and Elizabeth Fox. Back row, left to right: Oliver Fox, Richard Fox, William Cabell Moore (Elizabeth's husband) and Rector Fox. Mother Fox is seated in the middle. Not shown are Everett Fox and Minnie, who was notoriously shy of the camera (courtesy John Fox, Jr., Museum).

1954. John's brother Rector is buried in the same cemetery, only a stone's throw away from Fritzi.

Over the years, after John's death in 1919, the other members of his family came to join him and his father and brother James on that little rise of ground in Paris Cemetery. Everett was brought there in December 1924, and a year later their mother came home. In 1937 Richard came, in 1954 Oliver began his rest from his labors, and finally Minnie, John's most ardent protector before and after his death, found her way there in the same year as Oliver. Only Horace, who did not think John liked the hot, dirty work around the mines in Jellico very much, is buried in Big Stone Gap, having died there, still working, in 1934.

Rector remained in New York and died there in 1931, and Elizabeth, the last of them all, died in 1970, and they both are buried a long way from Paris. For a family so close, death has separated them more than life ever did.

The old house that John Sr. built in Stony Point to replace the one that burned is also gone, and in its place a more modern brick home. Their neighbors are now more affluent than ever before, and the fields around the old home place grow horses instead of gardens and grass seed. On September 3, 1939, a little more than 20 years after John Jr.'s death, a memorial was erected by the state of Kentucky and unveiled in front of the old homesite. On the face of an old millstone, a bronze plaque was affixed, announcing to all that this was the birthplace of John Fox, Jr. In later years, the Kentucky Department of Highways erected a historical marker on the site. Other monuments are painfully and noticeably absent. Few slow down long enough to read anything on those plaques, and most of those who do only vaguely remember the name of John Fox, Jr.

In 1950, in Duncan Tavern in Paris, Kentucky, a small corner in the basement of the building was set aside and furnished with artifacts belonging to John Jr. that were donated by the surviving members of his family. In that memorial room is a desk where he supposedly wrote *The Trail of the Lonesome Pine* and *The Little Shepherd of Kingdom Come*. There are photographs of himself and his mother and father, as is a photo of the check for $262 that he received for *A Mountain Europa* and for which he was so proud.

After his death in 1919, John's popularity with the reading public quickly faded away to almost nothing. Over the years his books have been reprinted from time to time, and for a while a modest income from royalties kept up a trickle of money, but eventually even that vanished. His brother Horace was appointed as executor of his estate and sold off the remaining lots and a house that John still owned in the Gap. In the next few years, movie rights were also sold to *The Heart of the Hills* (1919) and *The Kentuckians* (1920), and Hollywood did its usual stereotypical rewriting of the novels.

It was probably not the worst will that was ever written, but the last will and testament of John Fox, Jr., came very close. When John died, he left an estate valued roughly at just over $100,000.

This photograph of Minerva Carr Fox was taken in 1919 just after her son John Jr.'s death in July (courtesy University of Kentucky).

After debts of some $22,000 were paid, including his funeral expenses, he had written that he wished the balance of his estate to be equally divided between his mother and sisters. Should any sister marry, or his mother die, their share was to go to the surviving sister, or should his mother survive Elizabeth and Minnie, then his mother would get everything. No one outside the family — the only exception being Horace's adopted daughter — was to get anything.

The will was the one he had made just before leaving for Manchuria in 1904 and was just ambiguous enough to cause confusion. John had stipulated that Mildred, Horace and May's adopted daughter, should receive enough to gain her a good education, and left it at that. Notations on the margin of the will asked that certain debts be paid, that in the intervening years between 1904 and 1919 had already been settled. It was more than Horace felt capable of sorting out, especially when his own daughter was involved, and he had to eventually ask the courts to settle the whole thing.

One of the last photographs of "Mother" Fox before her death on September 8, 1925. Age revealed her strong resemblance to her son John Jr. (courtesy of John Fox, Jr., Museum).

Elizabeth had married, thus relinquishing her share to Minnie and her mother and ending her involvement. Finally, after several months of depositions and attorney's fees throughout the spring and summer of 1920, it was decided that Mildred should receive $2,500 for her education, and the rest was divided equally between Minnie, who had not married, nor ever would marry, and her mother.

When Minnie died in 1962, it was almost all gone, and Elizabeth had to pay nearly $2,000 of her own money to settle her sister's debts and see that she was buried beside the rest of the family in the cemetery at Paris, Kentucky.

Chronology

1852: John Fox, Sr., and Catherine Rice are married on August 3.
1853: James Wallace Fox is born on June 11.
1856: Sidney Allan Fox is born on July 3.
1858: Everett Buford Fox is born on September 2.
1860: Catherine Fox dies on June 14 and is buried at Bethel, Kentucky, just north of Mt. Sterling, Kentucky.
1862: John Fox, Sr., remarries to Minerva Worth Carr, daughter of William and Elizabeth Clary Carr of Mayslick, Kentucky.
1862: John Fox, Jr., is born on December 16.
1864: Horace Ethelbert Fox is born on April 4.
1866: Oliver Edwin Fox is born on August 31.
1868: Minerva (Minnie) Carr Fox is born on March 27.
1871: Richard Talbott Fox is born on April 25.
1873: Rector Kerr Fox is born on February 19.
1876: Sarah Elizabeth Fox, John Jr.'s youngest sister and the last of the Fox children to born, is born on September 19.
1878: John Jr. goes to Lexington, Kentucky. Enters Kentucky University that fall after money problems prevented him from entering Harvard whose entrance exams he had passed. He is fifteen years old.
1880: After two years at Kentucky University (soon to be renamed Transylvania University), John Jr. takes Harvard exams again, passes, and enters Harvard in September, joining the sophomore class at the age of seventeen years and nine months.
1881: Family's first venture into Jellico coal mines. James is prominent in this venture, which will soon involve most of the boys in the family, including John Jr.
1882: Home from Harvard for the summer. Works with his brothers at the mines in Jellico. First exposure to the mountains. Returned to Harvard again in the fall.
1883: Graduated from Harvard cum laude on June 21, being the youngest member of his class. Went to New York looking for a job. Worked for the New York *Sun* until that fall. Entered Columbia Law School in September, but dropped out in December.
1884: Went to work for the New York *Times*. Worries about his future. Not sure what he wants to do.
1885: Returned to Paris, Kentucky, because of illness.
1885: Has "A Betrothal Ring" published in *Frank Leslie's Illustrated Magazine* in May, and "Deceiver's Ever" published in *New York Life* magazine for November 21.
1886: Worked in the mines at Jellico for a short time. Left over dispute with other workers. Went home to Paris where he helped his father teach in his father's school in Paris.

1888: James and Horace move their mining interests to Big Stone Gap, Virginia.

1889: John Jr. enters into business with James and Horace. Move to Big Stone Gap, Virginia, imminent.

1890: Entire Fox family moves to Big Stone Gap, Virginia.

1890: Sidney Allan Fox dies in Brooklyn, New York, on December 12. Buried in Brooklyn in Greenwood Cemetery.

1894: "Europa" a success. John Jr. goes on the speaking circuit. Meets James Whitcomb Riley, Bill Nye and others. Theodore Roosevelt writes praising his writing.

1894: "A Cumberland Vendetta" published in *Century* magazine for June, July and August.

1894: Writes "On Hell-fer-Sartin Creek," published November 24, 1894, in *Harper's Weekly*.

1894: Makes his second home in Frankfort, Kentucky, where he shares a room with Kentucky poet-artist Robert Burns Wilson at the corner of Ann and Clinton streets in Frankfort. At this time he is gathering information for his first novel *The Kentuckians* and possibly *The Little Shepherd of Kingdom Come*.

1895: *A Cumberland Vendetta and Other Stories* published. Included "A Mountain Europa" and "The Last Stetson," which had been published in *Harper's Weekly* for June 29, 1895, and was a sequel to "Cumberland Vendetta." "Fox Hunting in Kentucky" published by *Century* magazine in August, and "Courtin' on Cutshin" published by *Harper's Weekly* December 21, 1895.

1896: Publishes "Through the Gap" (*Harper's Weekly*, January 11, 1896), "The Senator's Last Trade" (*Harper's Weekly*, May 2, 1896), "The Passing of Abraham Shivers" (*Century* magazine, June, 1896), "Message in the Sand" (*Harper's Weekly*, June 20, 1896), "Grayson's Baby" (*Harper's Weekly*, May 9, 1896), "Preachin' on Kingdom Come" (*Harper's Weekly*, September 5, 1896), "A Purple Rhododendron" (no magazine publication), and "After Br'er Rabbit in the Blue Grass" (*Century* magazine, November 1896).

1897: *Hell-fer-Sartin and Other Stories* published by *Harper's* in June against the advice of *Harper's*. It is sold out by November.

1897: *The Kentuckians* published in December. It had been serialized in *Harper's Monthly* for the July through October 1897 issues.

1897: "Through the Bad Bend" published in *Harper's Weekly*, December 18, 1897.

1898: Continues on speaking circuits. "Br'er Coon in Old Kentucky" published in *Century* magazine for February 1898.

1898: John leaves Big Stone Gap for Tampa, Florida, on June 2, 1898, and after much wrangling with *Harper's*, goes to Cuba as correspondent.

1898: Back to Tampa, Florida, from Cuba. Quarantined July 27 either with malaria or yellow fever.

1898: In Louisville, Kentucky, by August 3. Too sick to go home.

1898: August 8: Arrives home in Big Stone Gap that morning from Louisville.

1898: Spanish-American War articles: "A Day in Atlanta," "Chickamauga," "Santiago and Caney," and "Truce." All were published by *Harper's Weekly* for May 21, May 7-14, June 11, July 23-August 6 and August 13, 1898. These articles were used in *Harper's Illustrated History of the Spanish-American War*. Also, "With the Rough Riders at Las Guásimas," July 16, 1898, "With the Troops for Santiago," July 16, 1898, and "Volunteers in the Bluegrass," June 18, 1898.

1900: William Goebel, governor-elect of Kentucky, is assassinated in Frankfort, Kentucky, on January 30, 1900. John attends the funeral and is soon sick again from standing in the rain during the funeral procession in Frankfort.

1900: Published *Crittenden*, his novel based on the Spanish-American War. Published by Scribner's who had become his publisher after his leaving *Harper's*.

1900: Publishes "Down the Kentucky on a Raft" for Scribner's magazine for June, "Man Hunting in the Pound" for *Outing* magazine for July and "To the Breaks of the Sandy" for Scribner's magazine for September.

1901: Wrote "The Southern Mountaineer" for Scribner's magazine for April-May, "The Hanging of Talton Hall" for *Outing* magazine for October, and "Christmas Eve on Lonesome" for *Ladies Home Journal* for December.

1901: Published *Bluegrass and Rhododendron* which included "The Southern Mountaineer," "The Kentucky Mountaineer," "Down the Kentucky on a Raft," "After Br'er Rabbit in the Bluegrass," "Through the Bad Bend," "Fox-Hunting in Kentucky," "To the Breaks of Sandy," "Br'er Coon in Ole Kentucky," "Civilizing the Cumberland," "Man Hunting in the Pound." "Christmas Eve on Lonesome" was not included in this volume.

1902: "The Army of the Callahan" published in Scribner's magazine for July.

1903: *The Little Shepherd of Kingdom Come* is published by Scribner's. It was dedicated to Currie Duke.

1903: "Christmas Night with Satan" published in Scribner's magazine for December.

1904: *Christmas Eve on Lonesome and Other Stories* published by Scribner's. Includes title story, plus "The Army of the Callahan," and "Christmas Night with Satan. Also "A Crisis for the Guard" and "The Pardon of Becky Day," written for the volume.

1904: Rejoined Richard Harding Davis as correspondent for Scribner's magazine covering the Russo-Japanese War.

1905: Upon returning from Japan and Manchuria, *Following the Sun Flag*, which included his dispatches to Scribner's from the Russo-Japanese War, is published.

1906: *A Knight of the Cumberland* is published by Scribner's. Was serialized in Scribner's magazine for September, October and November issues.

1908: *The Trail of the Lonesome Pine* is published by Scribner's. He is living more and more in New York during this time, and hobnobbing with Richard Harding Davis. Meets and courts Fritzi Scheff, Austrian immigrant and opera star. They marry at Mt. Kisco, New York, at the home of John's brother Rector on December 13.

1909: "Christmas Tree on Pigeon" published in *Collier's* magazine for December 11.

1910: "Christmas for Big Ame," "On Horseback to Kingdom Come," "On the Road to Hell-fer-Sartin," and "On the Trail of the Lonesome Pine" are published by Scribner's magazine for December, August, September and October.

1912: John Fox, Sr., dies on June 28.

1912: John and Fritzi separate.

1913: John and Fritzi's divorce becomes final on January 31.

1913: *The Heart of the Hills* is published in March by Scribner's. It had been serialized in Scribner's magazine for April through December 1912, and January through March 1913. Went to Palm Beach, Florida, and Aiken, South Carolina. Also went west for a hunting trip that fall. Writes "A Hunting Peep into Jackson's Hole" and "After Bear in Colorado" neither of which was ever published.

1914: Made his first trip to Europe (England) in February and March, and thoroughly enjoyed himself. Returned to New York where difficulties in dramatizing *The Little Shepherd of Kingdom Come* kept him away from home for several weeks.

1916: "Tarpon Fishing at Boca Grande" published in Scribner's magazine for February.

1917: "The Lord's Own Level," "Battle Prayer of Parson Small," "Compact of Christopher," "Angel from Viper," "Pope of the Big Sandy," "Marquise of Queensberry," "Goddess of Happy Valley" and "Courtship of Allaphair" all written in 1917 and published in Scribner's magazine for March, April, February, May, June, September, October and January. They were all gathered into one volume the same year entitled *In Happy Valley* and published by Scribner's.

1918: James Wallace Fox dies on January 14, 1918, in New York. Buried in Paris Cemetery, Paris, Kentucky.

1919: On fishing trip to Norton, Virginia, John Fox, Jr., becomes ill, quickly developing into pneumonia. Taken home to Big Stone Gap where he died on July 8, 1919. Taken back to Paris, Kentucky, on special L&N train number 34, and buried on the 11th in Paris Cemetery.

1920: His nearly finished novel *Erskine Dale: Pioneer* is finished by his sister Elizabeth and published in 1920 by Scribner's. It too was serialized in Scribner's magazine for the months of January through June.

1924: Everett B. Fox dies on December 1, 1924. Buried at Paris, Kentucky.

1925: Minerva Carr Fox, John Jr.'s mother, dies on September 8, 1925, and is buried at Paris, Kentucky.

1931: Rector Fox dies on November 29, 1931, and is buried at Vahalla, New York.

1934: Horace Ethelbert Fox dies on June 9 and is buried in Big Stone Gap, Virginia.

1937: Richard Talbott Fox dies on April 24 at Elizabeth's home in Washington, D.C., and is buried at Paris, Kentucky.

1946: Currie Duke Mathews dies in Virginia on March 2 of influenza while living with her daughter and son-in-law. She is cremated and buried in the Ware Church cemetery, near Mathews, Virginia.

1954: Oliver Edwin Fox dies on December 25 at Big Stone Gap, Virginia, and is buried in Paris, Kentucky.

1954: Fritzi Scheff is found dead in her apartment at 300 E. 79th Street, on April 8. She was 74. She was buried at the Actors Fund Cemetery in Kensico, New York. She left an estate of only $476.

1962: Minerva Carr Fox dies on November 8, 1962, at the age of 94 in Big Stone Gap, Virginia. She never married. Buried Paris Cemetery, Paris, Kentucky.

1970: Elizabeth Fox Moore dies on June 21, 1970, in Big Stone Gap, Virginia. Funeral at Berryville, Virginia. Buried in Washington, D.C., beside her husband Cabell Moore.

Notes

Introduction

1. Fox, John, Jr., "Through the Shadows of the Big Black," unpublished manuscript, U.K. Collection.
2. Creason, Joe, *Crossroads and Coffee Trees,* The Louisville *Courier-Journal* and the Louisville *Times,* Louisville, 1975, page 52.
3. Page, Thomas Nelson, "John Fox," *Scribner's,* December, 1919, pages 679–680. Although Page doesn't say so, it is just possible that the lady was Fritzi Scheff, and that Page is describing that first meeting between John and his future wife.
4. Letter to Currie Duke Mathews from John Fox, Jr., July 7, 1903. Courtesy of Frank Cabot.
5. *Ibid.,* November 3, 1903. Courtesy of Frank Cabot.
6. Page, page 676.
7. Fox, John, Jr., "A Story of Some Stories," unpublished manuscript, U.K. Collection.
8. Fox, John, Jr., "On Horseback to Kingdom Come," *Scribner's,* August, 1910.
9. "John Fox, Jr.," The Atlanta *Constitution,* January 2, 1898.

Chapter 1: Early Spring — 1916

1. Fox, John, Jr., Introduction to *The White Mice* by Richard Harding Davis, Charles Scribner's Sons, New York, 1916.

Chapter 2: A School Teacher's Son

1. Around 1910, John Sr. wrote his oldest son James about those first years: "You ask about dates and places of living prior to coming to Stony Point. I was married in Aug. '52 and that fall moved to Clark County and we went to housekeeping ... the next spring — 1853, we moved down to the house where you were born. I was teaching during this time at the Jefferson Seminary.... In the fall of that year we moved ... [and] ... I taught at the Sudduth school house over on the Mt. Sterling pike ... in the fall of that year, 1854 we moved to Montgomery County to the house in which your mother afterwards died. We lived in this house one year, and in the fall of 1855 ... we quit housekeeping and moved about ¾ of a mile south.... In the next spring, 1856 we moved over to the Donohue house ... where Sidney was born, July 3rd of that year. During this time I was teaching at Lulbugrud Seminary. In the summer of the same year, 1856, we moved to Sharpsburg, living at first in the house, opposite and south of the hotel, then for a while in the upper rooms of the Academy, but finding these too small, we moved to the house west of the hotel named above ... where Everett

was born. [September 2, 1858] Not well pleased in Sharpsburg we removed to Montgomery again in the fall of 1858, to the same house which we had occupied first in 1854..." (Letter to James from John, Sr., incomplete, U.K. Collection; no date; Probably 1910).

    2. Fox, John, Sr., Diary, June 14, 1860. U.K. Collection.
    3. *Ibid.*, June 15, 1860.
    4. *Ibid.*, January 28, 1860.
    5. *Ibid.*, February 2, 1862.
    6. John, Sr., his sons and Aunt Carrie had moved from Montgomery County, Kentucky, to Mayslick in Mason County the latter half of 1861, and after his marriage, he and his new wife, now pregnant, had moved to the house at Stony Point just south of Paris in Bourbon County in August of 1862, where John, Jr. was born.
    7. John Fox, Sr. Diary, December 16, 1862. U.K. Collection.
    8. *Ibid.*, December 17, 1862.
    9. Letter to John Fox, Sr. from Boaz Fox, December 27, 1862. U.K. Collection.
    10. John Patterson, one of John Fox, Jr.'s schoolmates at Harvard, wrote in 1907 that the senior John Fox, "...was a man of refined taste and old time culture, a dignified and kindly gentleman schoolmaster. The impression that ... [John, Jr.'s] mother (Minerva Carr Fox) makes on me at this distant day, when I come to reflect and to remember, is that of a woman of sweetness and wit..." (Patterson, John, *Library of Southern Literature,* Volume 4, The Martin & Holt Company, Atlanta, 1907, page 1683).
    11. Moore, Elizabeth Fox, "Two Early Kentucky Schoolmasters," The *Register* of the Kentucky Historical Society, Volume 45, No. 151, April, 1947, page 181.
    12. Moore, Elizabeth Fox, *Personal and Family Letters and Papers,* University of Kentucky Collection, page 5. Also in Holman, Harriet R., "John Fox, Jr.: Appraisal and Self-Appraisal," *Southern Literary Journal,* Spring, 1971, page 32.
    13. Letter to Minerva Carr Fox from John Fox, Sr., June 27, 1875. U.K. Collection.
    14. Moore, Elizabeth Fox, "Two Early Kentucky Schoolmasters," page 162.
    15. *Ibid.*, pages 162–163. On April 2, 1879, disaster struck the Fox household. The house where John Fox, Sr., was busy raising a family, and where John Fox, Jr., was born, burned to the ground, taking with it almost everything they owned, little though it was. Finances being what they were in the Fox household, the house was not insured, but with the usual Fox fortitude, the elder Fox, with the help of his sons, cleared away the debris, and on October 9, 1879, laid out the foundation for a new home, "and on April 17, 1880, one year and 13 days after the old house had burned, he moved to the new house, where he lived until August 1883 when he moved family and school to Paris."
    16. Letter to Horace Fox from John, Jr., March 21, 1880. U.K. Collection. Transylvania University was known as Kentucky University until 1908 when it changed its name to Transylvania University. For clarity, I have used this latter choice of names.

Chapter 3: Off to School

    1. Wright, John D., *Transylvania: Tutor to the West,* The University Press of Kentucky, Lexington, 1975, page 283.
    2. *Ibid.*, page 284.
    3. There were, however, exciting moments. John D. Wright, Jr., relates (*Transylvania: Tutor to the West,* pages 284–285) that Champ Clark, who eventually became a national figure in the United States House of Representatives, was involved in what is arguably the most famous incident at the old school at that time. It all started as a small argument, and like many arguments, escalated into a full-blown battle of words, with threats and posturing from both Clark and another student named Webb. Eventually, Clark's adversary called Clark a liar, and was smacked over the head with a piece of wooden planking for his insolence. So far so good. But when others attempted to restrain the two, Webb broke free and began striking Clark with awkward blows that seem to characterize such disagreements, but still doing considerable damage to the helpless Clark who was being held by his classmates, still attempting to break up the fight. Clark was able, finally, to throw off his restrainers, and what happened next was what got him expelled. "Under the head of the bed," Clark later wrote, "I had an old revolver, whose cylinder would not revolve except by hand manipulation, for which I had swapped a German grammar and a French grammar. I got that and fired at Webb." A

bystander knocked up his arm and deflected the shot so that it missed Webb's head by about an inch. In something of an understatement, Clark said, "That ended the fight." Clark was expelled despite his best efforts to plead his case to the president of the college. Clark was on track to receive class honors for the class of 1871, but his expulsion ended all that. Stepping forward to receive the honors, now that Clark was out of the way, was none other than James Lane Allen, and second place honors went to the son of the college president. Clark had the last laugh, so to speak; after leaving Transylvania University, he went to Bethany College in West Virginia, was admitted to the bar in 1875, elected to Congress in 1892 and elected again in 1896, and served there until his death in 1921. He also ran unsuccessfully for the Democratic nomination for President in 1912, losing out to Woodrow Wilson.

4. Letter to John Fox, Sr., from James Fox, September, 1879. U.K. Collection.
5. Letter to John Fox, Sr., from John Fox, Jr., March 21, 1880. U.K. Collection.
6. Letter to James Fox from John Fox, Jr., October 14, 1880. U.K. Collection.
7. *Ibid.*
8. Letter to John Fox, Sr., from John Fox, Jr., September 28, 1880. U.K. Collection.
9. Letter to James Fox from John Fox, Jr., October 14, 1880. U.K. Collection.
10. *Ibid.*
11. Letter to James Fox from John Fox, Jr., November 28, 1880. U.K. Collection.
12. Letter to Minerva Fox from John Fox, Jr., April 24, 1881. U.K. Collection.
13. Letter to James Fox from John Fox, Jr., May 15, 1881. U.K. Collection.
14. Fox, John, Sr., Diary, July 5, 1881. U.K. Collection.

Chapter 4: Boom and Bust, Part I

1. Klein, Maury, *History of the Louisville & Nashville Railroad*, The Macmillian Company, New York, 1972, page 281.
2. Letter to Oliver Fox from Horace Fox, August 1, 1882. U.K. Collection.
3. *Ibid.*
4. *Ibid.*
5. Letter to Minerva Fox from Horace, August 13, 1882. U.K. Collection.
6. *Ibid.*
7. Letter to Minerva Fox from John Fox, Jr., October 22, 1882. U.K. Collection.
8. Martin, Betty Fible, *John Fox Letters: 1883–1889.* Introduction. The University of Virginia. Barrett Collection. Although Fible's letters to John didn't often reveal it, his thoughts about his friend are interesting. While at Harvard, the "Kentucky boys" gravitated toward each other. Their exclusive group included, besides Fible, John Letcher Patterson, Edward William Stevens Tingle, Edwin Upshur Berryman, and Paul Shipman Drane. Fible's diary doesn't mention very much in the way of opinions about the others, but about John, Fible was more forthcoming. In an entry dated January 1, 1883, just after Fible wrote to his sister that Christmas, Fible wrote, "Ten fellows at my new club table: John W. Fox sits opposite me, senior, about 21— known throughout school for his laugh — all call him "Jack," slap him on the back when they meet him —from Stony Point, Ky.— handsome face, excellent form, being tumbler in Athletic Association — gets good marks in courses, owing however, to no inherent thoughtfulness, only quickness in appropriating ideas, whether erroneous or right, of others— has nothing in view after graduating — no money — will sail around for four or five years. I prophesy, acting variously as schoolteacher in country school, clerk in lawyers office, finally turn up in new suit of clothes and marry a woman with money enough to take care of both —fine fellow, though, honest."
9. Wingate, Charles, "A Story-Teller of the Mountains," *Critic*, July 24, 1897. The portrayal was of Mme. Perrichon in the French comedy *Papa Perrichon.*
10. Letter to Minerva Fox from John Fox, Jr., February 6, 1883. U.K. Collection.
11. Letter to John Fox, Jr., from Everett Fox, May 5, 1883. U.K. Collection.
12. Letter to John Fox, Jr., from James, May 3, 1883. U.K. Collection.
13. *Ibid.*, May 17, 1883.
14. Letter to Minerva Fox from John Fox, Jr., March 11, 1883. U.K. Collection.
15. Letter to John Fox, Jr., from Minerva Fox, May 13, 1883. U.K. Collection.
16. Letter to John Fox, Jr., from James Fox, May 3, 1883. U.K. Collection.

284   Notes — Chapters 4, 5

17. Letter to John Fox, Jr., from John Fox, Sr., June 24, 1883. U.K. Collection.
18. Davis, Charles Belmont, *Adventures and Letters of Richard Harding Davis,* Charles Scribner's Sons, New York, 1917, page 29.
19. *Ibid.*
20. *Ibid.*
21. Moore, Elizabeth Fox, *Personal and Family Letters and Papers,* U.K. Collection, page 30.
22. Letter to John Fox, Jr., from John Fox, Sr., June 25, 1883. U.K. Collection.

Chapter 5: New York Reporter

1. Letter to Micajah Fible from John Fox, Jr., July 13, 1883. Collected in Martin, Betty Fible, *John Fox Letters — 1883–1889,* The University of Virginia. Barrett Collection, pages 2–3.
2. Lubow, Arthur, *The Reporter Who Would Be King,* Charles Scribner's Sons, New York, 1992, page 45.
3. *Ibid.,* page 41.
4. Letter to John Fox, Jr., from John Fox, Sr., July 18, 1883. U.K. Collection.
5. Moore, Elizabeth Fox, *Personal and Family Letters and Papers,* U.K. Collection, page 33.
6. Letter to James Fox from John Fox, Sr., July 18, 1883. U.K. Collection.
7. Letter to John Fox, Jr., from James Fox, July 29, 1883. U.K. Collection.
8. Letter to John Fox, Jr., from John Fox, Sr., August 27, 1883. U.K. Collection.
9. Letter to Micajah Fible from John Fox, Jr., September 27, 1883. Collected in Martin, *John Fox Letters,* page 9.
10. Letter to John Fox, Jr., from John Fox, Sr., November 18, 1883. U.K. Collection.
11. Moore, Elizabeth Fox, *Personal and Family Letters and Papers,* University of Kentucky Collection, page 32.
12. *Ibid.,* page 32.
13. *Ibid.,* page 5.
14. *Ibid.,* page 33
15. John had graduated from Harvard in 1883 with Fletcher Ryer, and through his old classmate, had gotten a job tutoring Fletcher's younger brother which was a common thing for the more wealthy to do in any city, but especially in larger cities like New York.
16. Letter to James Fox from John Fox, Jr., November 23, 1883. U.K. Collection.
17. Letter to John Fox, Jr., from John Fox, Sr., December 23, 1883. U.K. Collection. The letter outlining the cause of his troubles in the Ryer household remains elusive, and as far as is known, John never mentioned in print at least, any further explanation of the complications from that situation, nor the reason for his leaving.
18. Moore, *Personal and Family Letters and Papers,* page 33.
19. Letter to Minerva Fox from John Fox, Jr., November 11, 1883. U.K. Collection. Note: The play apparently never got any further than the planning stages, despite John's enthusiasm for it.
20. Letter to Minerva Fox from John Fox, Jr., November 23, 1883. U.K. Collection.
21. *Ibid.*
22. *Ibid.,* January 11, 1884.
23. Letter to Micajah Fible from John Fox, Jr., January 30, 1884. Collected in Martin, *John Fox Letters,* pages 12–13.
24. *Ibid.,* March 14, 1884, page 15.
25. *Ibid.,* May 11, 1884, pages 18–19.
26. *Ibid.,* page 19.
27. *Ibid.* Probably September–October, 1884, page 22.
28. Letter to Minerva Fox from John Fox, Jr., June 5, 1884. U.K. Collection.
29. Moore, *Personal and Family Letters and Papers,* page 35.
30. Letter to John Fox, Jr., from Minerva Fox, October 13, 1884. U.K. Collection.
31. *Ibid.,* December 5, 1884.
32. Letter to John Fox, Jr., from John Fox, Sr., February 6, 1885. U.K. Collection.
33. *Ibid.*
34. Letter to Micajah Fible from John Fox, Jr., February 3, 1885. Quoted in Martin, *John Fox Letters,* page 27.

35. *Ibid.*
36. *Ibid.*, February 14, 1885, page 31.
37. Moore, *Personal and Family Letters and Papers*, U.K. Collection, page 6. Also in Holman, Harriet R., "John Fox, Jr.: Appraisal and Self-Appraisal," *Southern Literary Journal*, Spring, 1971, page 33.
38. Letter to Micajah Fible from John Fox, Jr., March 4, 1885. Collected in Martin, *John Fox Letters*, page 34.ø

Chapter 6: A Tale of Two Cities

1. Letter to John Fox, Jr., from James Fox, February 27, 1885. U.K. Collection.
2. *Ibid.*, August 25, 1885.
3. Letter to Micajah Fible from John Fox, Jr., May 24, 1885. Collected in Martin, Betty Fible, *John Fox Letters*, The University of Virginia. Barrett Collection, page 42.
4. Moore, Elizabeth Fox, *Personal and Family Letters and Papers*, U.K. Collection, page 6.
5. Letter to Micajah Fible from John Fox, Jr., March 14, 1885. Collected in Martin, *John Fox Letters*, page 60.
6. *Ibid.*
7. *Ibid.*
8. He did find time to write a travel book about their experiences. It is typewritten obviously on pages supplied, or pages that could be brought from the railway company, and later bound into a book format by the writer. It numbers almost 100 pages, and he entitled it *The Wanderers*, and never offered it for publication, as far as we know.
9. *The Wanderers*, page 3, Unpublished. U.K. Collection.
10. *Ibid.*, pages 4–5.
11. *Ibid.*, pages 21–22.
12. Letter to Micajah Fible from John Fox, Jr., June 2, 1886. Collected in Martin, *John Fox Letters*, page 64.
13. *Ibid.*
14. *Ibid.*
15. Evidently this was a false alarm since Currie Duke didn't marry until 1898, some 12 years later.
16. Letter to Micajah Fible from John Fox, Jr., June 24, 1886. Collected in Martin, *John Fox Letters*, page 68.
17. *Ibid.*, August 18, 1886, page 73.
18. Letter to John Fox, Jr., from John Fox, Sr., September 19, 1886. U.K. Collection.
19. *Ibid.*
20. Moore, *Personal and Family Letters and Papers*, pages 38–39.
21. Letter to Micajah Fible from John Fox, Jr., May 13, 1887. Collected in Martin, *John Fox Letters*, page 79.
22. *Ibid.*, June 30, 1887, page 82.
23. Letter to James Fox from John Fox, Jr., September 3, 1887. U.K. Collection.
24. Letter to Micajah Fible from John Fox, Jr., January 7, 1888. Collected in Martin, *John Fox Letters*, page 89.
25. Letter to Minerva Fox from John Fox, Jr., January 8, 1888. U.K. Collection.
26. Letter to Micajah Fible from John Fox, Jr., April 1, 1888. Collected in Martin, *John Fox Letters*, page 92.

Chapter 7: Boom and Bust, Part II: The Gap

1. Letter to James Fox from John Fox, Jr., December 18, 1888. U.K. Collection.
2. *Ibid.*, incomplete, February, 1889.
3. *Ibid.*
4. Letter to James Fox from John Fox, Jr., February 21, 1889. U.K. Collection.

5. It would be very easy to read something into letters that the writer never intended to be there, but the break between Micajah Fible and John Fox, Jr., if indeed there ever was such a break, was definitely sudden, especially in light of their close friendship of several years. A descendant of Fible's suggested that money, possibly gambling debts, might have been involved, but there is yet no proof to say one way or the other. John's last letter to Fible in 1889, has a lonesomeness to it that is impossible to deny, and again, may suggest nothing except that John was lonely at Christmas that year.

6. Letter to Micajah Fible from John Fox, Jr., December 13, 1889. U.K. Collection.

7. *Ibid.*

8. There is a one page, typed document that has found its way into the pages of an unpublished rough draft of a story entitled *The Passing Star*, where John made a somewhat cryptic and less than intelligible remark about Fible's disappearance: "Then M.—boyish helpless melancholy. Wanted to pat him on the shoulder to encourage him ... F's [obviously Fritzi Scheff, John's wife at this time] characterization of M.F. already expressed her belief M. no longer existence. Remark: Would be fun to show him that women are of some good, but I'd hate to be the one" ("The Passing Star." Unpublished. U.K. Collection). The document is full of abbreviations, partial sentences and appears to be notes for something he later intended to "straighten up," but apparently never did. If there were other pages besides this one misplaced page, they were either destroyed or lost. In any event, John never spoke extensively—at least in print—of his missing friend again, but often wondered to himself that Fible might turn up somewhere—somehow. We certainly cannot know from so little what John meant in those few jumbled lines stored inside the pages of an unpublished story, but it is fairly obvious that as late as 1909—nearly 20 years later—that he was still thinking, from time to time, of his old friend, and his disappearance.

9. Letter to James Fox from John Fox, Jr., January 23, 1889. U.K. Collection.

10. Letter to James Fox from John Fox, Sr., March 7, 1890. U.K. Collection.

11. Fox, John, Jr., *The Trail of the Lonesome Pine*, Charles Scribner's Sons, New York, 1908, page 182.

12. *Ibid.*, page 233.

13. Fox, John, Jr., *The Trail of the Lonesome Pine*, Charles Scribner's Sons, New York, 1908, pages 236–237.

14. Letter to James Fox from John Fox, Jr., January 29, 1890. U.K. Collection.

15. Letter to John Fox, Jr., from *Century* Magazine, January 30, 1890. U.K. Collection.

16. *Ibid.*

17. *Ibid.*

18. Moore, Elizabeth Fox, *Personal and Family Letters and Papers*, U.K. Collection, page 7. Also quoted in Holman, Harriet R., "John Fox, Jr.: Appraisal and Self-Appraisal," *Southern Literary Journal*, Spring, 1971, page 35. Note: The photograph of the check is now on display at Duncan Tavern, in Paris, Kentucky, along with other items donated by John's family after his death in 1919. He had worked on the story for two years in his spare time away from the family business. He did not know it yet, but *Century* magazine would keep it for an additional two years before finally publishing it. "During those four years," he said, "I did not write another line."

19. Letter to James Fox from John Fox, Sr., April 20, 1890. U.K. Collection.

20. Letter to James Fox from John Fox, Jr., May 20, 1890. U.K. Collection.

21. *Ibid.*

22. *Ibid.*

23. Letter to James Fox from John Fox, Jr., June 23, 1890. U.K. Collection.

24. *Ibid.*, June 30, 1890.

25. *Ibid.*, July 5, 1890.

26. *Ibid.*, July 29, 1890.

27. Letter to James Fox from John Fox, Jr., June 22, 1890. U.K. Collection.

28. *Ibid.*, July 25, 1890.

29. *Ibid.*, August 26, 1890.

30. *Ibid.*, September 17, 1890.

31. Fox, John, Sr., Diary, December 9, 1890. U.K. Collection.

32. *Ibid.*, December 11, 1890.

33. *Ibid.*

34. Letter to James Fox from John Fox, Jr., February, 1891. U.K. Collection.

35. *Ibid.*, July 6, 1891.

36. *Ibid.*
37. When John was first introduced to Madison Cawein, he was accompanied by James Whitcomb Riley and James Lane Allen. They found him working behind the cashier's desk of a local pool room. Riley, for one, remarked that it was certainly a strange place to find a poet, and James lane Allen felt much the same way, but was too much of a gentleman to say very much about it. John, on the other hand, was ecstatic at meeting the author of *Blooms of the Berry*, and could hardly restrain himself in praise of the poet. Granted, poetry, for most of us is an acquired taste, or in some cases, one suspects, is an affected taste. In any event, neither Allen nor Riley were overly impressed with Cawein, but John was excited enough to make up for both Allen's and Riley's disappointment.
38. Letter to Thomas Nelson Page from John Fox, Jr., October 16, 1891. University of Virginia. Barrett Collection. Also collected in Holman, Harriet R., *John Fox and Tom Page as They Were*, Field Research Projects, Miami, 1970, pages 12–13.
39. Letter to James Fox from John Fox, Jr., October 30, 1891. U.K. Collection.
40. The Methodist-Episcopal Hospital in Brooklyn did not keep records from 1891, and as a result, no actual knowledge of what his illness was that required an operation exists.
41. Letter to John Fox, Jr., from *Century* Magazine, November 19, 1891. Incomplete. U.K. Collection.
42. Letter to James Fox from John Fox, Jr., January 16, 1892. U.K. Collection.
43. Fox, John, Jr., *The Trail of the Lonesome Pine*, Charles Scribner's Sons, New York, 1908, pages 41–42.
44. *Ibid.*, pages 244–245.
45. *Ibid.*

Chapter 8: A Knight of the Cumberland

1. The *Mountain-Echo*, London, Kentucky, January 2, 1896.
2. *Ibid.*, November 29, 1889.
3. Allen, James Lane, *The Bluegrass Region*, Harper & Brothers Publishers, New York, 1899, page 96.
4. The *Sentinel-Echo*, London, Kentucky, June 16, 1893.
5. Campbell, John C., *The Southern Highlander and His Homeland*, The University Press of Kentucky, Lexington, 1969.
6. Fox, John, Jr., *The Trail of the Lonesome Pine*, Charles Scribner's Sons, New York, 1908, page 140.
7. *Ibid.*, page 141.
8. The Big Stone Gap *Post*, Big Stone Gap, Virginia, September 12, 1890.
9. Fox, John, Jr., *Bluegrass and Rhododendron*, Charles Scribner's Sons, New York, 1901, page 209.
10. *Ibid.*, page 218.
11. *Lonesome Pine*, pages 94–95.
12. Big Stone Gap *Post*, December 18, 1890.
13. Rutherford, Mildred Lewis, *The South in History and Literature*, The Franklin-Turner Company, Atlanta, 1907, page 602.
14. *Bluegrass and Rhododendron*, page 211.
15. *Ibid.*, page 210.
16. *Ibid.*
17. "A Talk with John Fox, Jr., Author and Correspondent," The *Sentinel*, Milwaukee, Wisconsin, December 10, 1900.
18. *Bluegrass and Rhododendron*, page 224.
19. *Ibid.*, pages 210–211.
20. *Lonesome Pine*, page 343.
21. *Bluegrass and Rhododendron*, page 278.
22. Fincher, Jack, "A Unique Theatre Retells the Tale of Backwoods Feud," The *Smithsonian*, December, 1981.
23. Big Stone Gap *Post*, June 29, 1893.
24. Webb, W.B., "Blood-Stained Pound Gap," The Kentucky *Explorer*, April, 1991, page 48.

Notes — Chapter 8

25. Scalf, Henry P., "The Red Fox was Snared by Devil John Wright," The Kentucky *Explorer*, August, 1991, page 59.
26. Ibid., page 62.
27. *Lonesome Pine*, page 335.
28. Tincher, *Smithsonian*, December, 1981.
29. *Bluegrass and Rhododendron*, page 284.
30. The Swedenborgian religious beliefs that Taylor professed, takes its name from its founder Emanuel Swedenborg (1688–1772), a Swedish scientist who turned to spiritual teaching, the basis of which are direct contact with angels and the spiritual world in dreams, visions, etc. Though somewhat obscure, branches of the church exist mainly in Great Britain, with smaller congregations in Europe and the United States. Perhaps one of the more famous adherents was John Chapman, who was better known as "Johnny Appleseed."
31. *Bluegrass and Rhododendron*, page 284.
32. Big Stone Gap *Post*, June 29, 1893.
33. Ibid.
34. Handwritten note, U.K. Collection.
35. *Lonesome Pine*, page 340.
36. Ibid., page 341.
37. The *Mountain-Echo*, July 5, 1895. "Bad" Tom Smith was taken to Vicco, Kentucky, and is buried in the cemetery of the Vicco Baptist Church. In a few lines that illustrates both mountain admiration and optimism, his epitaph on a relatively modern stone reads: Bad Tom Smith, Oct. 15, 1859, June 28, 1895. "Bad enough to be hanged, Not too bad for God to save."
38. Fox, John, Jr., "Death by the Rope for Desperado Hall," The New York *Herald*, September 3, 1892.
39. Ibid.
40. Ibid.
41. *Lonesome Pine*, pages 293–294.
42. Ibid., page 348.
43. Fox, "Death by the Rope," The New York *Herald*, September 3, 1892.
44. *Lonesome Pine*, pages 348–349.
45. Talton Hall was taken by his sister back to Kentucky and buried there in the little community of Dunham just outside of Jenkins, Kentucky. Another version has it that Hall asked "Devil" John Wright to take him back to Kentucky and bury him there. Either way, Hall is buried in the Wright Cemetery at Dunham, just across the narrow road that cuts through the cemetery, within sight of John Wright's grave. The old grave marker is broken and the part that held Hall's name is missing. A new marker, put up by Ben Caudill Camp of the Sons of Confederate Veterans suggests that Hall was a veteran in the 13th Kentucky Cavalry, C.S.A., enlisting when he was 12 years old! Faron Sparkman of Wise, Virginia related that "the legend was that he was already so good with a gun by the age of 12, that the Confederate Army was glad to have him! Although this is extremely young, we know of at least one other 1850 man that served in the 13th — Austin G. Combs — [who] was born on May 13, 1850 and enlisted in Company I of the 13th when he was 13."

The location of Taylor's grave was apparently lost or kept secret for many years after his death. Recently, the location was found and marked inside the cemetery at Wise, Virginia. It too was marked by the Sons of Confederate Veterans. This stone suggest that Taylor was himself a member of the 13th Kentucky Cavalry. Oddly enough, Taylor's grave lies at the foot of Enos Hilton's (Hilton's name was sometimes spelled "Hylton") grave, the man whose murder in Norton, Virginia, sent Talton Hall to the gallows. Apparently the location of Taylor's grave was known to be beside that of his daughter-in-law Hattie Salyer Taylor, and the stone was set there. Again, Mr. Sparkman, who is a member of the Sons of Confederate Veterans camp who set the stone, explains: "It is my understanding that Hattie Salyer Taylor was Doc Taylor's daughter-in-law. The Pound Historical Society had done a lot of research on the Taylor story and they first informed us that although Taylor was intentionally buried in an unmarked grave at the Wise City Cemetery, old area residents knew the location of his grave was nearest to the grave of Hattie. With that information we chose to place the Confederate marker where we did. It may be off a few feet, but I feel certain that it is very near to his exact burial site."

46. Fox, John, Jr., *On the Trail of the Lonesome Pine*, Scribner's, October, 1910.
47. *Bluegrass and Rhododendron*, page 268.
48. Big Stone Gap *Post*, October 26, 1893.

49. *Ibid.*
50. *Bluegrass and Rhododendron*, page 269.
51. Big Stone Gap *Post*, November 2, 1893.
52. *Ibid.*
53. *Ibid.*
54. The state of Virginia did not adopt the electric chair as a means of execution until 1908. By that time, even the legislators admitted to the gruesomeness of the practice of hanging. The distaste of such executions was exacerbated, said the March 11, 1908, edition of The Big Stone Gap *Post*, "by several bungled executions, including that of a Negro in a Virginia town who was hanged three times, the rope breaking twice." From 1908 onward, condemned prisoners would be taken out of the hands of local amateur executioners, and put to death by state officials in Richmond.
55. *Bluegrass and Rhododendron*, page 239.
56. *Ibid.*, page 246.
57. *Ibid.*, page 247.
58. *Ibid.*
59. *Ibid.*, page 248. Henan Fleming's name was spelled variously as Heenan, Henon and Henan. The Big Stone Gap *Post* referred to him as "Henry" at least once or twice.
60. Big Stone Gap *Post*, January 18, 1894.
61. *Ibid.*, February 1, 1894.
62. Handwritten receipt from H.H. Dotson & Sons, dated July 30, 1895, for one hat for Henan Fleming. The cost was noted at $3.00. U.K. Collection.
63. Big Stone Gap *Post*, February 22, 1894.

Chapter 9: Feuds and Romance

1. Moore, Elizabeth Fox, *Personal and Family Letters and Papers*, U.K. Collection, page 7.
2. *Ibid.* Note: John didn't forget about Easter Hicks. Some years later, he resurrected her in an unfinished play based on *A Cumberland Vendetta*, replacing Martha Lewallan, heroine of *Vendetta*.
3. Letter to John Fox, Jr., from Richard Fox, December, 1892. U.K. Collection.
4. Bangs, J.K., "Literary Notes," *Harper's Monthly Magazine*, December, 1899.
5. *Ibid.*
6. *Ibid.*
7. Moore, Elizabeth, *Personal and Family Letters and Papers*, page 2. Also collected in Holman, Harriet R., "John Fox, Jr.: Appraisal and Self-Appraisal," *Southern Literary Journal*, page 27.
8. *Bluegrass Today*, March 30, 1977. Also Moore, William Cabell address "John Fox, Jr.," October 21, 1957, page 2.
9. Moore, Elizabeth, *Personal and Family Letters and Papers*, page 2.
10. *Ibid.*
11. Fox, John, Jr., *The Trail of the Lonesome Pine*, Charles Scribner's Sons, New York, 1908, page 58. Also *Bluegrass and Rhododendron*, Charles Scribner's Sons, New York, 1901, page 39.
12. *Lonesome Pine*, pages 97–98.
13. *Ibid.*, page 362.
14. Fox, *The Kentuckians*, pages 31–32.
15. John Fox, Sr. Diary, December 25, 1893. U.K. Collection.
16. Thomas Nelson Page (1853–1922) was possibly the best known Southern writer of his time. He came from an old distinguished Virginia family; was trained as a lawyer, but gave up the law to devote himself to writing novels and short stories until 1910 when he began to feel that editors wanted nothing more from him except the same old type-cast stories. He then began to dabble in politics, and eventually served under Woodrow Wilson as ambassador to Italy. He and his second wife Florence, were John's greatest friends and supporters. They arrived back in the United States from Italy on July 8, 1919, the very day that John died in Big Stone Gap, Virginia.
17. Letter to John Fox, Jr., from Thomas Nelson Page, January, 1894. U.K. Collection. Incomplete. Also collected in Holman, Harriet R., *John Fox and Tom Page as They Were*, Field Research Projects, Miami, 1970, page 16.
18. Letter to John Fox, Jr., from James Lane Allen, January 26, 1894. U.K. Collection.
19. *Ibid.*

20. *Ibid.*
21. *Ibid.*
22. Advertisement for speaking engagements. U.K. Collection.
23. Letter to R.C. Coldwell from Thomas Nelson Page. January 22, 1894. U.K. Collection. Also collected in Holman, page 18.
24. Letter to James Fox from John Fox, Jr., March 12, 1894. U.K. Collection.
25. The *Kentuckian-Citizen* (Paris, Kentucky) quoted in The Big Stone Gap *Post*, February 22, 1894.
26. Letter to John Fox, Jr., from Theodore Roosevelt, June 4, 1894. U.K. Collection.
27. *Ibid.*, August 11, 1894. U.K. Collection.
28. *Ibid.*, June 28, 1894. U.K. Collection.
29. Moore, Elizabeth, *Personal and Family Letters and Papers*, page 7.
30. *Ibid.*, page 8.
31. *Ibid.*

Chapter 10: *The Whirl-Wind*

1. Titus, Warren I., *John Fox, Jr.*, Twayne Publishers, Inc., New York, 1971, page 26.
2. Dutton, Lawerance, "Literary Notes," *Harper's Monthly Magazine*, Volume 41, No. 544, November, 1895.
3. *A Cumberland Vendetta and Other Stories*, The *Bookman*, Volume 2, No. 5, January, 1896.
4. "Hell-fer-Sartin and Other Stories," The *Critic*, No. 812, September 11, 1897. Note: Many of his reader's Victorian sensibilities were upset at his choice of a title for the book. Madison Cawein called it a "beautiful book, with the terrible title" (Letter, July 1, 1897 U.K. Collection). But Cawein liked it otherwise except for an almost non-existent complaint of being "already acquainted with [the stories] having read them in different periodicals." It was a perfect example of what *Harper's* had tried to warn him about.
5. "Hell-fer-Sartin and Other Stories," The *Bookman*, Volume 6, No. 1, September, 1897.
6. Letter to John Fox, Jr., from Rector Fox, 1895, U.K. Collection. Another publication, *The Nation*, thought the title: *Hell-fer-Sartin and Other Stories,* a little too lurid for their tastes, but the stories "leave little to be desired." In this first burst of writing activity, John was not neglecting his nonfiction work, and wrote for *Century* a piece called "Fox Hunting in Kentucky" that would be published in *Century* magazine in August of 1895. It would not see book publication until 1901 in *Bluegrass and Rhododendron* along with its companion pieces "After Br'er Rabbit in the Blue Grass" (*Century*, November, 1896) and "Br'er Coon in Old Kentucky" (*Century*, February, 1898). He had found his niche, and was happier than he had ever been in his life.
7. Letter to James Fox from John Fox, Sr., December 12, 1894, U.K. Collection
8. *Ibid.*
9. Townsend, John Wilson, "John Fox Junior." Text of radio broadcast given from the University of Kentucky radio studios of WHAS, June 22, 1932. It was not John's first foray into improvisation where clothing was concerned. While at Harvard, he and Edwin Upshur Berryman, classmate at Transylvania University and Harvard, were invited to a high-class social function "...and were horrified to understand at the last moment that they were expected to wear black silk sox. They did not have any and so, borrowed two pairs of long silk stockings from Mr. Berryman's Aunt ... pulled them almost waist-high and went to the ball" (Letter to James Fox from John Fox, Sr., December 12, 1894, U.K. Collection). It was a wonder that they kept straight faces throughout the ball, but apparently, everything went off without a hitch, so to speak.
10. Ardery, W.B., "John Fox, Jr. Still Popular With the Over-50 Set of Kentuckians," The *Lexington-Herald*, March 11, 1973.
11. Kesterson, David B., *Bill Nye*, G.K. Hall & Company, Twayne Publishers, Boston, 1981, page 67.
12. Moore, William Cabell, "John Fox, Jr.." Address delivered on October 21, 1957, at the Club of Colonial Dames, Washington, D.C. DAR Address, October 21, 1957. The Kentucky Historical Society Collection, Frankfort, Kentucky.
13. Promotional letter. U.K. Collection.
14. Kephart, Horace, *Our Southern Highlanders*, The University of Tennessee Press, Knoxville, page 282.

15. *Ibid.*, page 350.
16. Revell, Peter, *James Whitcomb Riley*, Twayne Publishers, Inc., New York, 1970, pages 36–37.
17. Kesterson, *Bill Nye*, page 82.
18. Pond, Major James Burton, *Eccentricities of Genius*, G.W. Dillingham Company, New York, 1900, pages 523–524.
19. Letter to Thomas Nelson Page from J.B. Pond, January 30, 1895. U.K. Collection. Also collected in Holman, Harriet R., *John Fox and Tom Page As They Were*, Field Research Projects, Miami, 1970, page 22.
20. Letter to Minerva Fox from John Fox, Jr., December 23, 1894, U.K. Collection.
21. *Ibid.*
22. Fox, John, Jr., "The Hanging of Bad Tom Smith," *Harper's Weekly*, August 10, 1895.
23. *Ibid.*
24. The *Mountain-Echo*, May 31, 1895.
25. Wingate, Charles E.L., "A Story-Teller of the Mountaineer," The *Critic*, July 24, 1887.
26. Jillson, Willard Rouse, *Literary Haunts and Personalities of Old Frankfort*, The Kentucky Historical Society, Frankfort, 1941, pages 80–82.
27. "The Kentuckians," The *Critic*, January 1, 1898.

Chapter 11: *The Spanish-American War*

1. Tom Page was already at Scribner's, and his advice was something John always considered seriously, even beyond that of older friends and family. It is not unlikely that Page played a huge part in bringing John into the Scribner's corral.
2. Currie Duke did marry, and soon. She married Wilbur Knox Mathews on June 7, 1898, in Louisville, Kentucky, five days after John left the Gap for Tampa, Florida.
3. Letter to Charles Davis from Richard Harding Davis, May, 1898, University of Virginia. Barrett Collection. Also quoted in Lubow, Arthur, *The Reporter Who Would Be King*, Charles Scribner's Sons, New York, 1992.
4. Willis, Walter, *The Martial Spirit: A Study of Our War with Spain*, Houghton Mifflin Company, Boston, 1981, page 10.
5. Cuba did not abolish slavery until 1886, removing at last, one of the real reasons the South supported annexation of the island before the Civil War.
6. Willis, Walter, *The Martial Spirit*, page 8.
7. Letter to Bob (?) from John Fox, Jr., March 13, 1898. New York Public Library.
8. Letter to John Fox, Jr., from Thomas Nelson Page, April 8, 1898. Collected in Holman, Harriet R., *John Fox and Tom Page As They Were*, Field Research Projects, Miami, page 38. Tom Page was closer to the mark than he may have intended, for The *Mountain Echo* reported the middle of April that "it is rumored that the men of Clay County [Kentucky] are not interested in the Cuban War, but have one of their own." (The *Mountain-Echo* (London, Kentucky) No date given. Typed copy in the Laurel County Historical Society). But by the first of May, that same paper was carrying advertisements for applications for enlistment in Theodore Roosevelt's "'Special Corps, better known as 'Rough Riders,'" and the paper went on to say that, "Applicants must be between 18 and 45 years of age, and in addition to regular physical qualifications must be expert riders" (*ibid.*, May 6, 1898).
9. Letter to John Fox, Jr. from Theodore Roosevelt, May 3, 1898. U.K. Collection.
10. Roosevelt, Theodore, *An Autobiography*, The Macmillian Company, New York, 1914, page 209.
11. A major Civil War battle had been fought on those grounds in 1863 between Confederate Braxton Bragg and Union General William S. Rosecrans. If not for the heroic, stubborn stand of Union General George H. Thomas on the Union's left flank, Rosecrans' army would have likely been destroyed. Two months later, the Confederates were defeated at nearby Chattanooga. The new soldiers now encamped on the old battlefield were not unmindful of where they were, and nearly all, including John Fox, Jr., set aside some time for sight-seeing. Camp Thomas at Chickamauga, was named for that Union general who saved the day 35 years before.
12. Letter to John Fox, Jr., from *Harper's Magazine*, April 21, 1898. U.K. Collection.
13. Fox, John, Jr., "Chickamauga," *Harper's*, May 7, 1898.

14. *Ibid.* John's description of black and white soldiers mingling on the streets could give the wrong impression. Black soldiers did not serve in the same units with white soldiers. Each race had their own separate units, and would fight separately. It would not be until after World War II that the armed forces would be integrated.
15. *Ibid.*
16. *Ibid.*
17. Barton, Clara, *The Red Cross,* J.B. Lyon Company, Albany, New York, 1898, page 411.
18. Colonel Walter Reed of the U.S. Army Medical Corps was appointed in 1900, two years after the Spanish-American War ended (government moved slowly then, too), to head a commission in Cuba to study the cause of yellow fever, and found it was spread by bacteria introduced by the bite of mosquitoes, the same as malaria.
19. Milton, Joyce, *The Yellow Kids: Foreign Correspondents in the Heyday of Yellow Journalism,* Harper & Row, Publishers, New York, 1989, page 343.
20. Fox, John, Jr., "Chickamauga," *Harper's,* May 21, 1898.
21. *Ibid.* Their commanding officer very likely was Lieutenant John J. Pershing who ever-afterward would be known as "Black Jack" Pershing, and would gain more laurels in World War I. In the charge up San Juan Hill, Pershing's command, by some accounts, saved the day, but lost half of its officers and fully 20 percent of its men.
22. *Ibid.*
23. Letter to John Fox, Jr., from H.L. Nelson, April 25, 1898. U.K. Collection.
24. Fox, John, Jr., "Volunteers in the Bluegrass," *Harper's,* June 18, 1898. The "bear-grass" refers to that area in and around Louisville, Kentucky, and "blue-grass" refers, of course, to that area in and around Lexington, Kentucky.
25. Moore, Elizabeth Fox, *Personal and Family Letters and Papers.* U.K. Collection, page 48.
26. Pringle, Henry F., *Theodore Roosevelt,* Harcourt, Brace and Company, New York, 1981, page 237.
27. Azoy, *Charge!,* page 38.
28. *Ibid.*
29. Davis, Richard Harding, *The Cuban and Porto Rican Campaigns,* Charles Scribner's Sons, New York, 1904, page 50.
30. Whitney, Caspar, "Wanted—A War," *Harper's Weekly,* May 21, 1898. Davis had come already prepared with his "outrageous" outfit, having bought in London—from whence he had just returned—and New York, an impressive ersatz uniform that "included top boots, a canvas shooting jacket, a revolver and cartridge belt, a leather flask, and a jaunty cap" (Lubow, page 156). Davis, almost as well known for his immaculate dress as his journalistic skills—which were considerable— became the good natured butt of plenty of jokes and smiles over his dress, but took it all in great style. He was, after all, Richard Harding Davis. His fans would accept no less of him, he reasoned, and he enjoyed his status as a clothes horse immensely.
31. *Ibid.*
32. Davis, Richard Harding, *The Cuban and Porto Rican Campaigns,* Charles Scribner's Sons, New York, 1904, page 50.
33. Letter to family from John Fox, Jr., June 10, 1898. U.K. Collection.
34. Lubow, Arthur, *The Reporter Who Would Be King,* Charles Scribner's Sons, New York, 1992, page 163.
35. *Ibid.*
36. Letter to Charles Davis from Richard Harding Davis, University of Virginia. Barrett Collection. Also in Lubow, *Reporter Who Would Be King,* page 169.
37. Willis, *The Martial Spirit,* page 245.
38. Fox, John, Jr., "With the Troops for Santiago," *Harper's Weekly,* July 16, 1898.
39. Davis, *Campaigns,* page 86.
40. *Ibid.,* page 90.
41. *Ibid.*
42. Fox, "Troops for Santiago," *Harper's,* July 16, 1898.
43. Davis, *Campaigns,* page 94.
44. *Ibid.,* page 100.
45. Letter to family from John Fox, Jr., June 19, 1898. U.K. Collection.
46. *Ibid.*
47. Fox, "Troops for Santiago," *Harper's,* July 16, 1898. The town of Daiquiri was often spelled with a "B" in many of the dispatches from the correspondents in Cuba.

48. *Ibid.*
49. Letter to family from John Fox, Jr., June 22, 1898. U.K. Collection.
50. *Ibid.*
51. Davis, *Campaigns*, page 136. Also Lubow, *The Reporter Who Would Be King*, page 173.
52. Fox, John, Jr., "Las Guásimas," *Harper's Weekly*, July 30, 1898.
53. Roosevelt, *Autobiography*, page 240.

Chapter 12: The Horizontal Hail-Storm

1. Lubow, Arthur, *The Reporter Who Would Be King*, Charles Scribner's Sons, New York, 1992, page 249.
2. Roosevelt, Theodore, *An Autobiography*, The Macmillian Company, New York, 1914, page 241.
3. Stallman, R.W., *Stephen Crane: A Biography*, George Braziller, Inc., 1968, page 381.
4. Davis, Richard Harding, *The Cuban and Porto Rican Campaigns*, Charles Scribner's Sons, New York, 1904, page 132. Davis himself first wrote that the engagement resulted from the volunteers walking into an ambush of which they had been previously warned. He soon changed his mind, however, when he realized the disastrous public relations firestorm that he and other correspondents had unwittingly set off. "The first accounts of the fight of the Rough Riders and Guásimas," he wrote later and rather lamely, "came from correspondents three miles away at Siboney, who received their information from the wounded when they were carried to the rear, and from an officer who stampeded before the fight had fairly begun. These men declared they had been entrapped in an ambush, that Colonel Wood was dead, and that their comrades were being shot to pieces."
5. Davis, *Cuban and Porto Rican Campaigns*, page 144.
6. Freidel, Frank, *The Splendid Little War*, Bramhall House, New York, 1958, page 106.
7. *Ibid.*
8. Davis, *Campaigns*, page 170.
9. *Ibid.*
10. Fox, John, Jr., "Las Guásimas," *Harper's Weekly*, July 30, 1898.
11. Davis, *Campaigns*, page 153.
12. Fox, John, Jr., *With the Rough Riders, Harper's Weekly*, July 30, 1898.
13. Remington, Frederic, "With the 5th Corps," *Harper's*, November, 1898.
14. Samuels, Peggy and Harold, *Frederic Remington: A Biography*, Doubleday & Company, Inc., New York, 1982, page 279. Remington was pushing close to 300 pounds at this time, and no doubt the scarcity of food played heavily on his nerves. Also, it is reasonable to assume that his great weight did nothing to make him comfortable in the heat and humidity of the tropics.
15. Davis, *Campaigns*, page 153.
16. Remington, "With the 5th Corps," *Harper's*, November, 1898.
17. *Ibid.* It is a little strange that Remington was not invited into Roosevelt's company. Remington had illustrated Roosevelt's book *Ranch Life and the Hunting Trail* in 1887. Given also the fact that Remington was the greatest living illustrator of that area of the country that Roosevelt loved so well, cannot help but make one wonder at the circumstances that Remington found himself in when Roosevelt was so close by.
18. Lubow, *The Reporter Who Would Be King*, page 180.
19. Letter to family from John Fox, Jr., July 2, 1898, U.K. Collection.
20. *Ibid.*, June 29, 1898.
21. Creelman, James, "Creelman of the *Journal* Leads the Charge at El Caney," The New York *Journal*, July 1, 1898.
22. Fox, John, Jr., "Santiago and Caney," *Harper's Weekly*, August 6, 1898.
23. *Ibid.*
24. *Ibid.*
25. *Ibid.*
26. *Ibid.*
27. *Ibid.* Cosby's full name was Fortunatus Cosby and he survived his wound, and was able to write his mother from his hospital bed that "I am not suffering at all" (Freidel, *Splendid Little War*, page 183). Presumably, Cosby got all the peaches he wanted in the hospital, and when he got back home.

28. *Ibid.*
29. Freidel, *Splendid Little War*, page 184.
30. "Santiago and Caney," *Harper's*, August 6, 1898.
31. Roosevelt, Theodore, *An Autobiography*, page 247.
32. *Ibid.*
33. Davis, *Campaigns*, page 218.
34. Stallman, *Stephen Crane*, page 393.
35. Remington, Frederic, "With the 5th Corps," *Harper's*, November, 1898.
36. "Santiago and Caney," *Harper's*, August 6, 1898.
37. Crane, Stephen, *War Memories*, quoted in Stallman, page 284.
38. Davis, *Campaigns*, pages 219–220.
39. *Ibid.*, pages 278–279.
40. Letter to family from John Fox, Jr., July 5, 1898. U.K. Collection.
47. Davis, *Campaigns*, page 227.
48. "Santiago and Caney," *Harper's*, August 6, 1898.
49. *Ibid.*
41. Letter to family from John Fox, Jr., July 5, 1898. U.K. Collection.
42. After the war, when everyone was rushing into print with books about the war before interest died down, *Harper's* used John's articles in their *Illustrated History of the War with Spain*, which shows the regard they had for his writing. In later years, Frank Freidel would quote from John's dispatches in his *The Splendid Little War*.
43. Barton, Clara, *The Red Cross*, J.B. Lyon Company, Albany, New York, 1898, page 564.
44. Fox, John, Jr., "Truce," *Harper's Weekly*, August 13, 1898.
45. Carnes, Cecil, *Jimmy Hare: News Photographer*, The Macmillian Company, New York, 1940, page 70.
46. *Ibid.*, page 76.
47. "Truce," *Harper's*, August 13, 1898.
48. Roosevelt, *An Autobiography*, page 250.
49. "Truce," *Harper's*, August 13, 1898.
50. Letter to family from John Fox, Jr., July 5, 1898. U.K. Collection.
51. "Santiago and Caney," *Harper's*, July 30, 1898.
52. Letter to Frederic Remington from John Fox, Jr., July 28, 1898. Quoted in Splete, *Selected Letters*, page 227.
53. Fox, John, Sr., Diary, July 27, 1898. U.K. Collection. The real scare was that any fever was yellow fever, and that could not be allowed into the United States. Suspected victims were either held at Cuba until well or dead, and those who sickened on the way to Tampa Bay, were held on their transports until the doctors were certain they either did not have the dreaded "yellow jack," or in some cases, until they had died.
54. Page, Thomas Nelson, "John Fox," *Scribner's*, December, 1919.
55. Letter to Frederic Remington from John Fox, Jr., July 28, 1898. Quoted in Splete, *Selected Letters*, page 227.
56. Fox, John, Sr., Diary, August 3, 1898. U.K. Collection.
57. A couple of years later, John would say that he was thrown out of the Galt House in Louisville because the manager thought he was sick with yellow fever. Whether this was before or after he entered the hospital is uncertain, but it is likely to have been before (The *Sentinel*, Milwaukee, Wisconsin, December 10, 1900).
58. Fox, John, Sr., Diary, August 8, 1898. U.K. Collection.
59. Letter to Frederic Remington from John Fox, Jr., August 15, 1898. Quoted in Splete, *Selected Letters*, page 228.

Chapter 13: The Road to Recovery

1. Stallman, R.W., *Stephen Crane: A Biography*, George Braziller, Inc., New York, 1968, page 401.
2. Roosevelt would be re-elected in 1904 with a majority of votes approaching 3 million. He had ruled out a run for a third term early on, but ran again in 1912 as a third party candidate, losing

to Woodrow Wilson. He had apparently set the groundwork for another run in 1920, but died early in 1919.

3. Letter to Thomas Nelson Page from John Fox, Jr., November 1, 1898. Duke University Collection. Also collected in Holman, Harriet R., *John Fox and Tom Page As They Were,* Field Research Projects, Miami, 1970, page 40.

4. John is referring to the first time he went to Tampa after Harper's wire to him of around the first of June.

5. Letter to *Harper's* from John Fox, Jr., July 30, 1898 U.K. Collection.

6. Letter to John Fox, Jr., from H.L. Nelson *(Harper's),* August 8, 1898. U.K. Collection.

7. Ibid.

8. Apparently Whitney was not too upset over the controversy or, at least if he was, he did not let it interfere with his work, especially in light of his quick acceptance of *Harper's* assignment to Hawaii so soon after his tiff with the magazine.

9. Obviously, John was able to put the past aside, and convinced Currie's new husband to invest in another land scheme that James had cooked up. Mathews was only too willing to invest, but before the decade was out, he would be broke, having lost perhaps a million dollars or more of his inheritance in different bad luck schemes. When he died in 1910, he left his widow and two children practically penniless.

10. Letter to John Fox, Jr., from Richard Harding Davis. Undated. U.K. Collection.

11. The reason for Davis's demeanor during the wedding may have been because his wife, even before they were married, assured him that their marriage would be a platonic one. There would be no consummation of that marriage. Perhaps Davis thought time would change Cecil's mind, but apparently it never did. After a few years of misery, they were divorced. Cecil never remarried, but Davis would marry again to Bessie McCoy. John would have a minor falling out with Davis over that second marriage, having liked Cecil a great deal, but perhaps John did not know the circumstances of Davis's marriage to Cecil. It would not have been something that Davis would have been inclined to talk about, even to as close a friend as John.

12. Letter to James Fox from John Fox, Jr., July 28, 1899. U.K. Collection.

13. Ibid., August 9, 1899. John's anger at his brother's suggestion that he had wasted valuable time after coming home from Cuba, when he should have been finishing the novel to take advantage of the public's interest in the war, probably hit closer to home than John would have like to admit. He was always putting things off, and knew it himself. It was a characteristic that he was never able to rid himself of.

14. Letter to John Fox, Jr., from Caspar Whitney, August 2, 1899. U.K. Collection.

15. Letter to Scribner's from John Fox, Jr., July 26, 1899. University of Virginia. Barrett Collection.

16. Page, Thomas Nelson, "John Fox," Scribner's, December, 1919.

17. Fox, John, Jr., *The Heart of the Hills,* Charles Scribner's Sons, New York, 1912, page 168.

18. Letter to John Fox, Jr., from John Fox, Sr., February 13, 1900. U.K. Collection.

19. Pilcher, Louis, "John Fox, Jr. Spends A Day In Lexington," The Lexington *Leader,* March 3, 1900.

20. Letter to James Fox from John Fox, Jr., July 26, 1900. U.K. Collection.

21. Letter to John Fox, Jr., from Scribner's, September 19, 1900. U.K. Collection.

22. Letter to John Fox, Jr., from James Fox, November 19, 1900. U.K. Collection.

Chapter 14: The Crook of the Shepherd

1. Letter to Minerva Fox from John Fox, Jr., August 13, 1901. U.K. Collection. This would not be the last time that he was thought to be married when he wasn't. Perhaps the most notorious example was that of another John Fox, Jr., from New York who eloped against his father's wishes. Since the father was a well-to-do businessman in that town, the papers ate the scandal up. Years later, when Harold Green wrote his biography of the real John Fox, Jr., he confused the two, and placed this non-existent marriage in the pages of his book. That book, incidentally, provoked a strong reaction from the surviving members of the family—John had died 20 years before its publication—and they attempted to remove all copies of it from the bookstores, even threatening lawsuits. They were not entirely successful.

2. Moore, Elizabeth Fox, *Personal and Family Letters and Papers*. U.K. Collection, page 69.
3. By the first of August, 1901, Scribner's had sold nearly 10,000 copies of *Crittenden* for which John received 18¾ cents per copy, or nearly $2000.
4. "There has not," wrote *The Nation*, "to our knowledge, been written a more illuminating document upon Kentucky open-air life than Mr. Fox's little book of sketches." But they were a little concerned about the bloodiness of the stories. "The Kentucky mountaineer kills his enemies among the rhododendrons; the Blue-grass lad and lass kill their foxes. The pages are punctured with rifle and pistol shots, even as we learn that the tavern sign-boards are, in the mountain towns. These, however," said *The Nation*, "are familiar features in stories of the region." ("Still More Novels," The *Nation*, Volume 75, No. 1903, December 19, 1901.)
5. Letter to Scribner's from John Fox, Jr., September, 1901. University of Virginia. Barrett Collection.
6. *Ibid.*, November 20, 1901.
7. Throughout his association with Scribner's, they regularly advanced him money towards his writing and never complained about it. Occasionally, he asked for more, and they never failed to send him an advance. They treated him well, and he remained loyal to them for the rest of his life.
8. Letter to Minerva Fox from John Fox, Jr., February 1, 1903. U.K. Collection.
9. *Ibid.*, March 1, 1903.
10. Letter to John Fox, Jr., from Currie Duke Mathews, October 16, 1903. U.K. Collection.
11. Letter to Charles Scribner from John Fox, Jr., December 1, 1903. University of Virginia. Barrett Collection.
12. Moore, Elizabeth Fox, *Personal and Family Letters and Papers*, page 86.
13. *Ibid.*, page 85.
14. Fox, John, Jr., *The Little Shepherd of Kingdom Come*, Charles Scribner's Sons, New York, 1903, page 190.
15. Letter to Minerva Fox from John Fox, Jr., January 14, 1904. U.K. Collection.
16. *Ibid.*
17. Fox, John, Jr., *Crittenden*, Charles Scribner's Sons, New York, 1900, page 150.

Chapter 15: Cherry Blossom Correspondents

1. Fox, John, Jr., *Following the Sun Flag*, Charles Scribner's Sons, New York, 1905, page 120.
2. Letter to John Fox, Jr., from Florence Page, February 23, 1904. U.K. collection.
3. Letter to Minerva Fox from John Fox, Jr., February 21, 1904, U.K. collection.
4. Letter to Rebecca Davis from Richard Harding Davis. Quoted in *Adventures and Letters*, page 298.
5. Letter to Rebecca Davis from Richard Harding Davis, July 6, 1904. University of Virginia, Barrett Collection.
6. Fox, *Sun Flag*, page 9.
7. The daughter was named Barbara. After Tommie's death, her body was brought back to the United States, and buried in the family plot in Lexington (Kentucky) Cemetery. Her mother, Henrietta Hunt Morgan, sister to CSA General John Hunt Morgan, would die four years later on October 20, 1909, and Tommie's father Basil Duke would die at Currie's home in New York on September 16, 1916. All are buried side-by-side along with other members of the Morgan family. Just a few yards from their gravesites is the grave of another of John's friends, James Lane Allen.
8. Letter to Rebecca Davis from Richard Harding Davis, February, 1904. Quoted in *Adventures and Letters*, pages 299–300.
9. *Ibid.*, March 22, 1904.
10. Fox, *Sun Flag*, page 12.
11. Letter to Rebecca Davis from Richard Harding Davis, March 22, 1904. Quoted in *Adventures and Letters*, page 300.
12. Fox, *Sun Flag*, page 14.
13. *Ibid.*, page 15.
14. Letter to Rebecca Davis from Richard Harding Davis, May 22, 1904. Quoted in *Adventures and Letters*, page 304.

15. Letter to family from John Fox, Jr., March 27, 1904. U.K. collection.
16. Letter to Rebecca Davis from Richard Harding Davis, June 13, 1904. Quoted in *Adventures and Letters*, page 305.
17. In October, 1906, John received a letter from Japan that explained the Japanese position concerning foreign correspondents, and whether they should be allowed to see battles and battlefields. The letter was pretty convincing, but John was not too impressed. It was mentioned in the letter that one of the correspondents was deported because of his bias towards the Russians. "Moreover," the letter said, "a foreign war correspondent — whose name I need not mention, for I dare say everyone knows it — made every effort to bring some of his colleagues to the Russian side." It is purely supposition, but knowing Jack London as we do, it is a better than good chance that the writer was speaking of him (Letter to John Fox, Jr., October 30, 1906. U.K. collection).
18. Letter to Minerva Fox from John Fox, Jr., April 15, 1904. U.K. collection.
19. Moore, Elizabeth Fox, *Personal and Family Letters and Papers*, U.K. collection, page 61.
20. Fox, *Sun Flag*, page 50. It is likely that it was Richard Harding Davis who first coined the phrase.
21. Letter to family from John Fox, Jr., May 7, 1904. U.K. collection.
22. Fox, *Sun Flag*, pages 60–61.
23. Letter to Minerva Fox from John Fox, Jr., May 23, 1904. U.K. collection.
24. *Ibid.*, July 16, 1904.
25. Fox, *Sun Flag*, pages 72–73.
26. *Ibid.*, page 78.
27. *Ibid.*, pages 78–79.
28. *Ibid.*, page 80.
29. *Ibid.*, page 84.
30. Fox, *Sun Flag*, page 86.
31. *Ibid.*
32. Letter to family from John Fox, Jr., July 25, 1904. U.K. collection.
33. *Ibid.*
34. Fox, *Sun Flag*, page 97.
35. *Ibid.*, page 98.
36. *Ibid.*, page 101.
37. Letter to Rebecca Davis from Richard Harding Davis, July 31, 1904. Quoted in *Adventures and Letters*, page 309.
38. Fox, *Sun Flag*, page 90.
39. *Ibid.*, pages 104–105.
40. Fox, *Sun Flag*, page 107. Gutta-percha is a hard, plastic-like material that was used for everything from pistol and revolver handles to electric insulators.
41. *Ibid.*
42. *Ibid.*, page 113.
43. *Ibid.*, pages 113–114.
44. *Ibid.*, pages 117–118.
45. *Ibid.*, pages 117–118.
46. John says they were sick with beri-beri, or "the sleeping sickness." It is caused by continued deprivation of Vitamin B, and would likely be the result of the Japanese soldiers', or any soldiers' for that matter, unvarying and sometimes irregular diet.
47. Fox, *Sun Flag*, page 122.
48. *Ibid.*, page 125.
49. *Ibid.*, page 127.
50. *Ibid.*, pages 129–130.
51. Davis, Charles, *Adventures and Letters*, August 18, 1904, page 310.
52. Letter to Minerva Fox from John Fox, Jr., August 15, 1904. U.K. collection.
53. Fox, *Sun Flag*, page 134.
54. *Ibid.*, page 143.
55. Davis, Charles, *Adventures and Letters*, August 18, 1904, page 310.

Chapter 16: The Backward Trail

1. Fox, John, Jr., *Following the Sun Flag*, Charles Scribner's Sons, New York, 1905, page 152.

2. *Ibid.*, pages 154–155.
3. *Ibid.*, pages 161–162.
4. Davis, Richard Harding, *Notes of a War Correspondent*, Charles Scribner's Sons, New York, 1911, page 214.
5. *Ibid.*, pages 215–216.
6. Fox, *Sun Flag*, page 163.
7. Davis, *Notes,* pages 216–217
8. Fox, *Sun Flag*, page 169.
9. Melton Prior was an artist for *The Illustrated London News*.
10. W.H. Brill was reporter for the Associated Press.
11. Fox, *Sun Flag,* pages 169–170.
12. *Ibid.*, page 170.
13. Davis, *Notes*, page 225.
14. Fox, *Sun Flag*, page 177. It was too ridiculous for words, and he always suspected that his being a graduate of Harvard had much to do with his arrest by a Yale graduate. How the Japanese knew of his being a Harvard graduate, however, he neglected to explain, but it was as good an explanation as any.
15. *Ibid.*, pages 177–178.
16. Davis, *Notes,* page 229.
17. *Ibid.*, page 231.
18. Fox, *Sun Flag*, page 182.
19. *Ibid.*, page 183.
20. Davis, Charles Belmont, *Adventures and Letters of Richard Harding Davis*, Charles Scribner's Sons, New York, 1917, page 311.
21. *Ibid.*
22. Fox, *Sun Flag*, page 186.
23. *Ibid.*, page 187.
24. *Ibid.*, page ix.
25. Letter to Minerva Fox from John Fox, Jr., October 25, 1904. U.K. Collection.

Chapter 17: A Noble Profession

1. Letter to Minerva Fox from John Fox, Jr., January 8, 1908. U.K. Collection.
2. Letter to Thomas Nelson Page from John Fox, Jr., October 31, 1907. Duke University Collection. Also collection in Holman, Harriet R., *John Fox and Tom Page As They Were*, Field Research Projects, Miami, Florida, 1970, page 52.
3. Letter to John Fox, Sr., from John Fox, Jr., February 14, 1907. U.K. Collection.
4. *Ibid.*
5. "Writers Meet in Old Hotel," 1921. Newspaper clipping. No source given. U.K. Collection.
6. Letter to Minerva Fox from John Fox, Jr., July 29, 1907. U.K. Collection.
7. The song she sang, written by Victor Herbert, was the one she was most identified with all her life. However, she left special instructions before she died that the song would not be sung at her funeral. Apparently, she had grown to dislike the song after singing it uncounted times throughout her career.
8. Letter to Thomas Nelson Page from John Fox, Jr., November 3, 1908. Duke University Collection. Also collected in Holman, Harriet R., *John Fox and Tom Page As They Were*, Field Research Projects, Miami, 1970, page 55.
9. There has been since the book's publication in 1908 more than one story concerning the location of that famous literary landmark. Some have stated firmly that no such tree ever existed outside Fox's imagination. Others say that Fox heard of the tree from another, and used it in his story, but never saw it for himself. He tried, so the story goes, to have someone guide him to where the tree stood, but in the meantime, lightning had destroyed the tree, and the mountaineer guide had conveniently died. The story that rings truest, although there is no certainty here either, is one that is related in Everett Green's much maligned biography of John Fox, Jr., and repeated some years later by Thomas Clark in *The Filson Club Quarterly*. The hero of the story, Jack Hale, was apparently based on real personages Fox knew or had described to him. Thomas Clark states that "Jack Hale

was the composite character of James M. Hodge, a geologist, Judge Henry C. McDowell, and ... [R.C. Ballard Thruston, a photographer, geologist, and student of the mountains]...," and Clark goes on to say that June Tolliver "was the daughter of a mountaineer named Moore, and Hodge had undertaken to have her educated" (Clark, Thomas D., "Rogers Clark Ballard Thruston: Engineer, Historian, and Benevolent Kentuckian," The Filson Club *Quarterly*, Volume 58, No. 4, October, 1984). Neighboring mountaineers suspected that Hodge only wanted to educate her and them make her his wife, and were naturally suspicious of his motives. In any event, "June Tolliver" did not marry Hodge, but married another man named Palmer. It was reminiscent of the Miles Standish, Pricilla Mullens, John Alden menage à trois, but not exactly the same. Anyway, the Palmers set up housekeeping in the Moore cabin, and as luck would have it, there was a huge pine tree growing in their front yard. Ballard Thruston in a letter to Everett Green in 1939 said it was a unique tree, and the only one of its kind that he knew of in the area.

Soon, tourists began visiting the Moore home to admire two of the most famous characters of Fox's novel — June Tolliver née Moore and the "Lonesome Pine." It was enough to tax the patience of the most tolerant husband, and finally, taking all he could, the irate husband cut down the tree. It is not recorded what Mrs. Palmer thought of the whole thing.

In a similar description from an earlier edition of the Filson Club *Quarterly* (Tapp, Hamilton, "Rogers Clark Ballard Thruston: Good Kentuckian," The Filson Club *Quarterly*, Volume 21, No. 2, April, 1947), the author of that article concerning, again, Rogers Clark Ballard Thruston, related much the same story except the name of the original for June Tolliver was Morris, not Moore, and that the name for the book "was obtained from a pine tree that stood in the yard of a man named Moore." Mr. Moore's daughter, June, did not marry Hodge, but married a man named W.S. Palmer, and the story continued with the cutting down of the tree for the sake of peace and quiet from tourists and lovesick sightseers. Interestingly, a newspaper article from the Big Stone Gap *Post* dated July 23, 1919, suggested that Myrtle Wright, the daughter of "Devil" John Wright was the model for June, and the article said she seldom tired of letting everyone know it. Her father was supposedly the model for "Devil" Judd, and of course, "Bad" Rufe Tolliver was based on badman Talton Hall.

In later years, newspapers would, from time to time, speculate upon the existence of the tree, and a few would claim to know its whereabouts. One such newspaper made headlines by announcing that the Lonesome Pine was dying. They placed the tree almost exactly on the border between Kentucky and Virginia, and admitted that John had never seen it. "The tree is still living," said this anonymous reporter, "but its top branches, while still there, are dead. Plainly," he went on sadly, "its days are numbered, and the processes of nature are at work slowly doing what the wild storms of centuries have failed to accomplish" ("Lonesome Pine of John Fox, Jr. Fame Is Dying," Neon, Kentucky, *News*; no date, newspaper clipping. U.K. Collection).

Chapter 18: An Imperfect Union

1. Letter to Minerva Fox from John Fox, Jr., September 7, 1908. U.K. Collection.
2. Letter to James Fox from John Fox, Sr., November 10, 1908. U.K. Collection.
3. Moore, William Cable, "John Fox, Jr.." Address delivered on October 21, 1957, at the Club of Colonial Dames, Washington, D.C. The Kentucky Historical Society Collection, Frankfort, Kentucky.
4. Letter to James Fox from John Fox, Sr., December 1, 1908. U.K. Collection.
5. Ibid.
6. Fox, John, Sr., Diary, November 23, 1908. U.K. Collection.
7. Letter to Florence Page from John Fox, Jr., December 11, 1908. The University of Virginia. Barrett Collection.
8. Mt. Kisco was not unknown to John, since his brother owned a home there, and he had visited there often. Also, Richard Harding Davis lived nearby, and John had spent some time there as well.
9. "Fritzi Scheff Weds," The New York *Times*, December 14, 1908. It appears from the newspaper accounts, especially the New York papers, that they were more concerned with Fritzi Scheff than with her new husband, John Fox, Jr. Of course, it depended upon where the newspapers were printed. New York papers, where Fritzi was a local (if anything in New York City could be called local) the headline was "Fritzi Scheff Weds." In Kentucky, The *Lexington-Herald* headlined its notice

of the event as "Marriage of John Fox, Jr. Confirmed." The difference in the two were already becoming apparent from the very first, and would only be intensified over the relatively short duration of the marriage.

10. Moore, William Cabell, "John Fox, Jr."
11. Tapp, Hamilton, "Rogers Clark Ballard Thruston: Good Kentuckian," The Filson Club *Quarterly,* Volume 21, No. 2, April, 1947.
12. Townsend, John Wilson, "John Fox, Junior." Text of a radio broadcast given from the University of Kentucky radio studios of WHAS, June 22, 1932.
13. Rothert, Otto, *The Story of a Poet: Madison Cawein,* John P. Morton & Company, Inc., Louisville, 1921, pages 293–294.
14. Letter to Minerva Fox from John Fox, Jr., December 2, 1909. U.K. Collection.
15. Moore Address, October 21, 1957.
16. In addition to those stories and articles he published while married to Fritzi, he also apparently outlined what he intended to be another novel based on the feuds in Eastern Kentucky, and the part the state militia played in putting down those incidences of violence. The outline, in the Fox Family Papers at the University of Kentucky is just over 20 pages, typewritten. There are no editing marks, so he obviously never returned to the story. It is dated August 2, 1909.
17. Fox, John, Jr., "On Horseback to Kingdom Come," Scribner's, August, 1910.
18. *Ibid.,* page 176.
19. *Ibid.,* pages 177–178. The "little trouble" he is referring to is probably when he was almost mobbed in Berea, Kentucky, after speaking at the college there to a group of mountain students who took offense at his use of mountain dialect.
20. *Ibid.,* page 181.
21. *Ibid.,* page 182.
22. *Ibid.,* page 183.
23. *Ibid.,* page 185.
24. *Ibid.,* page 186.
25. Fox, John, Jr., "On the Road to Hell-fer-Sartin," Scribner's, September, 1910, page 350.
26. *Ibid.,* page 351.
27. *Ibid.,* page 354. Two years after John's visit to ex-sheriff Ed Callahan of Breathitt County, all of Ed's care and preparations against another shooting became academic. On May 3, 1912 Callahan was in his fortified store, and crossing the front room, he passed the front door for an instant. It was enough. He was shot from almost the same spot he had been shot two years before. This time it was more than a flesh wound, and John's careful host died of what might have been called in those days, of "natural causes" in "Bloody Breathitt."
28. *Ibid.,* page 356.
29. *Ibid.*
30. *Ibid.,* page 359.
31. *Ibid.*
32. *Ibid.,* page 361.
33. Fox, John, Jr., "On the Trail of the Lonesome Pine," *Scribner's,* October, 1910, pages 419–420. Local option refers to the propensity of certain Kentucky counties to allow the sale of legal alcohol, while surrounding counties may or may not do so. It is the voters of any particular county — their "option" as it were — that determines this. Naturally, and to the dismay of moonshiners, legal alcohol can usually be purchased more cheaply than locally produced moonshine. Sales of the illegal stuff still enjoy something of a moderate following in the mountains, as John and his companion witnessed in Letcher County that day.
34. *Ibid.,* page 422.
35. *Ibid.,* page 423.
36. *Ibid.,* page 424. There is a photograph still hanging behind John's desk at his home in Big Stone Gap of a bearded mountaineer that looks suspiciously like the 1940's and 1950's Western movie star George "Gabby" Hayes. A notation on the back of the photograph, however, identifies the picture as being of Hosey Bowling.
37. *Ibid.*
38. *Ibid.,* page 428.
39. *Ibid.,* page 429.

Chapter 19: The Eyes of a Father

1. Fox, John, Sr., Diary, February 18, 1912.
2. *Ibid.*, March 18, 1912.
3. *Ibid.*, April 28, 1912.
4. *Ibid.*, May 4, 1912.
5. *Ibid.*, May 4, 5, 8 and 9, 1912.
6. *Ibid.*, May 19 and 20, 1912.
7. Moore, Elizabeth Fox, *Personal and Family Papers and Letters,* U.K. Collection, page 75.
8. Letter to Minerva Fox from John Fox, Jr., July 15, 1912. U.K. Collection.
9. *Ibid.*, July 22, 1912.
10. The divorce was finally granted on January 31, 1913. Before the end of that year, Fritzi remarried to George Anderson, her leading man in one of her operettas, but that marriage too soon ended in divorce.
11. Fox, John, Jr., *Crittenden,* Charles Scribner's Sons, New York, 1900, page 80.
12. The reviews of the book were mostly favorable, though certainly not all, and of those that were, one cannot help but wonder if those reviewers were just being polite. It was, after all, a kinder and gentler age. *The Literary Digest* thought it would, and should, have the same success as *The Trail of the Lonesome Pine*; it did not. *The Dial* was less impressed, though not by much. "Mr. Fox seems to be only at his second best in this book," they said, "which means, no doubt, that he has worked out his richest veins, and has to fall back upon tailings. The story would be striking enough," they admitted, "if we were not all the time forced to contrast it with its predecessors" ("The Heart of the Hills," The *Dial,* Volume 54, No. 647, June 1, 1913). The New York *Times,* however, did not care for the book, and said so, right up front, saying the historical passages in the book dealt "too vaguely and anonymously with the actual events to produce the effect or possess the value of history, pad out the slender tale and provide the not too alert reader with excuse to fall asleep over it. For the arts of the historical novelist," they continued, having gotten their steam up, "Mr. Fox has no gift. He lacks skill to make the greater movement part and parcel of the dramatic development of the story of his own invention, to merge and fuse fact and fiction. With him the two are simply mixed together and strung along until book length is reached, and then served up for reading. Nor does a disposition to moralize and rhapsodize about the virtues and short-comings of the Kentuckians very much assist the general effect" (The New York *Times,* March 23, 1913). *Bookman* took the opposite position, while apologizing for not being more kind in the past to Fox's other books. "Mr. Fox," they said, "is one of the few true artists among the younger makers of American fiction" (Cooper, Frederick Taber, "Big Moments in Fiction and Some Recent Novels," The *Bookman,* August, 1913).
13. Eventually, *The Trail of the Lonesome Pine* would be made into a movie three times: in 1916, the result of which he was not satisfied with at all, once more in 1922, and again in 1936 starring youthful stars Fred MacMurray, Henry Fonda and Sylvia Sidney, and had John been alive when it was released, it is unlikely he would have been pleased with the result of that effort either. *The Little Shepherd of Kingdom Come* fared no better, and although he bargained with producers off and on until his death, the film version was not produced until 1928, and again in 1961. The latter effort, filmed in and around Lexington, Kentucky, caused something of a revival in local interest in John's novels, mainly from those trying to figure out upon whom the characters in the book were based, but that resurgence of interest in his works was fairly short lived. At any rate, both efforts at bringing the stories of Jack Hale and June Tolliver and Chad and Melissa and all the rest to the big screen, soon faded into cinematic history, and are seldom heard from anymore, nor even remembered except by a select few who make it their business to know these things.

Chapter 20: West and East — North and South

1. Fox, John, Jr., "A Hunting Peep into Jackson's Hole." Unpublished. U.K. Collection, page 1.
2. *Ibid.*, page 2.
3. *Ibid.*, page 3.
4. *Ibid.*, page 7.

5. *Ibid.*, page 11.
6. *Ibid.*, page 12.
7. *Ibid.*, page 13.
8. Fox, John, Jr., "After Bear in Colorado." Unpublished. U.K. Collection, page 5.
9. *Ibid.*, page 9.
10. *Ibid.*, page 13.
11. *Ibid.*, page 14.
12. Fox, "A Hunting Peep into Jackson's Hole," page 1.
13. Letter to Minerva Fox from John Fox, Jr., February 27, 1913. U.K. Collection.
14. *Ibid.*, March 15, 1914.
15. *Ibid.*, September 7, 1915.
16. Fox, John, Jr., "Tarpon Fishing at Boca Grande," Scribner's, February, 1916, page 214.
17. *Ibid.*, page 215.
18. *Ibid.*
19. *Ibid.*
20. *Ibid.*, pages 216–217.
21. Letter to Minerva Fox from John Fox, Jr., January 26, 1916. U.K. Collection.
22. Richard had married Elizabeth Genevieve McElroy on July 8, 1912 after his marriage to Cecil fell apart. Elizabeth usually went by her stage name of Bessie McCoy, and they were very happy. Tragedy, however, would haunt the Davis family after Richard's death. Bessie would try an abortive attempt at returning to the stage and in vaudeville until the 1920's when she retired from both. She would die at the age of 44 in France in 1931. Their only daughter married a frenchman; tried unsuccessfully to continue her father's career with her own writing, and ended her life in 1976.
23. Lubow, Arthur, *The Reporter Who Would Be King*, Charles Scribner's Sons, New York, 1992, page 332.
24. Moore, Elizabeth Fox, *Personal and Family Letters and Papers*, U.K. Collection, pages 72–73.

Chapter 21: Last Dance

1. He had been sick with influenza in March, and had not completely recovered his health. He was still weakened, but felt with the coming of spring, that he might soon be all right again. There is a possibility that John was sickened in March by the last vestiges of the great flu epidemic of 1918. That epidemic was worldwide, killing an estimated 20 to 100 million people. Records are incomplete at best. In America, nearly 700,000 died of the disease.
2. Crawford, Bruce, The Dearborn (Michigan) *Independent*. No date. Newspaper clipping. U.K. Collection.
3. "When John Fox Danced His Last Dance," *Literary Digest*, October 17, 1925.
4. According to the July 12, 1919 edition of the Lexington *Herald*, "Honorary pall bearers were Webb J. Willett, Norton, Va. [it was with Willett that he had planned to go fishing that day in Norton], J.M. Hodge, Big Stone Gap, Va., Maj. J.F. Bullitt, Philadelphia, Pa., R.E. Taggert, Big Stone Gap, Va., Victor Paynter, Norton, Va., and H.L. Miller, Big Stone Gap, Va. Active pall bearers were C.M. Thomas, Judge H. Clay Howard, William E. Simms, Charles H. Berryman, Upshur Berryman and Edward Bassett."
5. Moore, Elizabeth Fox, *Personal and Family Letters and Papers*. U.K. Collection, page 78.
6. *Ibid.*
7. Letter to Minna Page from Minerva (Minnie) Fox, August 3, 1919. U.K. Collection.
8. *Ibid.*
9. An undated newspaper clipping in one of the scrap-books kept by himself and his sisters said the room where he wrote, and was writing *Erskine Dale*, remained as he left it the day he left for Norton, and remained untouched long after he died. The last words of the unfinished novel, it said, remained in the typewriter, but this is likely untrue. "His unanswered letters are still on the desk, also a copy of Keats, his favorite poet, together with a pair of scissors, a magnifying glass and a wooden tobacco box," they wrote. The books he was using to write *Erskine Dale* remained, and the usual clutter that finds its way onto every desk, went undisturbed. In 1950, the desk along with much of the items left on it was donated by the surviving members of the Fox family to the museum at Duncan Tavern, in Paris, Kentucky, where they remain today.

Chapter 22: Surviving

1. Fox, John, Jr., *Erskine Dale: Pioneer,* Charles Scribner's Sons, New York, 1920, page 255.
2. Rohe, Alice, "Famous Author, Talks Entertainingly of Novel Building," The Denver *Times,* January 8, 1910.
3. *Ibid.*
4. Fox, John, Jr., *The Heart of the Hills,* Charles Scribner's Sons, New York, 1912, pages 167–168.
5. "John Fox, Jr. Decries Theory that Unknown Writers Have No Chance with Magazine Editor." No source given. Undated. U.K. Collection.
6. Letter to Minerva Fox from Oliver Fox, December 7, 1919. U.K. Collection.

Bibliography

Most of the primary sources for this book came in the way of letters and original manuscripts held at the Special Collections and Archives Service Center at the Margaret I. King Library at the University of Kentucky in Lexington. These sources are referred to as U.K. Collection.

The same is true of those materials found at the University of Virginia, and are found in the John Fox, Jr., Letters 1863–1919 and the Papers of Richard Harding Davis, both from the Clifton Waller Barrett Library, Special Collections Department at the University of Virginia in Charlottesville. These are referred to simply as from the Barrett Collection.

Magazines, Newspapers and Journals

"All Injured," The Big Stone Gap *Post,* February 1, 1894.
"Allen's 'Flute and Violin,'" The *Critic,* July 4, 1891, Vol. 16, No. 392.
Ardery, W.B., "John Fox, Jr. Still Popular with the Over-50 Set of Kentuckians," The Lexington *Herald-Leader,* March 11, 1973.
Bangs, John Kendrick, "Literary Notes," *Harper's Monthly Magazine,* December, 1899, Vol. C, No. DXCV.
Banks, Nancy Huston, "Four Novels of Some Importance," The *Bookman,* January 1901.
"Battle with Desperadoes," The Big Stone Gap *Post,* January 18, 1894.
"Body of E.L. Wentz," Big Stone Gap *Post,* May 12, 1904.
Brown, Richard C., "Looking Back," The Kentucky *Advocate,* Danville, Kentucky, May 17, 1992.
Clark, Thomas D., "John Fox, Jr.," The *Kentuckian-Citizen,* Paris, Kentucky, January 9, 1935.
_____, "Rogers Clark Ballard Thruston: Engineer, Historian, and Benevolent Kentuckian," The Filson Club *Quarterly,* Vol. 58, No. 4, October, 1984.
Clark, Ward, "Five Books of the Month," The *Bookman,* December 1908.
Cooper, Frederick Taber, "Big Moments in Fiction and Some Recent Novels," The *Bookman,* August 1913.
Creason, Joe, "Friend to the Mountaineer," The Louisville *Courier-Journal,* June 12, 1949.
_____, "Joe Ceason's Kentucky," The Louisville *Courier-Journal,* November 8, 1963.
_____, "Kentucky Mountain Folk Lost Their First and Best Literary Friend as John Fox, Jr., Died 38 Years Ago," The Louisville *Courier-Journal,* July 9, 1957.
_____, "The Lonesome Pine's Gone from the Trail," The Louisville *Courier-Journal,* February 22, 1948.

Bibliography 305

———, "31 Years After," The Louisville *Courier-Journal*, November 12, 1950.
Creelman, James, "Creelman of the *Journal* Leads the Charge at El Caney," The New York *Journal*, July 1, 1898.
———, "A Japanese Massacre," The New York *World*, December 12, 1894.
"Crime of Royalty Causes Man to Vanish," Newspaper clipping in Filson Club collection (Louisville, Kentucky). No source or date given. Probably Seattle *Times*, 1904.
"Crittenden," The *Critic*, February 1901, Vol. 38, No. 2.
"A Cumberland Vendetta & Other Stories," The *Bookman*, January 1896, Vol. 2, No. 5.
Davies, Acton, "A Light of Light Opera," *Cosmopolitan*, December, 1912, Vol. 54, No. 1.
"Doc Taylor," The Big Stone Gap *Post*, October 26, 1893.
"Dorchester Wins Fritzi Scheff Cup," Big Stone Gap *Post*, September 8, 1909.
Dutton, Laurence, "Literary Notes," *Harper's Monthly Magazine*, November, 1895, Vol. XCI, No. DXLVI, and January, 1898, Vol. XCVI, No. DLXXII.
"England's Foul Scandal," The Chicago *Tribune*, November 15, 1889.
"Erskine Dale, Pioneer," *Catholic World*, January, 1921, Vol. 62, No. 670.
"Erskine Dale, Pioneer," The New York *Times*, October 17, 1920.
"Erskine Dale, Pioneer," *Outlook*, October 20, 1920.
"Exposure Sure to Come," The Chicago *Tribune*, November 21, 1889
"False Reports of Sanitary Condition of Camp at Chickamauga Park," The Louisville *Courier-Journal*, June 7, 1898.
"A Few of the Season's Novels," *Review of Reviews*, June 1913.
Fincher, Jack, "A Unique Theatre Retells the Tale of Backwoods Feud," The *Smithsonian*, December, 1981.
"Former Star of Whom 'Kiss Me Again' Was Written, Fritzi Scheff, Dies at 74," The Louisville *Courier-Journal*, April 9, 1954.
Fox, John, Sr., Diaries, December 1890, The University of Kentucky, Lexington, Kentucky.
"Fox Family Troubles," The New York *Times*, January 29, 1901.
Fox, John Jr., "After Bear in Colorado," Unpublished. University of Kentucky Collection.
———, "Backward Trail of the Saxon," *Scribner's*, March, 1905.
———, "Chickamauga," *Harper's Weekly*, May 7 and May 14 and June 11, 1898.
———, "A Day in Atlanta," *Harper's Weekly*, May 21, 1898.
———, "Death by the Rope for Desperado Hall," The New York *Herald*, September 3, 1892.
———, "The Hanging of Bad Tom Smith," *Harper's Weekly Magazine*, August 10, 1895.
———, "Hardships of the Campaign," *Scribner's*, July, 1904.
———, "A Hunting Peep into Jackson's Hole," Unpublished. University of Kentucky Collection.
———, "Making for Manchuria," *Scribner's*, December, 1904.
———, "On Horseback to Kingdom Come," *Scribner's*, August, 1910.
———, "On the Road to Hell-fer-Sartin," *Scribner's*, September, 1910
———, "On the Trail of the Lonesome Pine," *Scribner's*, October, 1910.
———, "On the War Dragon's Trail," *Scribner's*, January, 1905.
———, "Owned Up to Six Murders," The New York *World*, September 6, 1892.
———, "Santiago and Caney," *Harper's Weekly*, July 23 and July 30 and August 6, 1898.
———, "Tarpon Fishing at Boca Grande," *Scribner's*, February, 1916.
———, "Trail of the Saxon," *Scribner's*, June, 1904.
———, "Truce," Harper's *Weekly*, August 13, 1898.
———, "Volunteers in the Bluegrass," *Harper's Weekly*, June 18, 1898.
———, "White Slaves of Haicheng," *Scribner's*, February, 1905.
———, "With the Rough Riders at Las Guásimas," Harper's *Weekly* July 30, 1898.
———, "With the Troops for Santiago," *Harper's Weekly*, July 16, 1898.
"Fritzi Scheff Base Ball Cup," The Big Stone Gap *Post*, July 14, 1909.
"Fritzi Scheff in "'The Prima Donna,'" The New York *Times*, December 1, 1908.
"Fritzi Scheff Left $476," The New York *Times*, February 4, 1955
"Fritzi Scheff, 74, Star in Operettas," The New York *Times*, April 9, 1954.
"Fritzi Scheff Signed with Independents," The Big Stone Gap *Post*, August 31, 1910.
"Fritzi Scheff Weds," The New York *Times*, December 14, 1908.

Bibliography

Frost, William G., "Our Contemporary Ancestors in the Southern Mountains," The *Atlantic Monthly*, Vol. 83, (1899), pages 311–319.
"Hall vs. Commonwealth," The *Southeastern Reporter*, Volume 15, West Publishing Company, St. Paul, 1892, pages 517–520.
"Hammond of London in Seattle," Washington, The Chicago *Tribune*, December 18, 1889.
Hearst, William Randolph, "The Heroic Capture of Caney Told By The *Journal's* Editor-in-Chief, New York *Journal*, July 4, 1898.
"The Heart of the Hills," The *Dial*, June 1, 1913, Vol. 54, No. 647.
"The Heart of the Hills," The *Nation*, May 29, 1913, Vol. 96, No. 2500.
"The Heart of the Hills," *Outlook*, March 29, 1913.
"Hell-fer-Sartin and Other Stories," The *Bookman*, September, 1897, Vol. 6, No. 1.
"Hell-fer-Sartin and Other Stories," The *Critic*, September 11, 1897, No. 812.
"Hell-fer-Sartin and Other Stories," The *Nation*, November 4, 1897, Vol. 65, No. 1688.
Hickerson, Joan, "Shepherd Film Stirs Historical Interest," The Lexington *Herald*, February 12, 1961.
"The Highlands—A Strange Region: Part II," The Kentucky *Explorer*, September, 1991. Reprint of Louisville *Courier-Journal* article, December, 1884.
Holman, Harriet R., "John Fox, Jr. : Appraisal and Self Appraisal," *Southern Literary Journal*, Spring, 1971.
Howells, William Dean, "Other Novels of the Season," The *Literary Digest*, April 5, 1913.
"In Happy Valley," The *Bookman*, November, 1917
"In Happy Valley," The *Dial*, November 8, 1917, Vol. 63, No. 753.
"In Happy Valley," The New York *Times*, September 30, 1917.
"In Kentucky's Mountains," The New York *Times*, March 23, 1913.
"In the Bluegrass," The *Critic*, July 2, 1892, Vol. 18, No. 541.
"Jail Burns," The Big Stone Gap *Post*, December 29, 1892.
"James Lane Allen," The *Critic*, April 1, 1893, Vol. 19, No. 580.
"John Fox Dies After Illness of Pneumonia," The Lexington *Herald*, Wednesday, July 9, 1909.
"John Fox, Jr.," The Atlanta *Constitution*, January 2, 1898.
"John Fox, Jr. and His Kentucky," The *Nation*, July 19, 1919, Vol. 109, No. 2820.
"John Fox, Jr., Died at His Home," The Big Stone Gap *Post*, July 9, 1919.
"John Fox, Jr., Dies in Virginia Home," The New York *Times*, July 9, L909.
"John Fox, Jr. Is Buried at Paris," The Lexington *Herald*, July 12, L909.
"John Fox, Jr., Noted Author Dies Suddenly," The Louisville *Courier-Journal*, July 9, 1909, and July 12, L909.
"John Fox, Jr. Returns," The New York *Times*, January 28, 1901.
"John Fox, Jr., with Us Again," The *Idea* (University of Kentucky), September 28, 1911.
"June Brides: Wedding of Mr. Mathews and Miss Duke," The Louisville *Courier-Journal*, June 8, 1898.
"The Kentuckians," The *Critic*, January 1, 1898.
"The Kentuckians," The *Dial*, February 1, 1898, Vol. 24, No. 279.
"Kin of Famed State Novelist Dies at Age 94," The Louisville *Courier-Journal* October 11, 1962.
Klotter, James C., "Feuds in Appalachia: An Overview," The Filson Club *Quarterly*, Vol. 56, No. 3, Louisville, Kentucky, July, 1982.
_____, "Kentucky in the Twentieth Century," The Filson Club *Quarterly*, Vol. 66, No. 3, July, 1992.
"A Knight of the Cumberland, *Outlook*, November 24, 1906.
"Knoxville Loses in Golf Tournament," The Big Stone Gap *Post*, July 31, 1912.
"A Light of Light Opera," The Big Stone Gap *Post*, November 27, 1912.
"London's Dens of Vice," The Chicago *Tribune*, November 27, 1889.
"London's Low Morals," The Chicago *Tribune*, December 1, 1889.
"Lonesome Pine of John Fox, Jr. Fame, Is Dying," Newspaper clipping. Fox Family Papers. No source or date given. University of Kentucky, Lexington.
"Marriage of John Fox, Jr. Confirmed," The Lexington *Herald*, December 17, 1908.
Marrosson, Issac F., "The South in Fiction," The *Bookman*, December 1910.

Martin, Betty Fible, "John Fox Letters—1883-1889," The University of Virginia, Barrett Collection.
Mason, Ester E., "Towering Pines, The Life of John Fox, Jr.," by Lieutenant Harold Everett Green, The Filson Club *Quarterly*, Vol. 18, No. 1, January, 1944.
"Minerva Carr Fox, Sister of Novelist," The New York *Times*, November 10, 1962.
"Miss Scheff Mass Attended by 300," The New York *Herald-Tribune*, April 13, 1954.
"Mr. Fox's New Story," The New York *Times*, November 24, 1906.
Moore, Elizabeth Fox, "Two Early Kentucky Schoolmasters," The *Register* of the Kentucky Historical Society, Volume 45, Number 151, April, 1947.
Moore, William Cabell, "John Fox, Jr.," Address delivered on October 21, 1957, at the Club of Colonial Dames, Washington, D.C. Kentucky Historical Society Collection, Frankfort, Kentucky.
"A Mountain Europa," The *Nation*, October 19, 1899, Vol. 69, No. 1790.
"Mountain Man," The Lexington *Herald*, March 4, 1900.
"Mystery Surrounds the Disappearance of Mr. E.L. Wentz As Deep As Ever," Big Stone Gap *Post*, November 5, 1903.
"The New Books," The *American Review of Reviews*, November 1908, Vol. 37.
"No Trace," The Big Stone Gap *Post*, October 29, 1903.
"Our Police," The Big Stone Gap *Post*, September 12, 1890.
Page, Thomas Nelson, "John Fox," Scribner's, Vol. 66, December, 1919.
"The Passing of the Unlawful Element in Wise County," The Big Stone Gap *Post*, February 22, 1894.
Pilcher, Louis, "John Fox, Jr. Spends a Day in Lexington," The Lexington *Leader*, March 3, 1900.
"Radio Address on John Fox, Jr.," The Bourbon *News*, Paris, Kentucky, November 30, 1934.
Repplier, Agnes, "A Sheaf of Autumn Fiction," *Outlook*, November 28, 1908.
Rohe, Alice, "Famous Author, Talks Entertainingly of Novel Building," The Denver *Times*, January 8, 1910.
Saad, Karen, "Fox Knew Intimately Mountaineers He Wrote About," *Blue Grass Today*, Lexington, Kentucky, March 30, 1977.
Scalf, Henry P., "The Red Fox Was Snared by Devil John Wright," The Kentucky *Explorer*, August, 1991.
Still More Novels, The *Nation*, December 19, 1901, Vol. 75, No. 1903.
"Talk Only of Scandal," The Chicago *Tribune*, November 24, 1889.
"Talk with "Dock" Taylor," The Big Stone Gap *Post*, June 29, 1893.
"A Talk with John Fox, Jr., Author and Correspondent," The *Sentinel*, Milwaukee, Wisconsin, December 10, 1900.
"Talt Hall On Trial," The Big Stone Gap *Post*, February 5, 1892.
Tapp, Hamilton, "Rogers Clark Ballard Thruston: Good Kentuckian," The Filson Club *Quarterly*, Volume 21, No. 2, Louisville, Kentucky, April, 1947.
"Taylor Hanged," The Big Stone Gap *Post*, November 2, 1893.
"Taylor vs. Commonwealth," The Southwestern *Reporter*, Volume 17, West Publishing Company, St. Paul, 1892, pages 812-813.
Thierman, Sue M., "Paris Will Honor the Memory of John Fox, Jr.," The Louisville *Courier-Journal*, December 16, 1962.
Townsend, John Wilson, "John Fox, Junior," Text of a radio broadcast given from the University of Kentucky radio studios of WHAS, June 22, 1932.
_____, "A History of Kentucky Literature Since 1913," The Filson Club *Quarterly*, Volume 13, No. 1, January, 1939.
_____, "Robert Burns Wilson," Text of a radio broadcast given from the University of Kentucky radio studios of WHAS, June 8, 1932.
"The Trail of the Lonesome Pine," The *Nation*, November 12, 1908, Vol. 87, No. 2263.
"The Trail of the Lonesome Pine," The New York *Times*, December 5, 1908, and October 17, 1908.
"The Trail of the Lonesome Pine," *Outlook*, October 17, 1908.

"Vice in Hideous Mein," The Chicago *Tribune*, November 17, 1889.
Webb, W.B., "Blood-Stained Pound Gap," The Kentucky *Explorer,* April, 1991.
"When John Fox Danced His Last Dance," The *Literary Digest*, October 17, 1925, 70–71.
Whitney, Caspar, "Wanted—A War," *Harper's Weekly Magazine,* May 21, 1898.
Wilson, Robert Burns, "Remember the Maine," New York *Herald*, April 17, 1898.
Wilson, Shannon H., "Window on the Mountains: Berea's Appalachia, 1870–1930," The Filson Club *Quarterly,* Vol. 64, No. 3, July, 1990.
Wingate, Charles E.L., "A Story-Teller of the Mountaineer," The *Critic*, July 24, 1897.
"Writers Meet in Old Hotel," Newspaper clipping. Fox Family Papers. No source or date given. University of Kentucky Collection.

Books

Allen, Douglas, *Frederic Remington and the Spanish-American War*, New York: Crown, 1971.
Allen, James Lane, *The Bluegrass Region,* New York: Harper & Brothers, 1899.
Azoy, A.C.M., *Charge! The Story of the Battle of San Juan Hill,* New York: Longmans, Green, 1961.
Barton, Clara, *The Red Cross*, Albany, New York: J.B. Lyon, 1898.
Brown, Charles H., *The Correspondent's War*, New York: Charles Scribner's Sons, 1967.
Campbell, John C., *The Southern Highlander and His Homeland*, Lexington: The University Press of Kentucky, 1969.
Carnes, Cecil, *Jimmy Hare: News Photographer*, New York: Macmillan, 1940.
Caudill, Harry M., *Night Comes to the Cumberland*, Boston: Little, Brown, 1962.
Chadwicke, Alice, *The Trail of the Lonesome Pine: A Play in Three Acts*, New York: Klaw and Erlanger, 1916. Also copyright 1937 by Samuel French.
Clark, Champ, *My Quarter Century of American Politics*, Two Volumes, New York: Harper & Brothers, 1920.
Creason, Joe, *Crossroads and Coffee Trees*, Louisville: The Louisville *Courier-Journal* and the Louisville *Times*, 1975.
Davis, Charles Belmont. *Adventures and Letters of Richard Harding Davis*, New York: Charles Scribner's Sons, 1917.
Davis, Elmer, *History of the New York Times: 1851–1921*, New York: The New York *Times*, 1921. (Republished 1971 by Scholarly Press, Inc., 22929 Industrial Drive East, St. Clair Shores, Michigan.)
Davis, Richard Harding, *The Cuban and Porto Rican Campaigns*, New York: Charles Scribner's Sons, 1904.
_____, *Notes of a War Correspondent*, New York: Charles Scribner's Sons, 1911.
_____, *The White Mice*, New York: Charles Scribner's Sons, 1919.
Dickey, John J., *Diary: Part I*, Barbourville, Kentucky: Knox County Genealogical Society, 1986.
Dierks, Jack Cameron, *A Leap to Arms: The Cuban Campaign of 1898*, Philadelphia: J.B. Lippincott, 1970.
Downey, Fairfax, *Richard Harding Davis: His Day*, New York: Charles Scribner's Sons, 1933.
Everett, Marshall, *Exciting Experiences in the Japanese-Russian War* (No publisher given) copyright Henry Neil, 1904.
Fox, John, Jr., *Bluegrass and Rhododendron*, New York: Charles Scribner's Sons, 1901.
_____, *Christmas Eve on Lonesome and Other Stories*, New York: Charles Scribner's Sons, 1901 and 1904.
_____, *Crittenden*, New York: Charles Scribner's Sons, 1900.
_____, *Erskine Dale: Pioneer,* New York: Charles Scribner's Sons, 1919.
_____, *Following the Sun Flag*, New York: Charles Scribner's Sons, 1905.
_____, *The Heart of the Hills*, New York: Charles Scribner's Sons, 1912.

_____, *Hell-fer-Sartin and Other Stories,* New York: Harper & Brothers, 1899. Also copyright Charles Scribner's Sons, New York, 1909.
_____, *The Kentuckians,* New York: Harper & Brothers, 1897. Also copyright Charles Scribner's Sons, New York, 1909.
_____, *A Knight of the Cumberland,* New York: Charles Scribner's Sons, 1906.
_____, *The Little Shepherd of Kingdom Come,* New York: Charles Scribner's Sons, 1903.
_____, *A Mountain Europa,* New York: Charles Scribner's Sons, 1912. Also contains *A Cumberland Vendetta* and *The Last Stetson. A Mountain Europa* also copyright 1897 by Harper & Brothers, New York.
_____, *The Trail of the Lonesome Pine,* New York: Charles Scribner's Sons, 1908.
Freidel, Frank. *The Splendid Little War,* New York: Bramhall House, 1958.
Hall, Wade, Introduction to *The Little Shepherd of Kingdom Come,* Lexington: The University Press of Kentucky, 1987.
Harkins, E.F., *Little Pilgrimages Among the Men Who Have Written Famous Books,* Boston: L.C. Page, 1903.
Hollyfield, Jeanne, Introduction to *A Purple Rhododendron and Other Stories,* by John Fox, Jr., Appalachia, Va.: Young Publications, 1967.
Holman, Harriet R., *John Fox and Tom Page as They Were,* Miami: Field Research Projects, 1970.
Hyde, Frederic Griswold, *American Literature and the Spanish-American War, A Study of the Work of Crane, Norris, Fox, and R.H. Davis,* University of Pennsylvania, 1963 (submitted for Ph.D in Philosophy).
Jillson, Willard Rouse, *Literary Haunts and Personalities of Old Frankfort,* Frankfort: The Kentucky Historical Society, 1941.
Kephart, Horace, *Our Southern Highlanders,* Knoxville: The University of Tennessee Press, 1976.
Kesterson, David B., *Bill Nye,* Boston: G.K. Hall & Twayne, 1981.
Klein, Maury, *History of the Louisville & Nashville Railroad,* New York: Macmillan, 1972.
Klotter, James C., *William Goebel: The Politics of Wrath,* Lexington: The University Press of Kentucky, 1977.
Kunitz, Stanley J., and Howard Haycraft, *Twentieth Century Authors,* New York: H.W. Wilson, 1942.
Langford, Gerald, *The Richard Harding Davis Years,* New York: Holt, Rinehart and Winston, 1961.
Lubow, Arthur, *The Reporter Who Would Be King,* New York: Charles Scribner's Sons, 1992.
Millis, Walter, *The Martial Spirit: A Study of Our War with Spain,* Boston: Houghton Mifflin, 1931.
Milton, Joyce, *The Yellow Kids: Foreign Correspondents in the Heyday of Yellow Journalism,* New York: Harper & Row, 1989.
Morison, Elting E. (editor), *The Letters of Theodore Roosevelt,* Vol. II, Cambridge: Harvard University Press, 1951.
Mott, Frank Luther, *American Journalism: A History 1690–1960,* Third Edition, Toronto, Ontario: Macmillan, 1969.
_____, *A History of American Magazines: 1885–1905,* Cambridge: Harvard University Press, 1957.
Nye, Frank Wilson, *Bill Nye: His Own Life Story,* New York: Century, 1926.
Patterson, John, *Library of Southern Literature,* Vol. 4, Atlanta: Martin and Holt, 1907.
Pearce, John Ed, *Days of Darkness: The Feuds of Eastern Kentucky,* Lexington: The University Press of Kentucky, 1994.
_____, Introduction to *The Trail of the Lonesome Pine,* Lexington: The University Press of Kentucky, 1984.
Pond, Major James Burton, *Eccentricities of Genius,* New York: G.W. Dillingham, 1900.
Pringle, Henry F., *Theodore Roosevelt,* New York: Harcourt, Brace, 1931.
Revell, Peter, *James Whitcomb Riley,* New York: Twayne, 1970.
Roosevelt, Theodore, *An Autobiography,* New York: Macmillan, 1914.

Rothert, Otto A., *The Story of a Poet: Madison Cawein*, Louisville: John P. Morton, 1921.
Samuels, Peggy and Harold, *Frederic Remington: A Biography*, New York: Doubleday, 1982.
Snyder, Louis L., and Richard B. Morris, *A Treasury of Great Reporting*, New York: Simon and Schuster, 1962.
Splete, Allen P., and Marilyn D., *Frederic Remington: Selected Letters*, New York: Abbeville Press, 1988.
Stallman, R.W., *Stephan Crane: A Biography*, New York: George Braziller, 1968.
Titus, Warren I., *John Fox, Jr.*, New York: Twayne, 1971.
Townsend, John Wilson, *Kentucky in American Letters 1784–1912*, 2 Vols., Cedar Rapids: The Torch Press, 1913.
Ward, William Smith, *A Literary History of Kentucky*, Knoxville, The University of Tennessee Press, 1988.
Wright, John D., *Transylvania: Tutor to the West*, Lexington: The University Press of Kentucky, 1975.

Index

"After Br'er Rabbit in the Blue-grass" 171
Alden, Henry M. 159, 162
Aldis, Owen 178
Aldrich, Thomas Bailey 175
Allen, Buckner 41
Allen, James Lane 72, 100, 101–103, 111–112, 177, 252
The American Lyceum 110
"The Angel from Viper" 257
Annapolis 137
Ardery, W.B. 110
Arkansas 153
"The Army of the Callahan" 177

"Backward Trail of the Saxon" 205
Baiquiri, Cuba *see* Daiquiri, Cuba
Ballou, Sidney (husband of Tommie Duke) 183
Bardeleben, Baron Fritz von 215
Barrymore, Ethel 162
Barton, Clara 127; at San Juan Hill 151
Bates 82
"The Battle Prayer of Parson Small" 257
Beach, Major William 140
Beecher, Henry Ward 110
Berea College 77, 112, 117
Berryman, Upshur 30, 50
"A Betrothal Ring" 49, 81
Big Stone Gap Post 79, 94, 95, 99
Big Stone Gap, Virginia 1; John Fox, Jr.'s first sight of 2, 26, 59–60; description of 63–64; sanitation and disease in 67–68, 75–76; frontier conditions of 77; crime in 77–78; formation of the Guard in 79–81, 95, 169, 205, 208, 210, 219–220, 223, 237–238, 239, 242, 243, 251, 260– 261, 262, 265, 267, 269, 273
Bigelow, Poultney 127, 132–133, 134, 157–156
Bluegrass and Rhododendron 86, 95, 171, 177, 217
Boer War 167, 182
Bookman 106, 107
Bowlin, Hosea 235–236
Bradley, William O'Connell 164
Branham, John 94
Breathitt County, Kentucky 104, 231
"Br'er Coon in Old Kentucky" 171
Bullitt, Joshua 79

Callahan, Ed 231
Campbell, John C. 77
Cambridge, Massachusetts *see* Harvard
Camp Thomas *see* Chickamauga
Cape Maisi, Cuba 136
Capron, Capt. Allyn 144–145, 148, 150
Carnes, Cecil 151
Carroll, Harry 245
Castine 136
Cawein, Madison 10, 72, 110, 111, 222
Century Magazine 64–65, 69–70, 72, 74, 81, 96, 97, 99, 102, 103–104, 107, 134, 158, 171
Chaffed, Brig. Gen. Ada 142
Chattanooga, Tennessee 124–125, 129
Chefoo, Manchuria, battle of 204
Chicago Tribune 61
Chickamauga, Georgia 124–125; sanitary conditions at 126–127; negro soldiers at 128

311

"Christmas Eve on Lonesome" 177, 209
Christmas Eve on Lonesome and Other Stories 177
"Christmas for Big Ame" 227
"Christmas Night with Satan" 177
"Christmas Tree on Pigeon" 227, 257
"Civilizing the Cumberland" 171, 217
Clark, Cecil (first wife of Richard Harding Davis) 160, 181, 182, 183, 186, 255
Clark, Mary 50
Clay, Henry, Jr. 119
Coldwell, R.C. 101
Collier's 227
Columbia Law School 30–31, 35–36, 39–41
"The Compact of Christopher" 257
Conrad, Joseph 110
Cosby, Arthur F. 146
"Courtin' on Cutshin" 106
"Courtship of Allaphair" 257
Crane, Stephen 133; at Las Guásimas 140, 148; at San Juan Hill 149, 151, 154, 155
Crawford, Bruce 263
"A Crisis for the Guard" 177
The Critic 116
Crittenden 156, 157, 160, 161, 162, 166, 168–169, 171, 173, 174, 178
A Cumberland Vendetta 70, 72–74, 98, 103–104, 105, 106, 114, 116, 270

Daiquiri, Cuba 136–138
Davis (school friend, not Richard Harding Davis) 30
Davis, Bessie McCoy (second wife of Richard Harding Davis) 9, 255–256
Davis-Bigelow-Whitney Controversy 157–160, 167
Davis, Charles 121, 134
Davis, Hope (daughter of Richard Harding Davis) 255, 256, 257, 260–261
Davis, Richard Harding 9, 109, 119, 121–122, 131, 133, 134–137, 139; death of 255–256, 272; at Las Guásimas 140–141, 142, 143, 148; marriage of 161, 162, 167, 181, 182, 183–184, 185–187, 191–192, 195–198, 199–205, 210, 211–212, 216, 223, 251, 252; at San Juan Hill 149, 154, 155, 158–160, 160
Dearborn Independent 263
"Deceiver's Ever" 49, 81
The Denver Times 267

Devil's Jump Branch, Leslie, County, Kentucky 231
"Down the Kentucky on a Raft" 171
Drane, Paul 30
Duke, Basil 52, 120–121, 183
Duke, Currie 3, 52–53, 120–121, 160, 175–176, 183; death of 272
Duke, Thomas Morgan "Tommie" (Currie's sister) 183
Duncan Tavern, Paris, Kentucky 274
Dunne, Finley Peter 162, 172, 174

Eccentricities of Genus (J.B. Pond) 113–114
El Caney, Cuba 137–138, 143, 144, 147, 148, 149
El Pozo, Cuba 143, 144, 148
Empress of China 181, 182, 189, 190
Erskine Dale: Pioneer 258, 261, 262, 266

Fible, Micajah 30, 38–39, 42–44, 46, 50, 51–52, 54, 56–57; disappearance of 61
Fidelio 212
Field, Minna (Thomas Page's stepdaughter) 215, 251
Fleming, Calvin 84–85, 92–93; death of 94, 233
Fleming, Henan 84–85, 92–93; capture of 94; trial of 95 233
Following the Sun Flag 2, 205, 207, 209
Fox, Boaz (father of John, Sr.) 14–15
Fox, Catherine Hill Rice (first wife of John, Sr.) 12; death of 12–13, 239, 269
Fox, Elizabeth (sister) *see* Fox, Sarah Elizabeth
Fox, Everett Buford (half-brother of John, Jr.) 12, 15, 31, 46, 238, 239, 240, 269, 273
Fox, Hilda Seccomb (wife of Rector) 170, 243
Fox, Horace Ethelbert (brother of John, Jr.) 15, 23, 25, 28–29, 46, 55, 62, 87, 108, 181, 220, 238, 242, 262, 267, 269, 273, 274, 275
Fox, James Wallace (half-brother of John, Jr.) 1, 10, 12, 15, 22, 25, 31–32, 38, 46, 55, 58–59, 64, 71, 108, 161–162, 163, 166, 167–168, 169, 170, 208–209, 219, 220, 237, 238, 239, 242, 243; death of 258, 260, 261, 264, 269, 273
Fox, John William, Jr. 1; arrested as a spy in Manchuria 203; arrives in

Index

Hawaii 182–183; arrives back in Tampa 153; at Atlanta, Georgia 129; back home after the Russo-Japanese War 205; in Big Stone Gap 62–63,154; birth of 14; break with *Harper's* 161–162; with Capron's Battery 144–145; at Charleston, South Carolina 254; courtship of 214–216; courtship of Currie Duke 120–121; death of 264; enters Transylvania University 17–19; first meeting with Fritzi Scheff 214; first visit to Jellico 28; in Frankfort, Kentucky 116–117; with Frederic Remington in Cuba 142; goes to Chickamauga for *Harper's* 124–126; goes to England 251–252; goes to Florida 253–254; goes West with the Kemp family 50–51; graduation from Harvard 31–32; as guard for Talton Hall 83; at Harvard University 20–22, 23, 29–33; hunting the Fleming brothers 92–94; hunting trip in the West 247–250; illness of 46–48; with James Whitcomb Riley 110; at Jekyll Island, Georgia 175; in Jellico 55–56; joins the Southern Lyceum Bureau 103, 108; at Las Guásimas 140–141; last illness 263–264; leaves Tampa, Florida, for Cuba 135; at Lexington, Kentucky 129–130; in Louisville, Kentucky 154; marriage to Fritzi Scheff 220–221; at mother and father's anniversary 238–239; at the *New York Sun* 36–38; at the *New York Times* 42–44, 47; rejoins the speaking circuit 178; rumors of drinking and jealousy 223; separation and divorce 243; sickness of 7–8; on the speaking circuits 113–114; at Tampa, Florida 132; at Thomasville, Georgia 254; troubles with eyesight 60; at wedding of Rector Fox 170; at wedding of Richard Harding Davis 161
Fox, John William, Sr. (father of John, Jr.): celebrates anniversary 238–239; death of 240, 269; first marriage of 12; physical description of 15, 16–17, 22, 33–34, 38–39, 45–46, 54, 59–60, 62, 65–66, 68, 70, 100, 107, 166, 208, 219, 220; second marriage of 13; sickness of 239–240
Fox, Louise Buckminster Cole (wife of Richard Fox) 170

Fox, May (wife of Horace Fox) 209, 220, 238, 242, 262, 275
Fox, Mildred (daughter of Horace Fox) 181, 275
Fox, Minerva Carr "Minnie" (sister of John, Jr.) 15, as traveling companion to Fritzi Scheff 223, 238, 239, 242–243, 250, 262, 265, 267, 273, 274
Fox, Minerva Worth Carr (mother of John, Jr.) 13, 45, 71, 238, 239, 240, 250
Fox, Oliver Edwin (brother of John, Jr.) 15, 28–29, 238, 240, 242, 267, 269, 273
Fox, Polly (Mary Combs Kelly, wife of Sidney Fox) 70, 101, 170, 238
Fox, Rector Kerr 15, 65, 71–72, 101, 107, 170, 181, 220, 238, 240, 242–243, 250, 267, 272, 273
Fox, Richard Talbott (brother of John, Jr.) 15, 65, 71–72, 101, 170 , 220, 238, 240, 242–243, 250, 267, 273
Fox, Sarah Elizabeth (sister of John, Jr.) 15, 219, 220, 238, 240, 241, 242, 243, 250, 252, 258, 265, 266, 267, 273, 274
Fox, Sidney Allan (half-brother of John, Jr.) 12, 15, 20, 46, 50, 56, 66, 68; death of 70, 240, 269
"Fox Hunting in Kentucky" 171
Frank Leslie's Illustrated Weekly 49, 81
Frazier, Allie 7
French Lick Springs Hotel (Indiana) 210
Fujina Hotel, Japan 186

Garth Fund 29
Gibson, Dana 109
Gibson, Preston 215, 251
Gibson, R.E. Lee 222
Gladesville (now Wise), Virginia 81, 85, 89–91
Glenwood Springs, Colorado 249, 250
"The Goddess of Happy Valley" 257
Goebel, William 163–165; assassination of 165; funeral of 165–166, 271
Goebel Election Law 164
Gordon Keith (Page) 177
"Grayson's Baby" 107
Grimes' Battery (Capt. George Grimes) 145, 148

Haicheng, Manchuria 196, 199, 202
Hall, Ed 86, 94–95

Hall, Talton 82–83, 85, 87; hanging of 88–91, 108, 115, 228, 233–234
"The Hanging of Talton Hall" 171, 217
"Hardships of the Campaign" 187, 205
Hare, Jimmy 148, 151, 153
Harper's Weekly Magazine 44, 104–105, 106, 107, 116, 117, 119, 120, 124, 128–130, 132, 141, 150–151, 153, 157, 160, 161, 163, 166–167, 168, 207
Harvard University 20, 49, 98
Hawkins, Gen. Hamilton 149, 150
Hay, John 181
Hearst, William Randolph 184
The Heart of the Hills 164, 244, 245, 245, 246, 257, 267, 274
Heijo Maru 191
Hell-fer-Certain, Leslie County, Kentucky 244, 271
"Hell-fer-Sartin" 96, 107, 230
Hell-fer-Sartin and Other Stories 177, 231, 233, 234
Hilton [Hylton], Enos 82–83, 84, 234

Imperial Hotel, Tokyo, Japan 211
In Happy Valley 255
Iroquois 135–136, 141, 153

Jackson, Kentucky 89, 230, 231–232, 233, 236
Jackson Hole, Wyoming 247, 249
Jellico, Tennessee 25–27, 49, 57–58, 98, 108, 167, 261, 270, 273
Jillson, Willard Rouse 116

Kaiping, Manchuria 196
Kemp, Arthur 50
Kemp, George 50
Kemp Family 53, 247
Kent, Virginia 50
The Kentuckians 104, 116, 117–118, 119, 210, 274
"The Kentucky Mountaineer" 171
Kentucky University *see* Transylvania University
Kephart, Horace 112
Kesterson, David 110
Kinchau, Manchuria 193
Kingdom Come Creek, Letcher County, Kentucky 228
Kingdom Come, Letcher County, Kentucky 227–229, 230, 231, 233, 244
Kipling, Rudyard 110

A Knight of the Cumberland 210
Kōbe, Japan 190

Ladies' Home Journal 177
Las Guásimas, Cuba 137–138, 139–140, 142, 144, 147–148, 151
"The Last Stetson" 99, 106
Lawton, Brig. Gen. Henry 142; at San Juan Hill 149
Lexington Herald 110
Lexington Leader 166
Liao-Yang 197; battle of 202, 204
Lindsey, Maria 116
"Lingering in Tokyo" 205
The Little Shepherd of Kingdom Come 3, 5, 156, 162, 167, 168, 171, 173, 175–177, 207, 208, 211, 212, 216, 226, 227, 228, 229, 245, 252, 271, 274
London, Jack 184–185, 192, 205, 272
"The Lord's Own Level" 257
Louisville Courier-Journal 166
Louisville Medical College 83
Lubow, Arthur 133
Lynch, George 203–204

MacDonald, Ballard 245
Mademoiselle Modiste 214
Maine 120, 123
"Making for Manchuria" 205
"Man Hunting in the Pound" 95, 171
Manassas, Battle of 125
Marion, Massachusetts 160
"The Marquise of Queensberry" 257
Mathews, Wilbur Knox 3, 160, 175–176
McClure's 148
McKinley, William 124, 156
"Message in the Sand" 107
The Mettle of the Pasture (Allen) 177
Middlesboro, Kentucky 27
Miles, Gen. Nelson 134
Mitchell, Weir 175
Miyanoshita, Japan 186
Mobile, Alabama 124, 129
Moji, Japan 191
Moore, William Cabell 111, 219–220, 221–222, 225–226, 238, 242–243, 250, 252
Morgan, John Hunt 52, 120–121
Morgan, J. Pierpont 175
Mt. Kisco, New York 220–221, 242, 255, 256
"A Mountain Europa" 64, 69–70, 72–73,

81, 96, 97, 100, 101, 102, 105, 106, 108, 211, 274
Mullins, Ira 85
Mullins, John Harrison 85
Mullins family 84, 233–234
Mullins Massacre 84–85, 95

Nagasaki, Japan 190–191, 205
Nansham, Manchuria, Battle of 190
Nashimoto, Prince Morimasa 197
Nelson, H.L. 159–160
New Chang, Manchuria 202–203
New Orleans 136
New Orleans, Louisiana 124, 129
New York Herald 89, 123, 157
New York Life 49, 81
New York Sun 34, 35, 40, 49, 54
New York Times 39, 42, 54, 244
New York World 140
Norris, Frank 155–156
Norton, Virginia 86
Nye, Bill 105, 110, 111, 113

Oku, General 191–193, 197, 201
Olympic 252
"On Hell-fer-Sartin Creek" 104–105, 230
"On Horseback to Kingdom Come" 227, 230
"On the Road to Hell-fer-Sartin" 227, 233
"On the Trail of the Lonesome Pine" 227
"On the War Dragon's Trail" 205
Our Southern Highlanders (Horace Kephart) 112
Outing Magazine 171

Page, Florence 156, 178, 181, 265, 272
Page, Thomas Nelson 2, 8, 72, 74, 101, 103, 109, 113, 114, 119, 123, 153, 156, 157, 160, 161, 162–163, 167, 171, 175, 177, 178, 181, 210, 215, 216, 220, 251, 252, 259, 265, 272
Pa-lien-tan, Manchuria 195
Papa Perrichon 31
"The Pardon of Becky Day" 177
Paris Kentuckian Citizen 103
Paris, Kentucky 45, 57
"The Passing of Abraham Shivers" 107
Phoenix Hotel (Lexington, Kentucky) 166
Pond, Major J.B. 101, 111, 113–114
"The Pope of Big Sandy" 257
Pound Gap, Virginia 84, 88, 95

Port Arthur, Manchuria 179–180, 184, 189, 191–193, 205
"Preachin' on Kingdom Come" 107
Prior, Melton 202–204
Proctor, John 25
Proctor, town of 28
Proctor Coal Company 58, 269
Puerto Rico 155

Red Ash, town of 28, 57–58
Red Fox *see* Taylor, Marshall Benton
"The Red Fox of the Mountains" 171, 217
"Remember the Maine" (Robert Burns Wilson) 123
Remington, Frederic 109, 125, 141, 142, 143; at San Juan Hill 148–149, 153, 154, 157, 167
Rice, Carrie 13
Riley, James Whitcomb 110, 111, 113, 119, 166; death of 257
Rohe, Alice 267
Roosevelt, Theodore 101, 104, 109, 119, 121–122, 123, 124, 138–139, 142, 144, 148; at San Juan Hill 149–150, 152, 156, 162, 165, 167, 181, 185
Rough Riders 124, 134, 138, 140, 144, 148; at San Juan Hill 149
Russo-Japanese War 2; beginnings of 179–180, 211, 231, 251, 272
Ryer Family 40–41

San Juan Hill 143, 144, 146–147, 148; aftermath of the battle 151–152; battle of 146–150
Santiago, Cuba 136, 137, 140, 143, 147; surrender of 152, 155
Scheff, Fritzi 2, 5, 212–215, 216, 218–219; in Big Stone Gap, Virginia 219–220; marriage to John Fox, Jr. 220–221, 236, 237, 238, 239, 242–247, 251, 252, 261, 266, 272
Scribner, Charles 171, 176, 187
Scribner's, Charles & Sons 44, 162, 166, 175, 176, 178, 181, 187, 189, 205, 209–210, 212, 216, 227, 266
Scribner's Magazine 173, 175, 205, 227, 235, 256, 257
Sea of Japan 190–191
Seelbach Hotel, Louisville, Kentucky 222
Seguranca 135, 138
"The Senator's Last Trade" 107

Seventy-first New York 148
Shafter, Gen. William 134, 138
Shanghai, China 204–205
Shimonoseki Straits, Japan 190
Siboney, Cuba 137–138, 146, 151, 153
Sixteenth U.S. Infantry, at San Juan Hill 149
Sixth U.S. Infantry, at San Juan Hill 149
Slemp, Bascom 257
Smith, "Bad" Tom 89; hanging of 115–116
Southern Lyceum Bureau 101, 103, 156–157
"The Southern Mountaineer" 171
Spanish-American War 114, 121; and the insurrection of 1895 122, 208, 270
A Stage Tragedy 55
Stewart, J.D. (Mr. and Mrs.) 222
Stony Point Academy, description of 15
Stony Point, Kentucky 11, 22, 45, 114, 208, 242, 267, 273
Sumner, Brig. Gen. Samuel S. 148
Swindell, E.J. 94

Takeuchi 192–193
Talienwan, Manchuria 192
Tampa, Florida 118, 124, 129; conditions in 133–134, 153, 157, 158; description of 130–131, 132–133
"Tarpon Fishing at Boca Grande" 256
Tashikao, Manchuria 196
Taylor, Marshall Benton 82–84, 85–86; hanging of 91–92, 108, 115, 233, 235; trial of 86, 87, 90
Taylor, William S. 164
Tehlitzu, Manchuria, Battle of 190
32nd Michigan 134
Thoreau, Henry David 110
"Through the Gap" 107
Thurston, Ballard 222
Titus, Warren 106

"To the Breaks of Sandy" 171
Tokyo, Japan 183–184, 186, 189, 200, 205
Townsend, John Wilson 109–110, 222
The Trail of the Lonesome Pine 5, 74–75, 81, 83, 85, 90, 99, 205, 210–212, 216–217, 218–219, 226, 233, 235; stage play 239, 245, 252, 274
"The Trail of the Saxon" 187, 205
Trans-Siberian Railway 179–180
Transylvania University, description of 18–19, 98, 270
Turnure, Adele 50
Twain, Mark 110, 113, 119, 123

United States Civil Service Commission 40

Vixen 137

The Wanderers 51
Wheeler, Major Gen. Joseph 138; at Las Guásimas 140, 144, 153
"White Slaves of Haicheng" 205
Whitesburg, Kentucky 7
Whitney, Caspar 127, 131, 132, 146, 153, 154, 162, 167
Wilson, Joe 59
Wilson, Robert Burns 10, 72, 116, 119, 123, 160, 165, 254–255, 271
Wingate, Charles 30, 116
Wise County, Virginia 81, 83, 90, 95
Wister, Owen 109
Wompatuck 137
Wood, Gen. Leonard 144, 149
Wright, "Devil" John 85–86

Yalu River 180
Yohatong, Manchuria 196
Yokohama, Japan 183, 186, 189, 205
Young, Brig. Gen. Samuel 144

www.ingramcontent.com/pod-product-compliance
Lightning Source LLC
Chambersburg PA
CBHW020859020526
44116CB00029B/481